# ELEMENTS OF
# SOLITON THEORY

# ELEMENTS OF
# SOLITON THEORY

**G. L. LAMB, Jr.**
*University of Arizona*

A WILEY-INTERSCIENCE PUBLICATION

**JOHN WILEY & SONS**
New York • Chichester • Brisbane • Toronto • Singapore

*Library of Congress Cataloging in Publication Data:*

Lamb, George L, Jr.  1931-
  Elements of soliton theory.

  (Pure and applied mathematics)
  Bibliography: p.
  Includes index.
    1.    Differential  equations,  Partial—Numerical
solutions.    2.    Solitons.    3.    Scattering (Mathematics)    I.    Title.

QA377.L33    515.3'53    80-13373
ISBN    0-471-04559-4

Printed in the United States of America

10 9 8 7 6 5 4 3

To

LAUREN, LARRY, JOEY, and JOANNE

# PREFACE

This book is intended to be an elementary introduction to the theory of solitons, a topic that has provided a fascinating glimpse into the inner workings of certain nonlinear processes during the past decade. The background assumed of the reader is within that usually accumulated by a senior or beginning graduate student in physics or applied mathematics. Some knowledge of integration in the complex plane is presumed, and prior exposure to eigenvalue problems, preferably in the context of quantum theory, will be found helpful but not essential. Since the subject matter is concerned with the solution of nonlinear partial differential equations, some familiarity with the rudiments of linear partial differential equations is, of course, assumed. The applications presuppose some familiarity with hydrodynamics, electromagnetic theory, and the quantum theory of a two-level atom.

The subject is thus presented at an elementary level and concentrates on the background material and introductory concepts that have played a role in setting the stage for some of the current research trends in the field. Recent reformulations involving modern differential geometry and group theory, as well as the ingenious techniques devised by R. Hirota and the results on lattice solitons pioneered by M. Toda, have not been included.

The exposition is pedagogic rather than historical and the topics chosen for consideration are those that, in the author's opinion, convey the basic ideas of the subject in the simplest and most direct way. The analytical formulations are those that present themselves naturally to a physicist raised in an applied tradition. Such workers usually find the more severe procedures of the pure mathematician to be less rather than more demonstrative.

After an introductory chapter that gives a brief indication of the connection between a nonlinear partial differential equation that exhibits soliton behavior (the Korteweg–deVries equation) and a linear eigenvalue problem (for the Schrödinger equation), the next two chapters provide an elementary account of one-dimensional scattering theory and inverse scattering methods. The Korteweg–deVries equation is then treated by inverse scattering techniques in Chapter 4. Chapter 5 provides a corresponding introduction to the other most common soliton equations. Chapters 6 and 7 present some examples of how soliton equations arise in various physical contexts. Chapter

8 introduces the subject of Bäcklund transformations and finally, in Chapter 9, the recently popular topic of soliton perturbation theory is considered.

The presentation is largely self-contained so that reference to the original literature should be unnecessary. A number of references to additional background material have been included as well as references to expositions that either extend or complement the presentation given here. They are not intended to document either priority or high points in the historical development of the subject. The reader interested in extensive bibliographies may consult the article "The soliton: a new concept in applied science" by A. C. Scott, F. Y. F. Chu, and D. W. McLaughlin, *Proc. IEEE* **63**, 1443–1483 (1973) as well as the volume *Solitons* (Springer Topics in Modern Physics Series) edited by R. K. Bullough and P. J. Caudrey (Springer-Verlag, Heidelberg, 1980).

Certain facets of soliton theory may be traced back quite directly to research in nineteenth-century mathematics. It has been this writer's experience that reference to the research of our predecessors can be especially rewarding when investigating soliton theory. The writings of A. R. Forsyth have been found to be particularly appropriate in this regard. Some aspects of soliton theory seem to provide fulfillment to the closing paragraph in the sixth volume of Forsyth's *Theory of Differential Equations*, where he writes:

> My desire has been to give a continuous exposition of those portions of the subject which . . . bear some promise of leading into paths of research that will be trodden by investigators in days yet to come.

I wish to record my appreciation to F. A. Otter, Jr., for a comment concerning the occurrence in dislocation theory of what is now known as the sine-Gordon equation. A solution technique used in this work was based upon Bäcklund transformations. The transfer of these results to coherent optical pulse propagation, where the sine-Gordon equation also arises, led to my consideration of what are now known as the solitons of coherent optics.

My thanks are due to F. A. E. Pirani for a very careful reading of many chapters of this work as well as to M. G. Forest and P. R. Schlazer, who have also read many sections. By their assistance several errors and obscurities have been eliminated, and the volume has been rendered less imperfect than it would otherwise have been.

I am also indebted to W. E. Ferguson, Jr., for providing the numerical solutions of the Korteweg–deVries equation that appear in Chapter 4 and to M. O. Scully and F. A. Hopf for the numerical results indicated in Figure 7.6. The computer assistance offered to me by L. A. Appelbaum and R. C. Dillon during the preparation of the other pulse profile figures is also appreciated. Finally, I am grateful to my wife, Joan, for many hours spent in typing.

Any corrections or suggestions for improvements with which my readers may favor me will be greatly appreciated.

G. L. LAMB, JR.

*Tuscon, Arizona*
*July 1980*

# CONTENTS

**1. Introduction**     **1**

   1.1   A Sturm–Liouville Equation,   2
   1.2   The Korteweg–deVries Equation,   4
         *Single-Soliton Solution of the Korteweg–deVries Equation,*   7
   1.3   Multisoliton Solutions as Bargmann Potentials,   8
         *Linear Bargmann Potential—The Single Soliton,*   8
         *Quadratic Bargmann Potential—Interaction of Two Solitons,*   9
   1.4   A Physical System Leading to the Korteweg–deVries
         Equation,   13
   1.5   Extensions to Other Nonlinear Equations,   20
   1.6   A Preview,   21

**2. Topics in One-Dimensional Scattering Theory**     **22**

   2.1   Waves on a String,   22
         *Energy Flow on a String,*   23
   2.2   Scattering by an Oscillator,   26
   2.3   The Elastically Braced String,   29
   2.4   The Schrödinger Equation,   33
   2.5   Scattering by a $\mathrm{sech}^2$ Potential,   34
   2.6   Associated Sturm–Liouville Equations,   38
   2.7   Two-Dimensional Waves in an Inhomogeneous Medium,   41
   2.8   A General Approach to Scattering,   46
         *Fundamental Solutions,*   47
         *Wronskian Relations,*   47
         *Poles of the Transmission Coefficient,*   50
         *Relation between Transmission and Reflection Coefficients,*   55
         *Asymptotic Solution,*   57
         *Number of Poles of the Transmission Coefficient,*   59
   2.9   Truncated Potentials,   60
   2.10 Scattering of Pulses—Marchenko Equations,   62

2.11 Two-Component Scattering,   67

   *A Time-Dependent Problem*,   67
   *Fundamental Solutions*,   69
   *Wronskian Relations*,   71
   *Poles of the Transmission Coefficient*,   73
   *An Example*,   75
   *Asymptotic Solution*,   78
   *Truncated Potentials*,   78

2.12 Relation between One- and Two-Component Equations—
   Riccati Equations,   80


**3.  Inverse Scattering in One Dimension**                                  **84**

3.1 Relation between the Potential and the Functions
   $A_R(x,y)$ and $A_L(x,y)$,   84
   *Example—Repulsive Delta Function Potential*,   86

3.2 The Presence of Bound States,   87
   *Example—Attractive Delta Function Potential*,   88

3.3 Reflectionless Potentials,   90
   *Example—The Case $N = 2$*,   92

3.4 Reflection Coefficient a Rational Function of $k$,   93

3.5 Bargmann Potentials,   96
   *Linear Case*,   97
   *Quadratic Case*,   97

3.6 Two-Component Inverse Method for Real Potentials,   99

3.7 Reflectionless Potentials for Two-Component Systems,   103

3.8 Reflection Coefficient for Two-Component System—A
   Rational Function of $k$,   105

3.9 Two-Component System with a Complex Potential,   107
   *Asymptotic Solution*,   111


**4.  The Korteweg–deVries Equation**                                        **113**

4.1 Steady-State Solution,   113

4.2 Results of Numerical Solutions,   115

4.3 Inverse Scattering and the Korteweg–deVries Equation,   118

4.4 Multisoliton Solutions,   121
   *Example—The Two-Soliton Solution (N = 2)*,   122

4.5 Conserved Quantities,   126

4.6 The Initial Pulse Profile $\delta'(x)$—Similarity Solution,   128

4.7 Alternative Approach to the Linear Equations for the
   Korteweg–deVries Equation,   131

5.  **Some Evolution Equations Related to a Two-Component Linear System**                                                                        133

    5.1  Modified Korteweg–deVries Equation,   134
    *The Linear Equations*,   135
    *Solution by Inverse Scattering*,   136
    *Breather Solution*,   138
    *Alternative Approach to the Linear Equations*,   142
    5.2  Sine–Gordon Equation,   143
    *Some Simple Solutions*,   144
    *Energy Considerations*,   150
    *Solution by Inverse Scattering*,   151
    *Two-Soliton Solution*,   153
    *$\pi$ Pulse and the Similarity Solution*,   153
    5.3  Cubic Schrödinger Equation,   155
    *Linear Equations*,   156
    *Solution by Inverse Scattering*,   156
    5.4  A General Class of Soluble Nonlinear Evolution Equations,   160

6.  **Applications I**                                                                               169

    6.1  Shallow Water Waves and the Korteweg–deVries Equation,   169
    6.2  Shallow Water Waves and the Cubic Schrödinger Equation,   174
    6.3  Ion Plasma Waves and the Korteweg–deVries Equation,   178
    6.4  Classical Model of One-Dimensional Dislocation Theory—Sine–Gordon Equation,   182
    6.5  Choice of Expansion Parameters,   186

7.  **Applications II**                                                                              190

    A SOLITON ON A VORTEX FILAMENT,   190
    7.1  Self-Induction of a Vortex,   191
    7.2  Motion of the Filament,   194
    7.3  Shape of the Single-Soliton Filament,   198
    7.4  Other Soliton Equations,   200

    COHERENT OPTICAL PULSE PROPAGATION,   204
    7.5  Description of the Electromagnetic Field,   205
    7.6  The Two-Level Atom,   206
    7.7  Equations of the Model,   209

7.8   Stationary Atoms—Sine–Gordon Limit,   210
7.9   Moving Atoms and the Area Theorem,   214
7.10  Solution by an Inverse Method,   216
7.11  Propagation in an Amplifier,   221
7.12  The Two-Component Method,   227
       *Conserved Quantities*,   234
7.13  Level Degeneracy,   238


8.  **Bäcklund Transformations**                                           **243**

8.1   Baçklund Transformation for the Korteweg–deVries
       Equation,   243
       *Validity of the Theorem of Permutability*,   247
8.2   Bäcklund Transformations for Some Other
       Evolution Equations,   248
       *Example—The Sine–Gordon Equation*,   249
8.3   More General Bäcklund Transformations,   252
       *Example—Liouville's Equation*,   254


9.  **Perturbation Theory**                                                 **259**

THE KORTEWEG–DEVRIES EQUATION,   259
9.1   Basic Equations,   259
9.2   Perturbation of the Single-Soliton Solution,   262
       *Perturbation of Bound-State Parameters*,   263
       *Perturbations in the Continuous Spectrum*,   264
       *A Simple Procedure*,   269
THE CUBIC SCHRÖDINGER EQUATION,   271
9.3   Basic Equations,   271
9.4   Damping of the Single Soliton,   275

**References**                                                               **279**

**Index**                                                                    **285**

# ELEMENTS OF
# SOLITON THEORY

# CHAPTER 1

## Introduction

A remarkable development in our understanding of a certain class of nonlinear partial differential equations known as evolution equations has taken place in the past decade. The key to our present knowledge of these equations is the realization that they possess a special type of elementary solution. These special solutions take the form of localized disturbances, or pulses, that retain their shape even after interaction among themselves, and thus act somewhat like particles. This independence among elementary solutions is a well-known effect in processes governed by linear partial differential equations where a linear superposition principle applies but was quite unexpected when first observed in processes governed by nonlinear partial differential equations. These localized disturbances have come to be known as *solitons*.

Although the partial differential equations that govern the motion of solitons are nonlinear, they are closely related to certain linear ordinary differential equations known as Sturm–Liouville equations. A study of solitons should thus be prefaced by a summary of the relevant topics in Sturm–Liouville theory. These considerations are developed in Chapters 2 and 3.

Before taking up these preliminaries, however, we shall give a cursory introduction to certain essential elements of soliton theory in this initial chapter. First, by asking the right question regarding an ordinary differential equation of Sturm–Liouville type, we shall be led to a consideration of one of the nonlinear partial differential equations that has soliton solutions. The equation that arises is known as the Korteweg–deVries equation (Korteweg and deVries, 1895). It occurs in a number of physical problems, mostly in hydrodynamics. At our present level of understanding, there appears to be no basis for expecting so fundamental a relationship between a Sturm–Liouville equation and the Korteweg–deVries equation. Secondly, to underscore the close relation between solitons and linear ordinary differential equations, a formula that expresses the interaction between two solitons is constructed by adapting a technique devised by Bargmann (1949) for obtaining a certain class of potentials for the Schrödinger equation, which is an example of a Sturm–Liouville equation. In a somewhat loose manner of speaking, one can say the the analytical expressions that describe multisoliton interactions are merely Bargmann potentials. The particle nature of the soliton is evident when this two-soliton solution is examined. Finally, we shall consider how the

Korteweg–deVries equation arises in a simple example of nonlinear dispersive wave propagation.

## 1.1   A STURM–LIOUVILLE EQUATION

The differential equation

$$\frac{d^2y}{dx^2} + [\lambda - U(x)]y = 0, \qquad a \leqslant x \leqslant b \qquad (1.1.1)$$

plus boundary conditions imposed at point $x = a$ and $b$ (either or both of which may be at infinity) is of frequent occurrence in applied mathematics. Such an equation is a simple example of a Sturm–Liouville equation (Ince, 1926). Equation 1.1.1 has been most thoroughly studied in the context of quantum theory, where it is known as a Schrödinger equation. This nomenclature sometimes persists even in applications of the equation to classical physics, such as wave propagation in inhomogeneous media.

For a given function $U(x)$, which would be the potential for a problem in quantum theory, imposition of the boundary conditions can lead to only certain specific values of the constant $\lambda$ (the eigenvalues $\lambda_j$) for which the equation will have a nonzero solution [the eigenfunction $y_j(x)$]. The determination of the dependence of the solution $y$ on the parameter $\lambda$ and the dependence of the eigenvalues $\lambda_j$ on the boundary conditions is known as a Sturm–Liouville problem.

One of the simplest examples of such an eigenvalue problem is obtained by setting $U(x) = 0$ and imposing the boundary conditions $y(a) = y(b) = 0$. Solution of the resulting equation $y'' + \lambda y = 0$ subject to the prescribed boundary conditions shows that the eigenfunctions are $y_j(x) = \sin[(\lambda_j)^{1/2}x], j = 1, 2, 3, \ldots$, with the eigenvalues $\lambda_j = [j\pi/(b-a)]^2$. As the length of the system $b - a$ increases, the $\lambda_j$ become more closely spaced and in the limit $b - a \rightarrow \infty$ we obtain the continuous range of eigenvalues $0 < \lambda < \infty$. When the foregoing equation arises in the study of vibrating systems, each eigenfunction $y_j(x)$ represents the shape of a normal mode of the system. An example would be a uniform string that is vibrating in free space and confined between fixed ends located at points $x = a$ and $b$. Since the eigenvalues $\lambda_j$ are related to the resonant frequencies of vibration of the system, it is customary to refer to them as a spectrum of eigenvalues. For the case considered here, say that of a homogeneous string, the entire length of the system takes part in the vibration of each normal mode. However, it is possible to construct inhomogeneous systems, the inhomogeneity represented by the function $U(x)$, for which the vibration is confined to only a portion of the system. The vibration is then confined by the inhomogeneity rather than by the boundaries. An example would be a vibrating string that is partly embedded in elastic surroundings. This system will be discussed at length in Chapter 2.

In the subsequent development we shall always be concerned with systems of infinite extent. Hence, any localized solutions will always be due to inhomogeneities. Such localized solutions have been perhaps most thoroughly studied in quantum theory, where they are used to describe the discrete energy levels of atomic systems.

There are relatively few functions $U(x)$ for which the corresponding ordinary differential equation (1.1.1) may be solved in terms of the standard transcendental functions. As an example (which will be considered in detail in Chapter 2) the choice $U(x) = -2\,\mathrm{sech}^2 x$ plus the boundary conditions $y(\pm\infty) = 0$ leads to the single eigenvalue $\lambda_1 = -1$ with the associated eigenfunction $y_1 = \mathrm{sech}\; x$. That is, $y_1 = \mathrm{sech}\; x$ is the solution of the equation $y_1'' + (-1 + 2\,\mathrm{sech}^2 x)y_1 = 0$ that vanishes as $x \to \pm\infty$. In quantum theory, the interpretation of this result is that a particle is confined by a potential well having a shape proportional to $\mathrm{sech}^2 x$ while the single value of $\lambda$ is proportional to the energy that the particle confined by this well can possess. As a classical interpretation of the same equation, we may consider the channeling of a wave in a medium having the depth-dependent refractive index $n^2 = 1 + 2\,\mathrm{sech}^2 x$, where depth is measured from the location of the maximum value of $n(x)$. The inhomogeneity establishes a waveguide in the medium. A wave can be confined to the depth about which the refractive index takes on its maximum value. When the sign of the potential is reversed so that $U(x) = 2\,\mathrm{sech}^2 x$, the potential is repulsive and no bound state occurs. Similarly, when the refractive index is given by $n^2 = 1 - 2\,\mathrm{sech}^2 x$, waves tend to emanate away from the region of refractive index variation and no channeling effect takes place. This example will be developed further in Chapter 2.

In addition to the discrete negative values $\lambda_j$, with their associated localized wave functions $y_j(x)$, equations such as (1.1.1) can also possess a continuous range of solutions for positive values of $\lambda$ when $b - a$ becomes infinite. In the quantum case the physical interpretation of such solutions is that of the scattering of an incident particle with energy proportional to $\lambda$ by some obstacle that is characterized by the potential $U(x)$. In the one-dimensional problems being considered here, the presence of the scatterer usually manifests itself in terms of reflection and transmission of the incident wave. [The fact that certain choices for the potential $U(x)$ can result in perfect transmission with no reflected wave will play a central role in later considerations.] Particles with any positive energy may be incident upon the scattering center, of course, and hence we expect the continuous range of positive eigenvalues $0 < \lambda < \infty$. In the example of the inhomogeneous medium mentioned above, the situation corresponding to $\lambda > 0$ is the reflection and transmission of a wave of arbitrarily high frequency that is incident upon the inhomogeneous layer from the outside. In Chapter 2 we will show that, for the potential function $U(x) = -2\,\mathrm{sech}^2 x$ given above, the scattering solutions are made up of linear combinations of the functions $y_\pm = e^{\pm i\sqrt{\lambda}\,x}(i\sqrt{\lambda} \mp \tanh x)$. Determination of the solution of a Schrödinger equation when the potential function $U(x)$ is specified is frequently referred to as solving a scattering problem.

## 1.2   THE KORTEWEG–deVRIES EQUATION

If the function $U(x)$ in (1.1.1) should contain a parameter, say $\alpha$, so that $U = U(x, \alpha)$, then variation of the shape of the potential by variation of $\alpha$ could be expected to lead to some corresponding variation in the eigenvalues $\lambda_j$; that is, we would expect the values of the $\lambda_j$ to depend upon $\alpha$. It is perhaps not too unnatural to ask whether or not there are potential functions $U(x, \alpha)$ for which the $\lambda_j$ remain unchanged as the parameter $\alpha$ is varied. One rather trivial example suggests itself immediately. Replacement of any $U(x)$ by $U(x + \alpha)$ merely translates the potential or refractive index inhomogeneity along the $x$ axis. This merely changes the location of the confined particle or depth of the sound channel and has no effect upon the bound-state energy or the frequency of the confined wave. Such a variation of $\alpha$ thus has no effect upon the eigenvalues $\lambda_j$. For comparison with future results, it should be noted that functions $U(x + \alpha)$ satisfy the linear partial differential equation $U_\alpha - U_x = 0$. We shall find that there are other more interesting possibilities that lead to *nonlinear* partial differential equations. In particular, functions $U(x, \alpha)$ that satisfy the nonlinear partial differential equation

$$U_\alpha + UU_x + U_{xxx} = 0 \qquad (1.2.1)$$

will also be shown to leave the eigenvalues invariant. When the parameter $\alpha$ is interpreted as time (the associated Sturm–Liouville equation is, of course, not the time-dependent Schrödinger equation), then (1.2.1) is the Korteweg–deVries equation.

Finding solutions to the Korteweg–deVries equation can thus be related to the determination of parameter-dependent potentials in a Sturm–Liouville equation, and vice versa. The scattering problem mentioned in Section 1.1 was concerned with the determination of a wave function $y$ when the potential $U$ was specified. In the present situation we are concerned with determining the potential when certain information about the wave function is specified (in a manner that will be considered in detail in Chapter 4). Determination of a potential from information about the wave function is appropriately referred to as an *inverse* scattering problem.

By using a method devised by P. Lax (1968), we may see quite easily that the Korteweg–deVries equation is one of an infinite number of equations that govern the variation in the potential of a Schrödinger equation in such a way that the eigenvalues remain constant. To see this, it is convenient to write the Schrödinger or Sturm–Liouville equation in the form

$$Ly = \lambda y \qquad (1.2.2)$$

where $L = D^2 - u(x,t)$ and $D = d/dx$. A time derivative of this equation yields

$$(Ly)_t = Ly_t + L_t y = \lambda_t y + \lambda y_t \qquad (1.2.3)$$

Since $(Ly)_t = y_{xxt} - uy_t - u_t y = Ly_t - u_t y$, we see that $L_t = -u_t$. We are interested in imposing a time variation on $u$ and hence $y$, such that $\lambda_t = 0$. Let us consider the possibility that the time dependence of $y$ may be expressed in the form $y_t = By$ where $B$ is some linear differential operator (not necessarily unique) that must be determined. The spatial variation of $y$ is, of course, given by (1.2.2). Equation 1.2.3 may now be written

$$(-u_t + [L,B])y = \lambda_t y \qquad (1.2.4)$$

where $[L,B] \equiv LB - BL$. We note that $\lambda$ will be a constant so that $\lambda_t = 0$ provided that $B$ is chosen to satisfy the equation $-u_t + [L,B] = 0$. In general, this would be an operator equation. However as we shall presently see, certain restrictions on the form of $B$ can yield an expression for $[L,B]$ that is devoid of differential operators and contains merely $u$ and its spatial derivatives. In such cases we shall have constructed a partial differential equation for $u(x,t)$ which, when satisfied, will imply $\lambda_t = 0$; that is, the eigenvalues remain constant in time.

As a first example of a differential operator $B$ that can lead to constant eigenvalues, let us consider $B_1 = aD$, where $a$ is initially allowed to be a function of $u$ and its spatial derivatives. Then, from the definition of $L$,

$$\begin{aligned}
[L, B_1]y &= (LB_1 - B_1 L)y \\
&= (D^2 - u)(ay_x) - aD(y_{xx} - u_y) \\
&= 2a_x D^2 y + a_{xx} Dy + au_x y \qquad (1.2.5)
\end{aligned}$$

If the coefficients of $D^2 y$ and $Dy$ in this last expression vanish, that is, if $a$ is a constant, then $[L, B_1] = au_x$ and (1.2.4) becomes

$$(u_t - au_x)y = -\lambda_t y \qquad (1.2.6)$$

Therefore, $\lambda_t$ will be zero and $\lambda$ thus constant in time provided that $u$ satisfies the partial differential equation $u_t - au_x = 0$. Since the solution of this equation is any function of $x + at$, we see that any potential of the form $u(x + at)$ will leave $\lambda$ unchanged in time. This somewhat uninteresting example in which the parameter $at$ merely translates the potential along the $x$ axis at a velocity $-a$ has already been alluded to in the preceding discussion.

To obtain a more interesting example we might try $B_2 = aD^2 + fD + g$, where $f$ and $g$ are, in general, functions of $u$ and its spatial derivatives and $a$ is again a constant, as in the previous example. However, a simple calculation shows that no extension of the previous result is obtained. We are merely led to the same linear partial differential equation for $u$.

If we proceed a step further and consider $B_3 = aD^3 + fD + g$, we find that

$$[L, B_3]y = (2f_x + 3au_x)D^2y + (f_{xx} + 2g_x + 3au_{xx})Dy$$

$$+ (g_{xx} + au_{xxx} + fu_x)y \qquad (1.2.7)$$

A new partial differential equation for $u$ now results when we again require that the coefficients of $D^2y$ and $Dy$ vanish. The vanishing of these coefficients yields simple differential relations that are readily integrated. We find that $f = -\frac{3}{2}au + c_1$ and $g = -\frac{3}{4}au_x + c_2$, where $c_1$ and $c_2$ are arbitrary functions of time that arise from integration. Then from (1.2.7),

$$[L, B_3]y = \left[\tfrac{1}{4}a(u_{xxx} - 6uu_x) + c_1u_x\right]y \qquad (1.2.8)$$

The partial differential equation satisfied by $u$ again follows from the relation $-u_t + [L, B_3] = 0$. The constant $a$ may be set equal to $-4$ to simplify the coefficients in the resulting equation. Also, the function $c_1(t)$ may be set equal to zero since it may be eliminated in the final equation for $u$ by merely transforming to new independent variables given by $dx' = dx + c_1(t)dt$ and $dt' = dt$. The new equation for $u$ is thus found to be

$$u_t - 6uu_x + u_{xxx} = 0 \qquad (1.2.9)$$

If $u$ is governed by this equation, the left-hand side of (1.2.4) will vanish and hence we again obtain $\lambda_t = 0$. Except for the factor of $-6$ (which could be eliminated by setting $u = -\frac{1}{6}U$) this is the nonlinear partial differential equation that was given in (1.2.1). It is one of the standard forms of the Korteweg–deVries equation. Thus, if the potential in a Schrödinger equation evolves according to the Korteweg–deVries equation, the eigenvalue parameter $\lambda$ remains constant.

Finally, since the functions $f$ and $g$ in the operator $B_3$ are now known, the time dependence of the solution $y$ is also known. It is given by

$$y_t = B_3y = (-4D^3 + 6uD + 3u_x)y \qquad (1.2.10)$$

The function $c_2(t)$ has also been set equal to zero since it may be eliminated by introducing a new dependent variable $\bar{y} = y \exp(\int c_2 dt)$. It should be noted that both the spatial variation of $y$,

$$y_{xx} - uy = \lambda y \qquad (1.2.11)$$

and the temporal variation given by (1.2.10) are expressed in terms of linear differential equations.

As might be expected at this point, an infinite sequence of higher-order equations, characterized by the odd linear operators $B_5, B_7, \ldots$, may be constructed (Lax, 1968; Gardner et al., 1974). However, these higher-order evolution equations do not seem to arise in physical applications at present and will not be considered here. Instead, we shall examine two simple solutions of the Korteweg–deVries equation (1.2.9).

## Single-Soliton Solution of the Korteweg–deVries Equation

The simplest solution of the Korteweg–deVries equation is the steady-state solution which is obtained by looking for a solution in the form $u(x - ct)$. The solution thus represents a disturbance that moves in the positive $x$ direction at a constant velocity $c$. It will be shown in Chapter 4 that a steady-state pulse solution of the Korteweg–deVries equation (1.2.9) is

$$u = -\frac{c}{2} \operatorname{sech}^2\left[ \frac{\sqrt{c}}{2}(x - ct) \right] \qquad (1.2.12)$$

This solution exhibits a common feature of nonlinear waves in that the amplitude and velocity of the pulse are related. Larger-amplitude pulses move more rapidly and also are narrower in width. A simple integration shows that the width and amplitude of the pulse are related in such a way that

$$\int_{-\infty}^{\infty} dx \sqrt{|u|} = \pi \qquad (1.2.13)$$

The solution given in (1.2.12), which represents a localized disturbance that is symmetric about its midpoint, is the single-soliton solution of the Korteweg–deVries equation. However, the true soliton nature of this expression is not yet evident. The essential element of the soliton is that the analytical form above is preserved, except for a phase shift, after the interaction of two or more such pulses. To see this preservation of form upon interaction, we must consider a more complicated solution than the mere steady-state result quoted above. Procedures for generating multisoliton solutions based upon inverse scattering techniques will be described in later chapters. To give a preview of these more general results, we shall here obtain the two-soliton solution by a simple method that predates the more sophisticated inverse scattering techniques. The method is one devised by Bargmann (1949) for application to the radial Schrödinger equation but which may be applied equally well in the present instance where the range of the independent variable is the entire $x$ axis. The close connection between potentials and multisoliton solutions is brought out quite clearly and simply by this method.

## 1.3  MULTISOLITON SOLUTIONS AS BARGMANN POTENTIALS

As a potential in a Schrödinger equation, the expression $2\operatorname{sech}^2 x$, referred to above as an expression for a soliton, is known as an Eckart potential (Eckart, 1930). It may be obtained by using the method devised by Bargmann. Here we shall merely outline the procedure and then use it to obtain the two-soliton solution as well. The method will be described at greater length in Chapter 3.

### Linear Bargmann Potential—The Single Soliton

The starting point of Bargmann's method is the assumption that there exist potentials for the Schrödinger equation

$$y'' + (k^2 - u)y = 0 \tag{1.3.1}$$

such that the solution may be written in the form

$$y = e^{ikx}F(k, x) \tag{1.3.2}$$

where $F(k,x)$ is a polynomial in $k$. If $F(k,x)$ is of zero order in $k$ and independent of $x$, we merely obtain the obvious possibility $u=0$. The simplest nontrivial example is the linear form

$$y_1 = e^{ikx}\left[2k + ia(x)\right] \tag{1.3.3}$$

where the numerical factors are introduced for later convenience. Introducing this form of the solution into the Schrödinger equation (1.3.1) and separating terms according to powers of $k$, we obtain

$$a' = -u, \quad a'' = ua \tag{1.3.4}$$

Elimination of $u$ and integration yields

$$a' + \tfrac{1}{2}a^2 = 2\mu^2 \tag{1.3.5}$$

where $2\mu^2$ is the constant of integration. The substitution

$$a = \frac{2w'}{w} \tag{1.3.6}$$

leads to the linear equation

$$w'' - \mu^2 w = 0 \tag{1.3.7}$$

with the solution

$$w = \alpha e^{\mu x} + \beta e^{-\mu x} \tag{1.3.8}$$

According to the first of (1.3.4) and (1.3.6),

$$u = -2(lnw)''$$ (1.3.9)

With the form of $w$ given in (1.3.8), this yields

$$u = -2\mu^2 \operatorname{sech}^2(\mu x - \phi)$$ (1.3.10)

where $\phi = \frac{1}{2}\ln(\beta/\alpha)$.

Furthermore, this result may now be made to satisfy the Korteweg–deVries equation by simply allowing the integration constants $\alpha$ and $\beta$ and hence $\phi$ to be functions of time. Substitution of (1.3.10) into the Korteweg–deVries equation (1.2.9) shows that

$$\phi_t = 4\mu^3$$ (1.3.11)

and hence

$$u = -2\mu^2 \operatorname{sech}^2(\mu x - 4\mu^3 t)$$ (1.3.12)

A constant of integration that fixes the origin of coordinates has been ignored. If we set $\mu = \frac{1}{2}\sqrt{c}$, this form of $u$ agrees with the steady-state solution given in (1.2.11). The linear case in Bargmann's procedure thus yields the single-soliton solution of the Korteweg–deVries equation.

## Quadratic Bargmann Potential—Interaction of Two Solitons

The method may be pursued a step further by considering the quadratic Bargmann potential. This extension will be found to provide the desired analytical expression that describes the interaction between two solitons. We begin with the assumed form

$$y_2 = e^{ikx}\left[4k^2 + 2ika(x) + b(x)\right]$$ (1.3.13)

Substitution into the Schrödinger equation yields the three conditions $u = -a'$, $ua = a'' + b'$, and $ub = b''$. A detailed consideration of these equations will be postponed until Chapter 3. For the present we merely note that with $a = 2w'/w$ as before, and by a minor extension of the analysis outlined above for the linear case, $w$ is now found to satisfy a fourth-order linear differential equation with constant coefficients. The solution may be written

$$w = \sigma(\alpha e^{\rho x} + \alpha^{-1} e^{-\rho x}) + \rho(\beta e^{\sigma x} + \beta^{-1} e^{-\sigma x})$$ (1.3.14)

where $\rho$ and $\sigma$ are constants $(\rho > \sigma > 0)$ and $\alpha$ and $\beta$ are integration constants that will again become functions of time when we require that the potential also satisfy the Korteweg–deVries equation. In Chapter 3 it will be shown that the eigenvalues depend only upon the two constants $\rho$ and $\sigma$ and are thus independent of time.

The solution given in (1.3.14) is more conveniently expressed in the form

$$w = 2\left[\sigma \cosh(\rho x - \phi) + \rho \cosh(\sigma x - \psi)\right] \qquad (1.3.15)$$

where $\exp(-\phi) = \alpha$ and $\exp(-\psi) = \beta$. Then we find that

$$a = 2\frac{w'}{w} = 2\rho\sigma \frac{\sinh(\rho x - \phi) + \sinh(\sigma x - \psi)}{\sigma \cosh(\rho x - \phi) + \rho \cosh(\sigma x - \psi)} \qquad (1.3.16)$$

Setting

$$\rho = p + q, \quad \sigma = p - q, \quad \phi = \theta + \chi, \quad \psi = \theta - \chi \qquad (1.3.17)$$

we obtain

$$a = 2(p^2 - q^2)(p \coth\zeta - q \tanh\eta)^{-1} \qquad (1.3.18)$$

where $\zeta = px - \theta$ and $\eta = qx - \chi$. The potential is then

$$u = -a' = -2(p^2 - q^2)\frac{p^2 \operatorname{csch}^2\zeta + q^2 \operatorname{sech}^2\eta}{(p \coth\zeta - q \tanh\eta)^2} \qquad (1.3.19)$$

For this expression to be a solution to the Korteweg–deVries equation as well, the function $\theta$ and $\chi$ must have appropriate time dependence. Direct substitution of the form for $u$ given in (1.3.19) into the Korteweg–deVries equation would, of course, be quite laborious. Instead, we may note that the expression takes on a much simpler form for values of $x$ and $t$ such that either $|\zeta| = |px - \theta|$ or $|\eta| = |qx - \chi|$ is much greater than unity. Since $\theta$ and $\chi$ contain no $x$ dependence, they may be determined for all $x$ in these limits.

For $|\zeta| \gg 1$ and $|\eta|$ less than unity, we readily obtain

$$u \approx -\frac{2q^2(p^2 - q^2)\operatorname{sech}^2\eta}{(\pm p - q \tanh\eta)^2} \qquad (1.3.20)$$

where the upper (lower) sign is used as $\zeta$ is large and positive (negative). This result may be rewritten as

$$u^{(q)} = -2q^2 \operatorname{sech}^2(qx - \chi \mp \Delta) \qquad (1.3.21)$$

where

$$\Delta = \tanh^{-1}\left(\frac{q}{p}\right) \qquad (1.3.22)$$

The superscript has been introduced to emphasize that the result is merely the single-soliton expression (1.3.12) with $q$ as the amplitude–velocity parameter. On the other hand, for $|\eta| \gg 1$, but $|\zeta|$ less than unity, a similar calculation yields

$$u^{(p)} = -2p^2 \operatorname{sech}^2(px - \theta \mp \Delta) \tag{1.3.23}$$

where the upper (lower) sign is used for $\eta$ large and positive (negative).

The time dependence in $\theta$ and $\chi$ may now be obtained quite easily by substituting (1.3.21) and (1.3.23) into the Korteweg–deVries equation (1.2.9). The results are the same as before for the linear case, namely

$$\theta_t = 4p^3, \qquad \chi_t = 4q^3 \tag{1.3.24}$$

Thus, there are values of $x$ and $t$ for which the quadratic Bargmann potential reduces to two individual soliton pulses. Since $p > q$, the pulse $u^{(p)}$ travels at a greater velocity than the pulse $u^{(q)}$. As $x \to \pm \infty$ the solution takes the form of two separate pulses as shown in Figure 1.1. We obtain a picture of a faster pulse overtaking, interacting with, and emerging ahead of a slower one. These results are perhaps better understood if we now consider the complete solution (1.3.19) for all $x$ and $t$. A numerical example that is frequently quoted (Zabusky, 1968; Gardner et al., 1974) is recovered here if we set $\rho = 3$ and $\sigma = 1$. Then $p = 2$ and $q = 1$ and $\theta = 32t$, $\chi = 4t$. The two-soliton solution (1.3.19) may then be transformed to the usually quoted form

$$u = -12 \frac{4\cosh(2x - 8t) + \cosh(4x - 64t) + 3}{[3\cosh(x - 28t) + \cosh(3x - 36t)]^2} \tag{1.3.25}$$

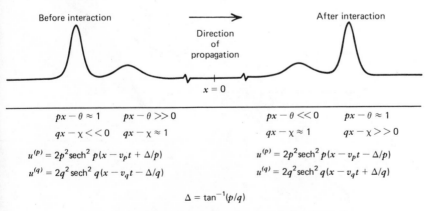

Before interaction

Direction
of
propagation

After interaction

$x = 0$

| $px - \theta \approx 1$ | $px - \theta \gg 0$ | | $px - \theta \ll 0$ | $px - \theta \approx 1$ |
| --- | --- | --- | --- | --- |
| $qx - \chi \ll 0$ | $qx - \chi \approx 1$ | | $qx - \chi \approx 1$ | $qx - \chi \gg 0$ |
| $u^{(p)} = 2p^2 \operatorname{sech}^2 p(x - v_p t + \Delta/p)$ | | | $u^{(p)} = 2p^2 \operatorname{sech}^2 p(x - v_p t - \Delta/p)$ | |
| $u^{(q)} = 2q^2 \operatorname{sech}^2 q(x - v_q t - \Delta/q)$ | | | $u^{(q)} = 2q^2 \operatorname{sech}^2 q(x - v_q t + \Delta/q)$ | |

$$\Delta = \tan^{-1}(p/q)$$

**Figure 1.1** Two isolated solitons before and after interaction.

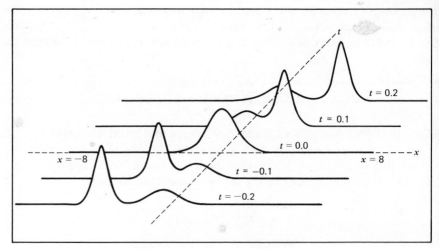

**Figure 1.2** Two-soliton interaction.

At $t=0$, when the two solitons completely overlap, this expression reduces to $u(x,0) = -6\,\text{sech}^2 x$. At earlier and later times, when the two solitons are well separated, (1.3.25) reduces to a sum of two expressions of the form given above in (1.3.21) and (1.3.25). The nonlinear interaction of the two solitons is apparent in that at $t=0$, when they overlap completely, the pulse amplitude has a magnitude of 6, which is *less* than that of the single well-separated pulse with $p=2$, which has an amplitude of 8. A graph of the two-soliton solution is shown in Figure 1.2. The pulses pass through each other in a completely

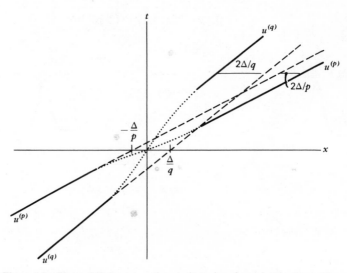

**Figure 1.3** Phase shift in space–time trajectories of two interacting solitons.

elastic fashion. The only vestige of their interaction is the phase shift of $2\Delta$. This shift in phase is clearly displayed when the trajectory of each pulse is viewed in an $x$–$t$ plane, as indicated in Figure 1.3.

We have now seen that a close relation may be established between solutions of the Korteweg–deVries equation and parameter-dependent potentials in the Schrödinger equation. Also, a simple technique for obtaining a certain class of potentials (Bargmann potentials) and solutions of the Schrödinger equation has been used to determine soliton solutions of the Korteweg–deVries equation. Before developing this and other solution techniques further, we shall indicate how the Korteweg–deVries equation arises in a simple physical situation.

## 1.4  A PHYSICAL SYSTEM LEADING TO THE KORTEWEG–deVRIES EQUATION

There are many examples to choose from, some of which will be described in Chapter 6. They all possess two features. First, they exhibit a characteristic hydrodynamic nonlinearity by containing a term of the form $u \, \partial u / \partial x$. Second, when only small-amplitude disturbances are considered, so that the above-mentioned nonlinear term can be neglected, single-frequency solutions having space–time dependence in the form $e^{i(kx - \omega t)}$ may be obtained. These solutions are found to impose a relation between wave number $k$ and frequency $\omega$, which, for small $k$ and $\omega$, may be approximated by $\omega = a_1 k + b_1 k^3$, where $a_1$ and $b_1$ are constants. The phase velocity of a wave is given by $c_p = \omega / k$, and if $b_1 = 0$ we see that $c_p$ is a constant. For $b_1 \neq 0$, however, waves with different values of $k$ will travel at different velocities and a pulse composed of a spectrum of such waves will change shape as it propagates. In particular, a localized pulse will tend to spread out. The wave motion is then said to be dispersive and the relation between $\omega$ and $k$ is referred to as the dispersion relation for the problem. In this case, the main portion of a pulse travels at the group velocity $v_g = \partial \omega / \partial k |_{k = k*}$, where $k*$ is an average of the values of $k$ that make up the pulse (Whitham, 1974, Chap. 11). Similar considerations apply to the equivalent relation $k = a_2 \omega + b_2 \omega^3$.

Unfortunately, all examples of nonlinear dispersive waves entail somewhat tedious perturbation calculations before the equation of interest, such as the Korteweg–deVries equation, is finally obtained. The physical system to be considered here is no exception, although the perturbation analysis is perhaps slightly less extensive than that of most other examples. Also, the experimental situation is readily visualized. We shall consider an incompressible fluid that is confined within an infinitely long circular cylinder. The walls of the cylinder are composed of elastic rings as shown in Figure 1.4. A localized pressure increase in the fluid (only axial variations are considered) causes a radially symmetric expansion of the elastic rings in the region of the pressure increase. Since the rings are uncoupled (but close enough so that the fluid does not leak out between them), no elastic waves are propagated along the cylindrical surface in the axial direction. Such a boundary is sometimes referred to as being locally reacting. The equations governing the fluid motion are those of conservation of mass and momentum. If the axial length of each ring is small compared to lengths of interest, the conservation of mass may be written

$$\frac{\partial A}{\partial t} + \frac{\partial}{\partial x}(Av) = 0 \qquad (1.4.1)$$

where $v$ is the velocity of the fluid and $A(x,t)$ is the cross-sectional area of the cylinder. The area is seen to play a role analogous to that of the density in a compressible fluid. For the conservation of momentum we may write

$$\frac{\partial v}{\partial t} + v\frac{\partial v}{\partial x} = -\frac{1}{\rho_0}\frac{\partial p}{\partial x} \qquad (1.4.2)$$

where $\rho_0$ is the constant density of the fluid and $p(x,t)$ is the fluid pressure. A third equation relating the area of the ring to the pressure must also be obtained. Since, as noted above, the area is analogous to a density, the equation that we seek plays the role of an equation of state.

From Figure 1.5 we can see that application of Newton's second law to a section of the elastic ring leads to the equation

$$\rho_R h l\, ds \frac{\partial^2 r(x,t)}{\partial t^2} = p(x,t)l\,ds - 2T_c h l \sin\left(\frac{d\theta}{2}\right) \qquad (1.4.3)$$

where $r(x,t)$ is the radius of the ring, $\rho_R$ is the density of the ring, and $h$ and $l$ are the ring thickness and axial length, respectively. For small angles $d\theta$, we may replace the sine of the angle by its argument and also set $d\theta = ds/a$, where $a$ is the equilibrium radius of the ring. Finally, the circumferential tension $T_c$ in the ring is related to the extension of the ring through Young's modulus $E$ by the usual relation

$$E = \frac{\text{stress}}{\text{strain}} = \frac{T_c}{2\pi(r-a)/2\pi a} \qquad (1.4.4)$$

$A(x,t)$

$x$

Figure 1.4  Soliton on tube composed of elastic rings.

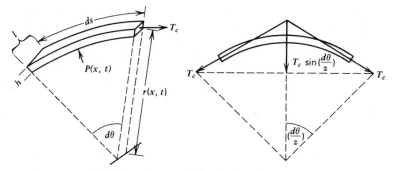

**Figure 1.5**  Forces acting on segment of elastic ring.

Noting that $A = \pi r^2$ so that $\Delta A = A - \pi a^2 \approx 2\pi r \, \Delta r$, we find that (1.4.3) may be written

$$\frac{2\pi a}{\rho_R h} p(x,t) = \frac{E}{\rho_R a^2} \Delta A + \frac{\partial^2 \Delta A}{\partial t^2} \tag{1.4.5}$$

The quantity $(E/\rho_R a^2)^{1/2}$ is the angular frequency for radial vibration of the ring (Love, 1927, p. 454). Introducing dimensionless quantities through the definitions $A = \pi a^2 \tilde{A}$, $p = p_0 \tilde{p}$, and $t = Tt'$, where $p_0 = Eh/2a$ and $T^2 = \rho_R a^2/E$, we obtain

$$\tilde{p} = (\tilde{A} - 1) + \frac{\partial^2 \tilde{A}}{\partial t'^2} \tag{1.4.6}$$

Also, introducing the length $L = (\rho_R ah/2\rho_0)^{1/2}$ and setting $x = Lx'$ and $v = L\tilde{v}/t$, we have the dimensionless equations

$$\frac{\partial \tilde{A}}{\partial t'} + \frac{\partial}{\partial x'}(\tilde{A}\tilde{v}) = 0 \tag{1.4.7}$$

$$\frac{\partial \tilde{v}}{\partial t'} + \tilde{v}\frac{\partial \tilde{v}}{\partial t'} = -\frac{\partial \tilde{p}}{\partial x'} \tag{1.4.8}$$

First we investigate the dispersion relation for the linearized version of (1.4.6) to (1.4.8). The linear equations are

$$\frac{\partial \tilde{A}}{\partial t'} + \frac{\partial \tilde{v}}{\partial x'} = 0 \tag{1.4.9a}$$

$$\frac{\partial \tilde{v}}{\partial t'} = -\frac{\partial \tilde{p}}{\partial x'} \tag{1.4.9b}$$

$$\tilde{p} = \tilde{A} - 1 + \frac{\partial^2 \tilde{A}}{\partial t'^2} \tag{1.4.9c}$$

If we allow $\tilde{p}$, $\tilde{v}$, and $\tilde{A} - 1$ to vary as $\exp i(\kappa x' - \tilde{\omega} t')$, then (1.4.9) becomes a system of homogeneous algebraic equations in the amplitudes of $\tilde{p}$, $\tilde{v}$, and $\tilde{A} - 1$. Nonvanishing solutions of this system of equations are obtained in the usual way by requiring that the determinant of the system vanish. We find that

$$\tilde{\omega}^2 = \frac{\kappa^2}{1 + \kappa^2} \tag{1.4.10}$$

which is the dispersion relation for this problem. For small $\tilde{\omega}$ and $\kappa$, this may be written in either of the two approximate forms $\kappa = \tilde{\omega} + \frac{1}{2}\tilde{\omega}^3$ or $\tilde{\omega} = \kappa - \frac{1}{2}\kappa^3$. We thus obtain the cubic terms in the dispersion relation that were referred to previously. It should be noted that the all-important cubic term is due to the elastic response of the cylinder wall. If the time derivative did not appear in (1.4.5), $\kappa$ would be proportional to $\tilde{\omega}$. In the units being employed here, the group velocity of the wave would then be equal to unity and no dispersion would occur.

When $\tilde{\omega}$ is expressed in terms of $\kappa$, the exponential factor introduced above takes the form $\exp\{i[\kappa x' - (\kappa - \frac{1}{2}\kappa^3)t']\}$. (The form that results when $\kappa$ is given in terms of $\tilde{\omega}$ could also be considered.) In developing a perturbation expansion of the nonlinear equations it will be useful to incorporate the linear dispersion at the outset. This may be done by introducing the new independent variables $\xi = \kappa(x' - t')$ and $\eta = \kappa^3 t'$. Then the differential operators become

$$\frac{\partial}{\partial x'} = \kappa \frac{\partial}{\partial \xi}, \qquad \frac{\partial}{\partial t'} = -\kappa \frac{\partial}{\partial \xi} + \kappa^3 \frac{\partial}{\partial \eta} \tag{1.4.11}$$

When the nonlinear equations (1.4.6) to (1.4.8) are written in terms of $\eta$ and $\xi$, we obtain

$$-\frac{\partial \tilde{A}}{\partial \xi} + \kappa^2 \frac{\partial \tilde{A}}{\partial \eta} + \frac{\partial}{\partial \xi}(\tilde{A}\tilde{v}) = 0 \tag{1.4.12a}$$

$$-\frac{\partial \tilde{v}}{\partial \xi} + \kappa^2 \frac{\partial v}{\partial \eta} + \tilde{v}\frac{\partial \tilde{v}}{\partial \xi} = -\frac{\partial \tilde{p}}{\partial \xi} \tag{1.4.12b}$$

$$\tilde{p} = \tilde{A} - 1 + \kappa^2 \frac{\partial^2 \tilde{A}}{\partial \xi^2} - 2\kappa^4 \frac{\partial^2 \tilde{A}}{\partial \xi \partial \eta} + \kappa^6 \frac{\partial^3 \tilde{A}}{\partial \eta^2} \tag{1.4.12c}$$

We now introduce the perturbation expansions

$$\tilde{p} = \kappa^2 p_1 + \kappa^4 p_2 + \cdots \tag{1.4.13a}$$

$$\tilde{A} = 1 + \kappa^2 A_1 + \kappa^4 A_2 + \cdots \tag{1.4.13b}$$

$$\tilde{v} = \kappa^2 v_1 + \kappa^4 v_2 + \cdots \tag{1.4.13c}$$

into (1.4.12) and require that the equations be satisfied for each power of $\kappa^2$. The terms in (1.4.12) that are proportional to $\kappa^2$ are

$$-\frac{\partial A_1}{\partial \xi} + \frac{\partial v_1}{\partial \xi} = 0 \qquad (1.4.14a)$$

$$-\frac{\partial v_1}{\partial \xi} + \frac{\partial p_1}{\partial \xi} = 0 \qquad (1.4.14b)$$

$$p_1 = A_1 \qquad (1.4.14c)$$

from which we may infer that $p_1 = A_1 = v_1 + \varphi(\eta)$, where $\varphi(\eta)$ is arbitrary. The terms in (1.4.13) that are proportional to $\kappa^4$ are

$$-\frac{\partial A_2}{\partial \xi} + \frac{\partial A_1}{\partial \eta} + \frac{\partial}{\partial \xi}(A_1 v_1) + \frac{\partial v_2}{\partial \xi} = 0 \qquad (1.4.15a)$$

$$-\frac{\partial v_2}{\partial \xi} + \frac{\partial v_1}{\partial \eta} + v_1 \frac{\partial v_1}{\partial \xi} = -\frac{\partial p_1}{\partial \xi} \qquad (1.4.15b)$$

$$p_2 = A_2 + \frac{\partial^2 A_1}{\partial \xi^2} \qquad (1.4.15c)$$

These equations are easily combined to eliminate $A_2$, $v_2$, and $p_2$. The result is

$$\frac{\partial v_1}{\partial \eta} + \frac{3}{2} v_1 \frac{\partial v_1}{\partial \xi} + \frac{1}{2} \frac{\partial^3 v_1}{\partial \xi^3} + \varphi \frac{\partial v_1}{\partial \xi} + \frac{\partial \varphi}{\partial \eta} = 0 \qquad (1.4.16)$$

The last two terms in this equation may be suppressed by the transformations $\bar{v} = v_1 + \frac{1}{2}\varphi(\eta)$ and $\tau = \eta, \zeta = \xi + \frac{3}{2}\int \varphi \, d\eta$. We then obtain the Korteweg–deVries equation in the form

$$\frac{\partial \bar{v}}{\partial \tau} + \frac{3}{2} \bar{v} \frac{\partial \bar{v}}{\partial \zeta} + \frac{1}{2} \frac{\partial^3 \bar{v}}{\partial \zeta^3} = 0 \qquad (1.4.17)$$

As a result of our previous consideration of this equation, we may expect that the physical system under investigation can support a disturbance in the form of a soliton. From (1.2.11), which is the single-soliton solution of the standard form of the Korteweg–deVries equation (1.2.9), we readily find that the corresponding solution of (1.4.17) is

$$\bar{v} = c \operatorname{sech}^2 \left[ \frac{1}{2}\sqrt{c} \left( \zeta - \frac{1}{2}c\tau \right) \right] \qquad (1.4.18)$$

Let us now return to dimensional variables. In particular, consider the radial displacement of the rings since this would be readily observable.

The change in area is $\Delta A = 2\pi a(r - a) = \pi a^2 \kappa^2 \bar{v}$. If the maximum amplitude of the soliton is now set equal to $r_0$ (i.e., we define $\frac{1}{2}ac\kappa^2 = r_0$), then we obtain

$$r - a = r_0 \operatorname{sech}^2\left[(x - Vt)/W\right] \tag{1.4.19}$$

Here, $W$, which may be considered the half-width of the soliton amplitude, is

$$W = L\sqrt{\frac{2a}{r_0}} = a\sqrt{\frac{h\rho_R}{r_0\rho_0}} \tag{1.4.20}$$

The soliton velocity $V$ that occurs in (1.4.19) is conveniently expressed in terms of $V_0$, the velocity of low-frequency linear waves in the system. We find that

$$V = V_0\left(1 + \frac{r_0}{a}\right) \tag{1.4.21}$$

where, as shown immediately below, $V_0^2 = Eh/2\rho_0 a$. Also the Korteweg–deVries equation takes the form

$$\frac{\partial v}{\partial t} + V_0\frac{\partial v}{\partial x} + \frac{3}{2}v\frac{\partial v}{\partial x} + \frac{1}{2}V_0 L^2\frac{\partial^3 v}{\partial x^3} = 0 \tag{1.4.22}$$

To examine the low-frequency limit ($\omega \ll T^{-1}$) of linear waves in the system, we ignore the time derivative term in (1.4.5) so that $p \simeq (Eh/2\pi a^3)\Delta A$. Also, linearization of (1.4.1) and (1.4.2) yields

$$\frac{\partial \Delta A}{\partial t} + \pi a^2\frac{\partial v}{\partial x} = 0 \tag{1.4.23}$$

and

$$\frac{\partial v}{\partial t} = -\frac{1}{\rho_0}\frac{\partial p}{\partial x} = -\frac{Eh}{2\pi\rho_0 a^3}\frac{\partial \Delta A}{\partial x} \tag{1.4.24}$$

Eliminating $v$ between these two equations, we obtain the linear wave equation

$$\frac{\partial^2 \Delta A}{\partial x^2} - \frac{1}{V_0^2}\frac{\partial^2 \Delta A}{\partial t^2} = 0 \tag{1.4.25}$$

with $V_0$ given by the expression employed in (1.4.21).

From (1.4.21) we see that if $r_0 \ll a$, which is necessary to justify the linearizations employed in treating the vibration of the ring in (1.4.3), the soliton velocity is slightly larger than the linear velocity $V_0$. Also, for a density ratio $\rho_R/\rho_0=8$ (appropriate for steel and water) and a soliton amplitude to ring thickness ratio $r_0/h=2$, we find that the soliton width $2W$ is approximately equal to twice the diameter of the ring.

The existence of a steady-state solution of the Korteweg–deVries equation may be expected on the basis of simple intuitive considerations. This solution, and thus the soliton itself, may be interpreted as describing a balance between two competing effects. One effect results from the nonlinearity of the equation. The nonlinearity is the type that tends to bring about a cresting and breaking of wave profiles such as those on the surface of water. The other effect is the spreading out of wave profiles that occurs whenever dispersion is present.

To see this competition in more detail, let us consider the Korteweg–deVries equation (1.4.22) in a coordinate system that moves with the velocity $V_0$. Then

$$\frac{\partial v}{\partial t} + \frac{3}{2}v\frac{\partial v}{\partial x} + \beta\frac{\partial^3 v}{\partial x^3} = 0 \qquad (1.4.26)$$

where $\beta = \frac{1}{2}V_0L^2$. The steady-state solution is

$$v = 8\left(\frac{\beta}{W^2}\right)\text{sech}^2\left[\frac{x-4(\beta/W^2)t}{W}\right] \qquad (1.4.27)$$

To examine the effect of the nonlinear term in (1.4.26), let us consider only the simpler nonlinear equation $v_t + \frac{3}{2}vv_x = 0$. The term $\frac{3}{2}v$ plays the role of a wave velocity. Since this velocity depends upon the solution itself, we may expect that portions of the wave profile at which $v$ is large will move more rapidly than portions of the wave near the edge of the profile where $v$ approaches zero. For the steady-state solution given in (1.4.27), the largest velocity should correspond to the maximum value of $v$ and hence be given by $\frac{3}{2}\cdot 8(\beta/W^2)$. The time $t_{NL}$ required for the peak to travel at this velocity for a distance equal to the half-width of the pulse, $W$, and thus give rise to a breaking of the wave, is $t_{NL} = \frac{2}{3}\cdot\frac{1}{8}(W^2/\beta)\cdot W \approx W^3/\beta$ if we neglect numerical factors.

Let us now examine the effect of dispersion in (1.4.26) by ignoring the nonlinear term. We then have the linear equation $v_t + \beta v_{xxx} = 0$. It has traveling wave solutions of the form $\exp(ikx + i\beta k^3 t)$. The largest wave numbers $k$ that contribute to a pulse of width $2W$ are approximately $k = 2\pi/2W \approx 1/W$. Such wave numbers will lead to a change in phase of $\pi$, and thus bring about considerable diffusion of an initial wave form, in a time $t_D$ given by $\beta(1/W)^3 t_D = \pi$. That is, $t_D \approx W^3/\beta$, which is on the same time scale as $t_{NL}$. Thus, for pulses of amplitude equal to that of the soliton, $\beta/W^2$, the two effects are comparable and the establishment of a steady pulse profile due to a balance between the two effects is not surprising.

## 1.5 EXTENSIONS TO OTHER NONLINEAR EQUATIONS

In this introductory chapter we have briefly considered the relation between a parametric-dependent potential in a Schrödinger equation. and a nonlinear evolution equation. The potential was considered to be a real function of both position and the parameter. When parametric variation of the potential was carried out in such a way that the Korteweg–deVries equation was satisfied the eigenvalues (energy levels) were found to remain constant. In subsequent chapters other nonlinear evolution equations will be considered that will be related in a similar way to other linear eigenvalue problems. These linear equations are equivalent to a Schrödinger equation in which the potential is complex, although this is not always the most convenient form in which to analyze the linear equations. In particular, we shall encounter the equation

$$y_{xx} + \left(\zeta^2 + u^2 + iu_x\right)y = 0 \tag{1.5.1}$$

The function $u$ will be found to be related to nonlinear evolution equation that have come to be known as the modified Korteweg–deVries equation $u_t + u^2 u_x + u_{xxx} = 0$ and the sine-Gordon equation $\sigma_{xt} = \sin \sigma$, where $\sigma = \frac{1}{2}u_x$.

It is easily shown that (1.5.1) results upon the elimination of $z$ between the two first-order equations

$$\begin{aligned} y_x + iuy &= \zeta z \\ z_x - iuz &= -\zeta y \end{aligned} \tag{1.5.2}$$

The first-order equations that have been studied most extensively in soliton theory are obtained from (1.5.2) by introducing the transformation $n_1 = \frac{1}{2}(y + iz)$ and $in_2 = \frac{1}{2}(y - iz)$. The functions $n_1$ and $n_2$ are found to satisfy

$$\begin{aligned} n_{1x} + i\zeta n_1 &= un_2 \\ n_{2x} - i\zeta n_2 &= -un_1 \end{aligned} \tag{1.5.3}$$

which are of the form of (1.5.2) with $u$ and $\zeta$ interchanged. These linear equations, as well as the more general form

$$\begin{aligned} n_{1x} + i\zeta n_1 &= qn_2 \\ n_{2x} - i\zeta n_2 &= rn_1 \end{aligned} \tag{1.5.4}$$

where $q$ and $r$ may be complex, play a fundamental role in soliton theory (Ablowitz et al., 1974a). The choice $r = -q^*$ in (1.5.4) is of particular importance. It leads to a consideration of the nonlinear evolution equation $iq_t + q_{xx} + |q|^2 q = 0$, which is known as the cubic or nonlinear Schrödinger equation. These and other nonlinear evolution equations will be considered after relevant properties of the linear eigenvalue equations are summarized in Chapters 2 and 3.

## 1.6    A PREVIEW

We have now seen a simple mathematical description of a soliton as well as
the physical significance of the soliton in a highly idealized physical situation.
Succeeding chapters of this book present a more detailed study of the various
topics that have been touched upon here.

It should be clear at this stage that the solution of soliton equations is
closely related to the solution of certain linear ordinary differential equations
of Sturm–Liouville type. Chapter 2, especially Sections 2.8 and 2.11, provides
an elementary survey of the relevant topics in Sturm–Liouville theory. The
special aspect of this subject that is of interest in the study of solitons is
known as the inverse scattering problem, and an introduction to this topic is
presented in Chapter 3.

It is not until Chapter 4, where the Korteweg–deVries equation is considered, that we again consider solitons per se. With the mathematical preliminaries dispensed with, we shall then be able to proceed directly to the
special use of these topics for the description of solitons. The presentation in
Chapter 4 and succeeding chapters contains extensive reference to the preliminary developments in Chapters 2 and 3 so that the reader with even a
nominal familiarity with Sturm–Liouville theory and the inverse scattering
problem may proceed directly to Chapter 4 and refer back only as needed.

Chapter 5 is devoted to the three equations that are most commonly
associated with the two-component linear equations, namely the modified
Korteweg–deVries equation, the sine-Gordon equation and the cubic (or
nonlinear) Schrödinger equation.

In Chapters 6 and 7 some of the physical contexts in which solitons arise
are considered. Chapter 8 is an introduction to Bäcklund transformations, a
topic that provides an alternative approach to the construction of multisoliton
solutions. Finally, in Chapter 9 a perturbation theory for treating equations
that differ slightly from known soliton equations is developed. Perturbed
Korteweg–deVries and cubic Schrödinger equations are considered as examples.

## CHAPTER 2

# Topics in One-Dimensional
# Scattering Theory

In Chapter 1 it was shown that when variations in the potential of a
Schrödinger equation are characterized by a parameter $\alpha$ and the shape of the
potential $U(x, \alpha)$ is required to vary in such a way that the Korteweg–deVries
equation is satisfied, that is, that $U(x, \alpha)$ satisfy $U_\alpha + UU_x + U_{xxx} = 0$, then the
energy levels associated with the potential will remain unchanged as $\alpha$ varies.
The present chapter contains a summary of certain aspects of the solution of
Schrödinger equations. In particular we consider both the scattering of waves
and the localized solutions, or bound states, associated with certain potentials
that are frequently used in the Schrödinger equation.

Although a considerable body of knowledge on scattering theory is extant,
it is predominantly concerned with the radial coordinate in three-dimensional
quantum scattering. In that case the independent variable is confined to a
semiinfinite interval $0 < r < \infty$. In the situation of interest here, the somewhat
different problem of one-dimensional scattering on the infinite interval $-\infty
< x < \infty$ is of concern. Besides applications to one-dimensional models in
quantum theory, such scattering problems have a number of physical inter-
pretations in terms of classical wave propagation, and this point of view will
also be considered.

As noted at the end of Chapter 1, a two-component formulation of the
linear ordinary differential equations will sometimes be more convenient in
analyzing other nonlinear evolution equations. The two-component scattering
problem will also be treated.

Much of this chapter may be omitted by readers who are already familiar
with elementary wave propagation. Sections 2.8 and 2.11 contain the material
that is especially pertinent for soliton theory.

## 2.1 WAVES ON A STRING

Since the notion of reflection and transmission of waves in one-dimensional
scattering plays a central role in soliton theory, it will be helpful to review this
topic in some detail. The vibrating string is perhaps the simplest physical
context in which we can develop most of the concepts that arise in one-

22

dimensional scattering. The equation governing the motion of small-amplitude waves on a homogeneous string is (Morse, 1948)

$$\frac{\partial^2 y}{\partial x^2} - \frac{1}{c^2}\frac{\partial^2 y}{\partial t^2} = 0 \tag{2.1.1}$$

where $y(x,t)$ is the string displacement and $c$, the speed of waves on the string, is equal to the square root of the ratio of string tension $T_s$ to string density $\rho$. The general solution of (2.1.1) in terms of arbitrary wave profiles moving in opposite directions along the string is

$$y = f(x - ct) + g(x + ct) \tag{2.1.2}$$

If the disturbance takes place at the single frequency $\nu = \omega/2\pi$, then

$$y = a\cos(kx - \omega t + \alpha) + b\cos(kx + \omega t + \beta) \tag{2.1.3}$$

where the amplitudes $a$ and $b$ and phases $\alpha$ and $\beta$ are constants and $k = \omega/c$. It is convenient to consider this solution to be the real part of the expression

$$y = A e^{i(kx - \omega t)} + B e^{-i(kx + \omega t)} \tag{2.1.4}$$

where $A = a e^{i\alpha}$ and $B = b e^{-i\beta}$. The time-dependent factors for waves in both directions are now identical and may be ignored. Hence $e^{ikx}$ is an adequate representation for a wave traveling in the positive $x$ direction with unit amplitude and $e^{-ikx}$ for such a wave traveling in the negative $x$ direction as long as we assume that the time dependence is $e^{-i\omega t}$. Disturbances that are localized in space and time may be built up by Fourier analysis (Hildebrand, 1976, Section 5.15),

$$\varphi(x + ct) = \int_{-\infty}^{\infty} \frac{d\omega}{2\pi} \Phi(\omega) e^{-i\omega(t \pm x/c)} \tag{2.1.5}$$

where $\Phi^*(-\omega) = \Phi(\omega)$ to assure the reality of $\varphi$. Examples of the use of this technique will appear later in this chapter.

## Energy Flow on a String

The transport of energy is one of the most important physical quantities associated with wave motion on a string. In general, the energy associated with a section of string between $x = a$ and $x = b$ $(b > a)$ is composed of kinetic energy $E_k$ given by

$$E_k = \tfrac{1}{2}\rho \int_a^b dx (y_t)^2 \tag{2.1.6}$$

as well as potential energy $E_p$. The potential energy is equal to the work done in stretching the string at a tension $T_s$, hence

$$E_p = T_s \int_a^b dx \left[ \sqrt{1+(y_x)^2} - 1 \right] \approx \tfrac{1}{2} T_s \int_a^b dx (y_x)^2 \qquad (2.1.7)$$

where the approximation is valid for small displacements from equilibrium [i.e., $(y_x)^2 \ll 1$]. The total energy of the wave on the section of string $a \leqslant x \leqslant b$ is thus

$$E = \tfrac{1}{2} \rho \int_a^b dx (y_t^2 + c^2 y_x^2) \qquad (2.1.8)$$

Change in the energy of the string in this region is then given by

$$\frac{dE}{dt} = \rho \int_a^b dx (y_{tt} y_t + c^2 y_{xt} y_x)$$

$$= \rho c^2 \int_a^b dx (y_x y_t)_x \qquad (2.1.9)$$

where (2.1.1) has been employed. After performing the spatial integration, we see that the quantity $\rho c^2 y_x y_t$ has the physical interpretation (and hence dimensions) of energy flow. Energy flow $J(x,t)$ for a wave moving in the positive $x$ direction will be positive provided that we define

$$J(x,t) = -\rho c^2 y_x y_t \qquad (2.1.10)$$

For, setting $y = y_0 \cos(kx - \omega t)$, we obtain $J = \rho c y_0^2 \omega^2 \sin^2(kx - \omega t)$ and the time average of this quantity over a complete cycle $T = 1/\nu$ is

$$\langle J \rangle = \frac{1}{T} \int_0^T dt\, J(x,t)$$

$$= \tfrac{1}{2} \rho c \omega^2 y_0^2 \qquad (2.1.11)$$

which is a positive quantity. Equations 2.1.9 and 2.1.10 now yield

$$\frac{dE}{dt} = J_a - J_b \qquad (2.1.12)$$

This has the obvious physical interpretation that the energy change in the region between $a$ and $b$ is the difference between the influx at $a$ and the efflux at $b$.

Finally, if the energy between points $a$ and $b$ is expressed in terms of an energy density so that $E = \int_a^b dx\, \mathscr{E}(x,t)$, then from (2.1.9) and (2.1.10), we obtain

$$\frac{dE}{dt} = \int_a^b dx \frac{\partial \mathscr{E}}{\partial t} = -\int_a^b dx \frac{\partial J}{\partial x} \qquad (2.1.13)$$

This result must hold in any region $a < x < b$. The integrands themselves must therefore be equal and we obtain the conservation law,

$$\frac{\partial \mathcal{E}}{\partial t} + \frac{\partial J}{\partial x} = 0 \qquad (2.1.14)$$

Since this result holds for any interval $a < x < b$, it is sometimes referred to as a local conservation law.

It is convenient to be able to calculate the time average of the energy flow when complex notation is used. Since energy is a second-order quantity, some care must be used in doing this with complex notation. Setting $y = \text{Re}(fe^{-i\omega t})$, where Re signifies "real part of" and $f = f_r + if_i$, we see from (2.1.10) and (2.1.11) that

$$\langle J \rangle = -\frac{\rho c^2}{T} \int_0^T dt \, \text{Re} \left( -i\omega f e^{-i\omega t} \right) \text{Re} \left( \frac{\partial f}{\partial x} e^{-i\omega t} \right)$$

$$= \tfrac{1}{2}\rho c^2 \omega \left( -f_i \frac{\partial fr}{\partial x} + fr \frac{\partial fi}{\partial x} \right) \qquad (2.1.15)$$

This may be rewritten in the form

$$\langle J \rangle = \tfrac{1}{2}\rho c^2 \omega \, \text{Im} \left( y^* y_x \right) \qquad (2.1.16)$$

where Im signifies "imaginary part of." A similar expression will be encountered in Section 2.4 for describing particle flux associated with the Schrödinger equation.

The expression for the energy on the string in (2.1.8) may be obtained in a more elegant fashion by using a Lagrangian density (Morse and Feshbach, 1953, Chap. 3; Goldstein, 1950, Chap. 7–9). This will be a convenient method to use in later considerations of wave motion. We first note that the wave equation for the string may be obtained from the Lagrangian density $\mathcal{L}(y_t, y_x, y) = \tfrac{1}{2}\rho[(y_t)^2 - c^2(y_x)^2]$. By substituting this expression for $\mathcal{L}$ into the Euler–Lagrange equation

$$\frac{\partial}{\partial t}\left( \frac{\partial \mathcal{L}}{\partial y_t} \right) + \frac{\partial}{\partial x}\left( \frac{\partial \mathcal{L}}{\partial y_x} \right) - \frac{\partial \mathcal{L}}{\partial y} = 0 \qquad (2.1.17)$$

we immediately recover the wave equation (2.1.1). Of more interest to us here is the fact that the Hamiltonian density, which is an expression for the energy density of waves on the string, is given by $\mathcal{H} = y_t \, \partial \mathcal{L}/\partial y_t - \mathcal{L}$. With the Lagrangian density given above, we obtain $\mathcal{H} = \tfrac{1}{2}\rho[(y_t)^2 + c^2(y_x)^2]$. Then, since $E = \int \mathcal{H} \, dx$, we immediately recover (2.1.8).

As an example of energy flow on a string, the wave $y = e^{ikx} + R(k)e^{-ikx}$ represents a superposition of two waves, one with unit amplitude traveling in the positive $x$ direction and one with amplitude $R$ (which may be complex and thus contain phase information) traveling in the negative $x$ direction. This is the wave field that would be set up in the region $x < 0$ by a steady wave of unit amplitude that is incident from $x = -\infty$ and reflected with an amplitude $R$ from some inhomogeneity on the string at $x = 0$. From (2.1.16) the energy flow is readily found to be

$$\langle J \rangle = \tfrac{1}{2}\rho c \omega^2 (1 - |R|^2) \tag{2.1.18}$$

If $|R|^2 < 1$, then $J$ is positive and there is a net energy flow in the positive direction. (This energy could be either transmitted beyond the inhomogeneity to the region $x > 0$ or be absorbed by it.) If $|R|^2 = 1$, then as much energy is reflected as is incident and there is no net flow of energy in either direction. This would be the case for a wave that is completely reflected by a lossless termination (i.e., one that absorbs no energy). The most general form for $R$, the reflection coefficient, would then be $R = e^{i\delta}$, where $\delta$ is real; the termination could at most introduce a phase change in the reflected wave. Two different lossless terminations that introduce the same phase shift might be called phase equivalent. The one-dimensional scattering problems to be considered subsequently will provide information on the reflection coefficient and also the transmission coefficient that occurs when a wave is incident upon some obstacle or inhomogeneity on an infinite string.

## 2.2 SCATTERING BY AN OSCILLATOR

As a simple example of a scattering process, consider a mass–spring system that is attached to a string at $x = 0$ as shown in Figure 2.1$a$. A steady wave of angular frequency $\omega$ and unit amplitude that is incident from $x = -\infty$ is partly reflected back to $-\infty$ and partly transmitted beyond the oscillator to $x = +\infty$. Indicating the string displacement by $y_>$ for $x \geqslant 0$ and $y_<$ for $x \leqslant 0$, we may write

$$y_<(x,k) = e^{ikx} + R(k)e^{-ikx}, \qquad x \leqslant 0$$
$$y_>(x,k) = T(k)e^{ikx}, \qquad\qquad x \geqslant 0 \tag{2.2.1}$$

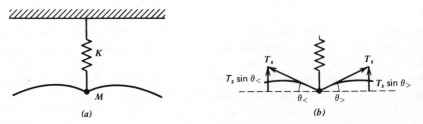

**Figure 2.1** ($a$) Mass–spring system attached to string. ($b$) Driving forces acting on mass due to vibration of string.

where again $k=\omega/c$. The string displacement is continuous at $x=0$ and equal to that of the oscillator. However, the slope of the string may be discontinuous at $x=0$. As shown in Figure 2.1$b$, the discontinuity in the slope at the origin provides the driving force for the oscillator. From Newton's second law, the displacement of the oscillator $\tilde{y}(t)$ is governed by

$$m\frac{d^2\tilde{y}}{dt^2} = -K_0\tilde{y} + T_s\left(\frac{\partial y_>}{\partial x}\bigg|_{x=0} - \frac{\partial y_<}{\partial x}\bigg|_{x=0}\right)e^{-i\omega t} \tag{2.2.2}$$

where $m$ and $K_0$ are the mass and spring constant of the oscillator, respectively. The oscillator displacement may be written $\tilde{y}=y(0,k)e^{-ikct}$, where $y(0,k)=y_<(0,k)=y_>(0,k)$. Consequently, $y(0,k)=1+R=T$. From (2.2.1) and (2.2.2), we obtain

$$y(0,k)=\frac{2i\kappa k}{k^2-k_0^2+2i\kappa k} \tag{2.2.3}$$

where $\kappa=\rho/m$ and $k_0c=\sqrt{K_0/m}$ is the resonant frequency of the oscillator. Also, the transmission and reflection coefficients are $T=y(0,k)$ given above and

$$R=T-1=\frac{k_0^2-k^2}{k^2-k_0^2+2i\kappa k} \tag{2.2.4}$$

If the frequency of the incident wave equals the resonant frequency of the oscillator then $R=0$ and $T=1$ while $R=-1$ and $T=0$ in the limit $m\to\infty$. In the limit $m\to 0$ we obtain $R=(-iK_0/2T_s)$ and $T=k/(k+iK_0/2T_s)$.

There are singularities in the expressions for the transmission and reflection coefficients at the zeros of the denominator in (2.2.3) or (2.2.4). They are located at the points

$$k=-i\kappa\pm\sqrt{k_0^2-\kappa^2} \tag{2.2.5}$$

They are in the lower half of the complex $k$ plane and are located either symmetrically about the imaginary axis (for $k_0>\kappa$) or on the imaginary axis (for $d_0<\kappa$). For $k_0=\kappa$ there is a single pole of second order at $k=-ik_0$.

Location of the zeros in the lower half-plane provides assurance that the string acts to dampen rather than amplify the motion of the oscillator. Indeed, (2.2.2) may by written in a form that corresponds to that of a driven damped oscillator. We may see this by noting that the expression for the driving force on the oscillator as given in (2.2.2) may be rewritten

$$T_s\left(\frac{\partial y_>}{\partial x}\bigg|_0 - \frac{\partial y_<}{\partial x}\bigg|_0\right)e^{-i\omega t}=2ikT_s(T-1)e^{-i\omega t} \tag{2.2.6}$$

Since $d\tilde{y}/dt = -i\omega T e^{-i\omega t}$ and $T_s = m\kappa c^2$, this driving force may be written as

$$-m\left(2\kappa c \frac{d\tilde{y}}{dt} + 2ik\kappa c^2 e^{-i\omega t}\right)$$

Hence the equation for the oscillator takes the form

$$\frac{d^2\tilde{y}}{dt^2} + 2\kappa c \frac{d\tilde{y}}{dt} + \omega_0^2 \tilde{y} = -2ik\kappa c^2 e^{-i\omega t} \qquad (2.2.7)$$

which is the equation for a driven damped oscillator. If the driving force on the oscillator were turned off, the amplitude of the oscillations would decay in a time given by $(\kappa c)^{-1}$. This is the time required for the energy stored in the oscillator to radiate away on the string.

The role played by the damping term may be seen quite simply by considering the scattering of an incident delta function pulse. The delta function may be represented as

$$\delta\left(t - \frac{x}{c}\right) = \int_{-\infty}^{\infty} \frac{d\omega}{2\pi} e^{-i\omega(t - x/c)} \qquad (2.2.8)$$

which is an example of the Fourier representation mentioned in Section 2.1. In this case, all frequencies contribute with equal amplitude so that $\Phi(\omega) = 1$. Since the problem is linear and solutions may be superposed, the scattering of the delta function pulse may be built up by merely integrating over the scattering of a sinusoidal wave at each frequency with equal amplitude. For the reflection of the delta function pulse, we then have

$$y_<(x,t) = \int_{-\infty}^{\infty} \frac{d\omega}{2\pi} y_<\left(x, \frac{\omega}{c}\right) e^{-i\omega t}$$

$$= \delta\left(t - \frac{x}{c}\right) + \int_{-\infty}^{\infty} \frac{d\omega}{2\pi} R\left(\frac{\omega}{c}\right) e^{-i\omega(t + x/c)} \qquad (2.2.9)$$

where (2.2.1) and (2.2.8) have been employed. Since $R \to 1$ as $|k| \to \infty$, the last integral in (2.2.9) contains an additional delta function. This may be extracted by writing $R = 1 - T$. When the remaining integral is evaluated by contour methods (see, for example, Hildebrand, 1976, Chapter 10), we obtain

$$y_<(x,t) = \delta\left(t - \frac{x}{c}\right) - \delta(\tau)$$

$$+ \frac{2\kappa c k_0}{\sqrt{k_0^2 - \kappa^2}} e^{-\kappa c \tau} \cos\left(c\tau\sqrt{k_0^2 - \kappa^2} + \alpha\right)\theta(\tau) \qquad (2.2.10)$$

where $\tau = t + x/c$, $\alpha = \tan^{-1}(\kappa/\sqrt{k_0^2 - \kappa^2})$, and $\theta(\tau)$ is the unit step function. Evaluating this result at $x = 0$, we see, as noted above, that the ringing of the oscillator decays in a time proportional to $(\kappa c)^{-1}$.

Use of the foregoing method to describe the scattering of waves by more complex dynamical systems would lead to rather tedious calculations. In the next section we will consider a more convenient way to describe scattering by some more elaborate systems.

## 2.3   THE ELASTICALLY BRACED STRING

If a portion of a string is embedded in a material that provides an elastic restoring force $K(x)$ per unit length of the string, the usual application of Newton's second law leads to an additional term $-K(x)y$ in (2.1.1). We obtain

$$\rho\frac{\partial^2 y}{\partial t^2} = Ts\frac{\partial^2 y}{\partial x^2} - K(x)y \tag{2.3.1}$$

We shall be interested in situations in which $K(x)$ is localized; that is, $K(x)$ either vanishes on portions of the string farther than some distance away from a central point or else decays exponentially away from a central point. A wave incident upon the elastic region may be expected to be partly reflected and partly transmitted. An obvious physical constraint upon the function $K(x)$ is that it must represent a restoring force density and hence cannot be negative.

Equation 2.3.1 may be rewritten

$$\frac{\partial^2 y}{\partial x^2} - \frac{1}{c^2}\frac{\partial^2 y}{\partial t^2} - \mu^2(x)y = 0 \tag{2.3.2}$$

where $\mu^2(x) \equiv K(x)/T_s$. Even if $\mu(x)$ is constant, the solution of this equation is not as simple as that for the equation governing the free string. The reason for this is not difficult to find. Considering the simple case in which $\mu$ has the constant value $\mu_0$, and looking for solutions at one frequency $\omega$ so that $y(x,t) = y(x,\omega)e^{-i\omega t}$, we find that $y(x,\omega)$ satisfies

$$\frac{d^2 y}{dx^2} + \left(\frac{\omega}{v}\right)^2 y = 0 \tag{2.3.3}$$

where

$$\frac{1}{v^2} = \frac{1}{c^2} - \left(\frac{\mu_0}{\omega}\right)^2 \tag{2.3.4}$$

The string displacement is then

$$y(x,t) = ae^{-i\omega(t-x/v)} + be^{-i\omega(t+x/v)} \tag{2.3.5}$$

and $v$ is seen to play the role of phase velocity for the wave. The wave motion is now dispersive, since by (2.3.4) the phase velocity $v$ of the wave is frequency-dependent. Distortionless propagation of the sort described by (2.1.2), the general solution of (2.1.1), no longer occurs.

In addition, the system now possesses a cutoff frequency since, according to (2.3.3) and (2.3.4), frequencies that are less than $\mu_0 c$ lead to an imaginary propagation constant. Such low-frequency disturbances do not propagate away as a wave at all but merely move the string up and down in phase. This leads to the possibility of localized wave motion being excited on a string with inhomogeneous elastic bracing. As a simple example consider

$$\mu^2(x) = \begin{cases} \mu_0^2, & |x| > a \\ 0, & |x| < a \end{cases} \tag{2.3.6}$$

The region $|x| < a$ could support a low-frequency standing wave that would be bounded by a nonpropagating disturbance that decays exponentially for $|x| > a$. Wave motion would then be localized to the region $|x| < a$. It should be noted that a bounded disturbance supported by a free string, that is,

$$\mu^2(x) = \begin{cases} 0, & |x| > a \\ -\mu_0^2, & |x| < a \end{cases} \tag{2.3.7}$$

is not permissible since, as noted above, $\mu^2 = K/T_s$ cannot be negative. Localized solutions resulting from negative values of $\mu^2(x)$ will be obtained when one-dimensional problems in quantum scattering and two-dimensional problems in classical wave propagation are taken up. Localized solutions will be considered after these topics have been introduced. We now consider the scattering of waves that are incident upon an inhomogeneity on the string.

As an example of the scattering of waves by an inhomogeneous elastic region, consider a string with a segment of constant elastic bracing so that

$$\mu^2 = \begin{cases} 0, & |x| > a \\ \mu_0^2, & |x| < a \end{cases} \tag{2.3.8}$$

If a wave of frequency $\omega$ and unit amplitude is incident from the negative $x$ direction, the wave motion in the three regions may be written

$$y_<(x) = e^{ikx} + R e^{-ikx}$$

$$y_>(x) = T e^{ikx}$$

$$y_i(x) = A e^{i\alpha x} + B e^{-i\alpha x} \tag{2.3.9}$$

where $k = \omega/c$ and $\alpha = \sqrt{k^2 - \mu_0^2}$. It is evident from the form of $\alpha$ that a wave does not propagate in the elastic region unless $k > \mu_0$. The four constants $A$,

$B$, $R$, and $T$ are determined by requiring continuity of string displacement and slope at $x = \pm a$. At $x = -a$ this yields

$$e^{-ika} + Re^{ika} = A\cos\alpha a - B\sin\alpha a$$

$$ik(e^{-ika} - Re^{ika}) = \alpha(A\sin\alpha a + B\cos\alpha a) \tag{2.3.10}$$

while at $x = a$, we find that

$$Te^{ika} = A\cos\alpha a + B\sin\alpha a$$

$$ikTe^{ika} = \alpha(-A\sin\alpha a + B\cos\alpha a) \tag{2.3.11}$$

Solution of these equations gives

$$Re^{2ika} = \frac{\mu_0^2}{D}\sin 2\alpha a$$

$$Te^{2ika} = \frac{2i\alpha k}{D} \tag{2.3.12}$$

where

$$D \equiv (k^2 + \alpha^2)\sin 2\alpha a + 2i\alpha k\cos 2\alpha a \tag{2.3.13}$$

As in the case of the simple mass–spring oscillator, perfect transmission ($R = 0$) can occur. This happens at the infinite number of frequencies that satisfy $\sin 2\alpha a = 0$ or $\alpha a = 0$, $\pi/2$, $\pi$, $3\pi/2,\ldots$, that is, whenever an integral number of half-wavelengths of the wave in the elastically braced medium fit into that region. As in the case of the simple mass–spring system, there are also zeros in the denominator of the expressions for $R$ and $T$. There are an infinite number of these as well. Determination of their location is facilitated by first employing the identity $\tan 2\theta = (\cot\theta - \tan\theta)^{-1}$ in the equation $D = 0$. This leads to

$$D = (k\tan\alpha a + i\alpha)(\alpha\tan\alpha a + ik) = 0 \tag{2.3.14}$$

and the zeros are obtained by setting each of these factors equal to zero separately. The first factor, which also arises in three-dimensional quantum scattering, has been thoroughly considered by Nussenzweig (1959). There are an infinite number of solutions that again all lie in the lower half of the $k$-plane. A similar procedure could be applied to the second factor. However, it will not be necessary for us to pursue the solution of these equations in any detail.

*Exercise 1*

Set $\tau^2 = t^2 - x^2/c^2$ in (2.3.2) with $\mu = \mu_0$ and show that $y$ satisfies Bessel's equation in the form $\tau\ddot{y} + \dot{y} + (\mu_0 c)^2\tau y = 0$, where the dot indicates a derivative with respect to $\tau$.

*Exercise 2*

Show that (2.3.12) implies energy conservation (i.e., $|R|^2 + |T|^2 = 1$).

*Exercise 3*

Show that in (2.3.12) the limit $a \to 0$, $\mu_0^2 \to \infty$ such that $2\mu_0^2 a = K_0/T_s$ yields $R = -i(K_0/2T_s)/(k + iK_0/2T_s)$ and $T = k/(k + iK/2T_s)$. This agrees with the previously quoted limiting forms of (2.2.3) and (2.2.4) in this limit.

*Exercise 4*

The limiting process employed in Ex. 3 is one that is frequently used in defining the delta function. The results of Ex. 3 may be obtained by initially taking $\mu_0^2(x)$ to be proportional to a delta function in the wave equation. Equation 2.3.3 then becomes

$$y'' + \left[ k^2 - \left( \frac{K_0}{T_s} \right) \delta(x) \right] y = 0$$

The amplitude of the delta function is the value of $2\mu_0^2 a$ as introduced in Ex. 3. Integration of this equation across the singularity at $x = 0$ yields

$$y'_>(0) - y'_<(0) = \left( \frac{K_0}{T_s} \right) y(0)$$

Since $y_<(x) = e^{ikx} + R e^{-ikx}$ and $y_>(x) = T e^{ikx}$, the discontinuity in the derivative yields $T(1 + iK_0c/T_s) = 1 - R$. Continuity of $y(x)$ at the origin also imposes $1 + R = T$. When these two equations are solved for $R$ and $T$, the results given in Ex. 3 are again recovered.

*Exercise 5*

Analyze the scattering by the potential $u(x) = A[\delta(x + a) - \delta(x)]$ by writing $y$ in the form used in (2.3.9). Show, by matching solutions across each delta function as in Ex. 4, that

$$R = \beta(\beta + i) \frac{1 - e^{-2ika}}{D}$$

$$T = D^{-1}$$

where $D = 1 + \beta^2(1 - e^{2ika})$ and $\beta = A/2k$.

Set $A = B/a$ and show that as $a \to 0$, so that $u(x) \to B\delta'(x)$, we obtain $R = -1$ and $T = 0$.

*Exercise 6*

Show that for $aa < < 1$, that is $k \approx \mu_0$, the second factor of (2.3.14) has the solutions $ka \approx -(i/2) \pm \sqrt{(\mu_0 a)^2 - \frac{1}{4}}$ .

## 2.4  THE SCHRÖDINGER EQUATION

Although the elastically braced string can provide considerable understanding of one-dimensional scattering, it has a severe restriction in that $K(x)$, the elastic restoring force density, cannot be negative. There is no corresponding restriction for one-dimensional scattering in quantum theory, however, since the scattering potential $V(x)$ may be either repulsive ($V > 0$) or attractive ($V < 0$). The case $V < 0$ introduces the possibility of localized disturbances or bound states.

The governing equation in quantum scattering theory is the Schrödinger equation. For one-dimensional time-independent problems it may be written (Schiff, 1949)

$$-\frac{\hbar^2}{2m}\frac{d^2\psi}{dx^2} + V(x)\psi = E\psi \qquad (2.4.1)$$

where $m$ is the mass of the particle, $E$ its total energy, and $\hbar$ is Planck's constant divided by $2\pi$. On setting $2mE/\hbar^2 = k^2$ and $2mV(x)/\hbar^2 = U(x)$, we obtain the standard equation that is used to describe steady-state one-dimensional waves in an inhomogeneous medium:

$$\frac{d^2\psi}{dx^2} + \left[k^2 - U(x)\right]\psi = 0 \qquad (2.4.2)$$

As shown in texts on quantum mechanics, the flux density associated with the particle is

$$\text{flux} = \frac{\hbar}{m}\,\text{Im}\left(\psi^*\frac{d\psi}{dx}\right) \qquad (2.4.3)$$

which should be compared with (2.1.16) for waves on a string.

Analysis of the quantum scattering problem that results for $U > 0$ will not be considered since it is the same as that for the string. For $U < 0$, bound-state or localized solutions of (2.4.2) will occur in addition to the scattering solutions. The simplest case to treat analytically is the attractive delta function potential. Setting $V(x) = V_0 a\delta(x)$ and following the procedure outlined in Ex. 4, we obtain

$$R = \frac{-iU_0 a/2}{k + iU_0 a/2}$$

$$T = \frac{k}{k + iU_0 a/2} \qquad (2.4.4)$$

where $U_0 = 2mV_0/\hbar^2$. Let us now write the scattering solution in the form that is appropriate for an incident wave of arbitrary amplitude $A$, that is,

$$y_< = Ae^{ikx} + ARe^{-ikx}$$
$$y_> = ATe^{ikx} \tag{2.4.5}$$

It is useful to observe that there could be nonvanishing solutions for $A = 0$ if $R^{-1}$ and $T^{-1}$ also vanished. From (2.4.4) it is clear that this happens if $k = -iU_0a/2$. Then $y_< \approx e^{-U_0ax/2}$ and $y_> \approx e^{U_0ax/2}$, which is a localized or bound-state solution when $U_0 < 0$. It should be noted that the pole of $R$ and $T$ at $k = -iU_0a/2$ is then in the upper half-plane. The solution obtained here could also be recovered by setting $y_> = e^{-\gamma x}$, $y_< = e^{\gamma x}$ and matching the solutions as in Ex. 4.

As a somewhat more elaborate example, consider the attractive square-well potential. The consideration of the repulsive square-well potential would be equivalent to that for an elastically braced string if we set $V(x) = \hbar^2\mu^2(x)/2m$ and use $\mu(x)$ as given in (2.3.8). Hence we need not consider this case. To analyze the role of the bound states that appear for the attractive potential

$$V(x) = \begin{cases} 0, & |x| > a \\ -V_0, & |x| < a \end{cases} \tag{2.4.6}$$

where $V_0 = |V_0|$, we first substitute $\mu_0^2 = -2mV_0/\hbar^2$ in the results of Section 2.3. We then obtain $\alpha = \sqrt{2m(E + V_0)/\hbar^2}$, which is real for negative values of $E$ as long as $E > -V_0$. The associated values of $k$ are the purely imaginary ones, $k = \sqrt{-2mE/\hbar^2} = i\kappa$.

The allowed values of $k$ may be obtained in either of two ways. We may proceed directly to the bound-state solutions $e^{-\kappa|x|}$ for $|x| > a$ and repeat the matching of solutions outlined in Section 2.3 or, as shown above for the delta function potential, we may locate the points in the upper half of the $k$ plane for which $R$ and $T$ have vanishing denominators. The former approach is the one employed in most introductory texts in quantum mechanics. For the latter approach we need merely set $\alpha = \sqrt{k^2 + 2mV_0/\hbar^2}$ in the expression for $D(k)$ given in (2.3.13). Again the factored forms as in (2.3.14) are more easily analyzed. A complete analysis of the first of these expressions has also been carried out by Nussenzweig (1959). In addition to a certain number of bound states (depending on the value of $V_0$) there are again an infinite number of complex zeros in the lower half of the complex $k$ plane as well as so-called antibound states on the imaginary axis in the lower half-plane.

## 2.5  SCATTERING BY A SECH$^2$ POTENTIAL

Solutions to the Schrödinger equation for potentials other than either the delta function or square-well potentials provide rather extensive exercises in the use of special functions. The case of the sech$^2$ potential will be considered here since it will play an important role in later developments. Also, the scattering process is expressed in terms of linear combinations of solutions in a way that appears later in a more general way.

Consider the attractive potential

$$V(x) = -V_0 \operatorname{sech}^2\left(\frac{x}{d}\right), \qquad V_0 > 0 \qquad (2.5.1)$$

Setting $\epsilon = 2md^2E/\hbar^2$, $v = 2md^2V_0/\hbar^2$, and $z = x/d$ in the Schrödinger equation (2.4.1), we obtain

$$\psi'' + (\epsilon + v\operatorname{sech}^2 z)\psi = 0 \qquad (2.5.2)$$

where the prime indicates differentiation with respect to $z$. This equation may be transformed to the hypergeometric equation (Morse and Feshbach, 1953, p. 1651). First, setting

$$\psi = A \operatorname{sech}^\beta z \cdot y(x) \qquad (2.5.3)$$

where $A$ is an arbitrary amplitude, we find that if $\beta^2$ is chosen to equal $-\epsilon$, then $y(z)$ satisfies

$$y'' - 2\beta \tanh z \cdot y' + (v - \beta^2 - \beta)\operatorname{sech}^2 z \cdot y = 0 \qquad (2.5.4)$$

The subsequent transformation $u = \frac{1}{2}(1 - \tanh z) = e^{-z}/(e^z + e^{-z})$ yields

$$u(1-u)\frac{d^2y}{du^2} + [c - (a+b+1)u]\frac{dy}{du} - aby = 0 \qquad (2.5.5)$$

with

$$c = 1 + \beta, \qquad a + b + 1 = 2(1 + \beta), \qquad ab = \beta^2 + \beta - v \qquad (2.5.6)$$

Equation 2.5.5 is the standard form for the hypergeometric equation. The solution that remains finite at $u = 0$ (i.e., as $z$ approaches $\infty$) is usually written (Abramowitz and Stegun, 1964)

$$y = F(a, b; c; u) = 1 + \frac{ab}{c}u + \cdots \qquad (2.5.7)$$

As $z \to \infty$ (i.e., as $u \to 0$), we have $y \to 1$. Thus (2.5.3) becomes

$$\psi \underset{z \to \infty}{\to} A2^\beta e^{-\beta z} \qquad (2.5.8)$$

For this to represent a plane wave $Ae^{ikx}$ going to $+\infty$ we must set $\beta = -ikd$. According to the previous definition of $\beta$, we then have $k = \sqrt{2mE/\hbar^2}$. Solving (2.5.6) for $a$, $b$, and $c$ and using the definition of $\beta$, we find that

$$a = \tfrac{1}{2} - ikd + \sqrt{v + \tfrac{1}{4}}$$

$$b = \tfrac{1}{2} - ikd - \sqrt{v + \tfrac{1}{4}} \qquad (2.5.9)$$

$$c = 1 - ikd$$

Although the solution $F(a,b;c;u)$ is valid for $0 \leqslant u \leqslant 1$ (i.e., $\infty > z > -\infty$), it is more readily understood in the region $u \to 1$ (i.e., $z \to -\infty$) by writing it as a linear combination of the two solutions that are expressed in terms of expansions about $u = 1$. The appropriate combination, which is one of the standard identities among hypergeometric functions, is (Abramowitz and Stegun, 1964, p. 559)

$$F(a,b;c;u) = \frac{\Gamma(c)\Gamma(c-a-b)}{\Gamma(c-a)\Gamma(c-b)} F(a,b;\ a+b-c+1;1-u)$$

$$+ (1-u)^{c-a-b} \frac{\Gamma(c)\Gamma(a+b-c)}{\Gamma(a)\Gamma(b)} F(c-a,c-b;c-a-b+1;1-u)$$

$$(2.5.10)$$

where $\Gamma$ refers to the gamma function (Abramowitz and Stegun, 1964, pp. 253 and 559). We now show that this relation actually provides the solution to the scattering problem. Noting that $(1-u)^{c-a-b} = e^{-2\beta z}$ and that according to (2.5.7), the hypergeometric functions with argument $1-u$ will approach unity as $u \to 1$, we find that (2.5.3) and (2.5.10) yield

$$\psi \underset{z \to -\infty}{\to} A2^{\beta} \left[ \frac{\Gamma(c)\Gamma(c-a-b)}{\Gamma(c-a)\Gamma(c-b)} e^{\beta z} + \frac{\Gamma(c)\Gamma(a+b-c)}{\Gamma(a)\Gamma(b)} e^{-\beta z} \right] \quad (2.5.11)$$

With $\beta = -ikd$ as noted above and writing (2.5.11) in the form

$$\psi \underset{z \to -\infty}{\to} A2^{\beta} \frac{\Gamma(c)\Gamma(a+b-c)}{\Gamma(a)\Gamma(b)} \left[ e^{ikx} + \frac{\Gamma(c-a-b)}{\Gamma(c-a)\Gamma(c-b)} \cdot \frac{\Gamma(a)\Gamma(b)}{\Gamma(a+b-c)} e^{-ikx} \right]$$

$$(2.5.12)$$

we can immediately determine the reflection coefficient and the incident wave amplitude. Returning to (2.5.8) and factoring out the same incident wave amplitude, the transmission coefficient is also determined. The results are

$$R = \frac{\Gamma(c-a-b)\Gamma(a)\Gamma(b)}{\Gamma(c-a)\Gamma(c-b)\Gamma(a+b-c)}$$

$$T = \frac{\Gamma(a)\Gamma(b)}{\Gamma(c)\Gamma(a+b-c)} \qquad\qquad (2.5.13)$$

Using the definitions of $a$, $b$, and $c$ given in (2.5.9) plus the identity $\Gamma(\frac{1}{2}-z)\Gamma(\frac{1}{2}+z) = \pi/\cos \pi z$, we find that $R$ is proportional to $\cos(\pi\sqrt{v+\frac{1}{4}}\,)$. The potential is thus reflectionless for $\sqrt{v+\frac{1}{4}} = n+\frac{1}{2}$, where $n=1,2,\dots$. It should be noted that this result holds for incident particles of any energy (i.e., classical waves of any frequency).

Also, both $R$ and $T$ have poles in the $k$ plane for $\frac{1}{2}-ikd\pm\sqrt{v+\frac{1}{4}} = -p$, where $p=0,1,2,\dots$. Those in the upper half of the complex $k$ plane are the positive solutions of

$$kd = i\left[\sqrt{v+\frac{1}{4}} - \left(p+\frac{1}{2}\right)\right] \qquad\qquad (2.5.14)$$

This imposes an upper limit on the value of $p$. If in addition we restrict consideration to reflectionless potentials, then $\sqrt{v+\frac{1}{4}} = n+\frac{1}{2}$ and $kd = i(n-p)$, where $p=0,1,\dots,n-1$.

In summary, reflectionless potentials and their eigenvalues are associated with the Schrödinger equation

$$\frac{d^2\psi}{dz^2} + \left[-(n-p)^2 + n(n+1)\operatorname{sech}^2 z\right]\psi = 0, \qquad \begin{matrix} n=1,2,\dots \\ p=0,1,\dots,n-1 \end{matrix} \qquad (2.5.15)$$

The eigenfunctions can be obtained as limiting cases of the hypergeometric function. An example is considered below in Ex. 9. However, a simpler technique for generating these eigenfunctions will be described in the next section.

*Exercise 7*

Show that $R$ and $T$ given in (2.5.13) satisfy $|R|^2 + |T|^2 = 1$.

*Exercise 8*

In the limit $V_0 \to \infty$, $d \to 0$ such that $V_0 d = $ constant, the potential given in (2.5.1) becomes $-2V_0\delta(x/d)$. Show that the reflection and transmission coefficients for the delta function potential are recovered from the corresponding limiting forms of $R$ and $T$ given in (2.5.13).

*Exercise 9*

For reflectionless potentials, so that $v = n(n+1)$, the transformation $u = \tanh z$ brings (2.5.2) to the standard form for the equation defining associated Legendre functions $P_\nu^\mu(u)$ (Abramowitz and Stegun, 1964, p. 331). The appropriate solution of the Schrödinger equation is then

$$\psi(z) \approx P_n^{ikd}(\tanh z)$$

Use the relations (Magnus and Oberhettinger, 1954, p. 63)

$$P_0^\mu(\cos\theta) = \frac{1}{\Gamma(1-\mu)}\left(\cot\frac{\theta}{2}\right)^\mu, \qquad P_{-\nu-1}^\mu(x) = P_{\nu}^\mu{}_{(x)}$$

and the recurrence relation

$$(2\nu+1)xP_\nu^\mu(x) = (\nu-\mu+1)P_{\nu+1}^\mu(x) + (\nu+\mu)P_{\nu-1}^\mu(x)$$

to show that for $n = 1$ the wave function reduces to

$$\psi \approx e^{ikx}\left[ikd - \tanh\left(\frac{x}{d}\right)\right]$$

## 2.6 ASSOCIATED STURM–LIOUVILLE EQUATIONS

In the last section, the Schrödinger equation for the potential $-V_0\operatorname{sech}^2(x/d)$ was solved in terms of hypergeometric functions. Certain values of the potential strength $V_0d^2$ led to reflectionless potentials. In these cases the solution can be expressed in terms of elementary functions. As indicated in Ex. 9, these solutions can be obtained by considering appropriate limiting forms of the general solution. However, since these elementary cases will subsequently be among our main concerns, it is desirable to have a more direct method for treating them. The equations that arise when the potential is reflectionless may be solved in a very simple manner by using a technique devised by Darboux (1882; 1915, Vol. II, p. 210; Ince, 1926, p. 132). Besides yielding solutions to certain specific equations, the method may be used to effect changes in potentials of arbitrary shape (Crum, 1955; Faddeev, 1963; Wadati et al., 1975).

Darboux's procedure was to investigate the relation between the two equations

$$y'' = \left[\lambda + u(x)\right]y \tag{2.6.1}$$

and

$$z'' = \left[\lambda + v(x)\right]z \tag{2.6.2}$$

when $z$ is related to $y$ through the linear combination

$$z = A(x,\lambda)y + B(x,\lambda)y' \qquad (2.6.3)$$

The analysis is especially simple, and adequate for our purposes, when $B(x,\lambda)$ is set equal to unity. Only this case will be considered. When (2.6.3) is substituted into (2.6.2), and the coefficients of $y$ and $y_x$ are separately equated to zero, we obtain

$$A_{xx} + u_x + A(u-v) = 0 \qquad (2.6.4)$$

$$2A_x + u - v = 0 \qquad (2.6.5)$$

Eliminating $(u-v)$ and integrating, we find that

$$A^2 - A_x - u = \tilde{\lambda} \qquad (2.6.6)$$

where $\tilde{\lambda}$ is the integration constant.

*Exercise 10*

Since there is no first derivative term in either (2.6.1) ot (2.6.2), the Wronskians for solutions of these equations must be constant. A relation between these Wronskians is imposed by (2.6.3). Show that when $B=1$, the constancy of the Wronskians leads to (2.6.6).

Equation 2.6.6 is a Riccati equation for $A$ and may be linearized by the substitution $A = -\tilde{y}'/\tilde{y}$ (cf. Section 2.12). Equation 2.6.6 then becomes

$$\tilde{y}'' = \left[\tilde{\lambda} + u(x)\right]\tilde{y} \qquad (2.6.7)$$

Thus, the significance of $\tilde{y}$ is that it is a particular solution of (2.6.1) when we set $\lambda = \tilde{\lambda}$.

From (2.6.5) we find that $v = u - 2(\ln\tilde{y})''$, so the equation for $z$ becomes

$$z'' = \left[\lambda + u - 2(\ln\tilde{y})''\right]z \qquad (2.6.8)$$

We thus obtain a Schrödinger equation in which the potential has been changed by the amount $\Delta u = -2(\ln\tilde{y})''$.

Also, the identity $2(\ln f)'' = f''/f - f(1/f)''$ yields

$$z'' = \left[\lambda - \tilde{\lambda} + \tilde{y}(1/\tilde{y})''\right]z \qquad (2.6.9)$$

It is sometimes convenient to note that for $\lambda = \tilde{\lambda}$, a particular solution of (2.6.9) is $z = 1/\tilde{y}$.

In summary, then, if $y$ is the general solution of the equation

$$y'' = [\lambda + u(x)] y \qquad (2.6.10)$$

and $\tilde{y}$ is any particular solution corresponding to the value $\lambda = \tilde{\lambda}$, then

$$z = y' - y\left(\frac{\tilde{y}'}{\tilde{y}}\right) \qquad (2.6.11)$$

is the general solution of the equation

$$z'' = \left[\lambda - \tilde{\lambda} + \tilde{y}\left(\frac{1}{\tilde{y}}\right)''\right] z \qquad (2.6.12)$$

or, equivalently,

$$z'' = [\lambda + u - 2(\ln \tilde{y})''] z \qquad (2.6.13)$$

As a simple example, consider (2.6.10) with $u = 0$. The general solution of the resulting equation is $y = A e^{\sqrt{\lambda}\, x} + B e^{-\sqrt{\lambda}\, x}$. As a particular solution, take $\lambda = 1$ and $A = B = \frac{1}{2}$ so that $\tilde{y} = \cosh x$. Since $2(\ln y)''$ equals $2\,\mathrm{sech}^2 x$ in this case, (2.6.11) and (2.6.13) give the general solution of

$$z'' = (\lambda - 2\,\mathrm{sech}^2 x) z \qquad (2.6.14)$$

in the form

$$z = A e^{\sqrt{\lambda}\, x}(\sqrt{\lambda} - \tanh x) - B e^{-\sqrt{\lambda}\, x}(\sqrt{\lambda} + \tanh x) \qquad (2.6.15)$$

Equation 2.6.14 is equivalent to (2.5.15) for the case $n = 1$. The solution obtained in Ex. 9 is of the form obtained here. The method may be extended to higher values of $n$ as indicated in the following example.

*Exercise 11*

Consider the equation $y_n'' = [\lambda - n(n+1)\,\mathrm{sech}^2 x] y_n$ and note that $y = \cosh^{n+1} x$ is a particular solution for $\lambda = (n+1)^2$. Show that the general solution for $n = 2$ is

$$y_2 = \left(\frac{d}{dx} - 2\tanh x\right)\left(\frac{d}{dx} - \tanh x\right) y_0$$

where $y_0 = A e^{\sqrt{\lambda}\, x} + B e^{-\sqrt{\lambda}\, x}$. In general, $y_p = O_p O_{p-1} \cdots O_1 y_0$ where $O_p = d/dx - p \tanh x$.

The method also provides a technique for changing a potential so as to add an additional eigenvalue (bound state). As a specific example, note that the equation

$$y_\lambda'' = [\lambda - n(n+1)\,\mathrm{sech}^2 x] y_\lambda \qquad (2.6.16)$$

has $n$ eigenvalues. According to (2.6.13), the increment in the potential $\Delta u = -2(\ln \tilde{y})''$ would increase the number of eigenvalues by one if we were to set $\tilde{y} = \cosh^{n+1} x$. In that case $\Delta u = -2(n+1) \operatorname{sech}^2 x$ and (2.6.13) would become

$$z'' = \left[ \lambda - (n+1)(n+2) \operatorname{sech}^2 x \right] z \tag{2.6.17}$$

which is the form of (2.6.16) for $n+1$ eigenvalues.

The form chosen above for $\tilde{y}^{-1}$ may be derived as follows. The equation with $n$ eigenvalues is written in the form

$$y''_{k,n} = \left[ k^2 - n(n+1) \operatorname{sech}^2 x \right] y_{k,n} \tag{2.6.18}$$

and the equation for $n+1$ eigenvalues as

$$z''_{k,n+1} = \left[ k^2 - \tilde{k}^2 + \tilde{y}^{-1} \left( \frac{1}{\tilde{y}} - 1 \right)'' \right] z_{k,n+1} \tag{2.6.19}$$

where $k = n+1$ is an eigenvalue of (2.6.19) but not of (2.6.18). Setting $k = \tilde{k} = n+1$ in (2.6.18), we obtain $z_{n+1,n+1} = \tilde{y}^{-1}$. But the equation for $z_{n+1,n+1}$ has the particular solution $\operatorname{sech}^{n+1} x$. Hence $\tilde{y}^{-1} = \cosh^{n+1} x$, as required. Similar considerations may be applied to arbitrary potentials (Wadati et al., 1975).

*Exercise 12*

The method may also be used to generate the bound-state solutions directly. Use the fact that $y''_{m,n} = [m^2 - n(n+1) \operatorname{sech}^2 x] y_{m,n}$ has solutions $y_{n,n} = \operatorname{sech}^n x$ and $y_{n,n-1} = \cosh^n x$ to show that

$$y_{n,n+1} = \left[ \frac{d}{dx} - (n+1) \tanh x \right] \operatorname{sech}^n x$$

$$y_{n,n+2} = \left[ \frac{d}{dx} - (n+2) \tanh x \right] y_{n,n+1}$$

For $n=2$ obtain the result $y_{1,2} = \tanh x \operatorname{sech} x$.

An alternative procedure that also provides a simple approach to the results of this section is the factorization method (Infeld and Hull, 1951; Morse and Feshbach, 1953, p. 729).

## 2.7 TWO-DIMENSIONAL WAVES IN AN INHOMOGENEOUS MEDIUM

Bound-state solutions can also appear quite naturally in the classical context of two- or three-dimensional wave propagation in an inhomogeneous medium. They appear whenever there is a region of the medium to which a wave may be confined by refraction. This can happen if there is a layer of the medium at which a minimum in the phase velocity of the wave occurs. An example from underwater sound propagation is shown in Figure 2.2. The effect is most easily understood in terms of a ray acoustic representation. For simplicity, the sound field is assumed to be generated by a source located at the depth of the minimum in the sound velocity. Rays that leave the source at a steep-enough angle with respect to the horizontal, such as $A$ in Figure 2.2, will radiate to $z = \pm\infty$ while rays at sufficiently small angles to the horizontal, such as $B$ and $C$, will be continually refracted back to the depth of the source and thus will be channeled so as to propagate in a horizontal direction.

For two-dimensional waves at a single frequency $\nu = \omega/2\pi$, the wave equation is

$$\frac{\partial^2 p}{\partial x^2} + \frac{\partial^2 p}{\partial z^2} + \left[\frac{\omega}{c(z)}\right]^2 p = 0 \tag{2.7.1}$$

where $p(x,z)$ is the sound pressure. This equation is valid provided the phase velocity of the wave $c(z)$ does not change too rapidly in a wavelength (Brekhovskikh, 1960, p. 171; Bergmann, 1946). It is convenient to write the phase velocity in the form

$$c^2(z) = \frac{c_0^2}{1 + \mu^2(z)} \tag{2.7.2}$$

where $\mu$ goes to zero as $z \to \pm\infty$. Then the standard separation of variables procedure leads to solutions of (2.7.1) in the form

$$p(x,z) = e^{ik_x x} v(z) \tag{2.7.3}$$

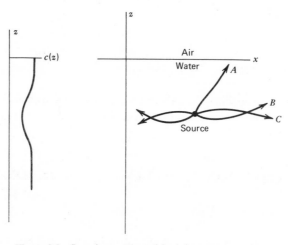

**Figure 2.2**  Sound rays trapped by inhomogeneous layer.

The function $v(z)$ is found to satisfy the Sturm–Liouville equation

$$\frac{d^2v}{dz^2} + \left[ k^2 - U(z) \right] v = 0 \qquad (2.7.4)$$

where $k^2 = (\omega/c_0)^2 - k_x^2$ and $U(z) = -[\omega\mu(z)/c_0]^2$. For $\mu^2 > 0$, this is equivalent to a Schrödinger equation for an attractive potential. Therefore, bound-state solutions, that is, solutions localized about the depth at which $\mu^2(z)$ has a maximum, can occur in this classical context. This happens at values of $k_x$ such that the constant $k^2$ is an eigenvalue of the Sturm–Liouville equation (2.7.4). It is clear on physical grounds that no such channeling can take place about a depth at which the sound velocity has a maximum (i.e., $\mu^2 < 0$).

It is frequently useful to express the solution to a partial differential equation as an expansion in terms of a set of eigenfunctions. For certain problems this set of eigenfunctions is not complete unless it includes two types of functions. One type may be interpreted as scattering solutions that propagate to infinity, whereas the other type has the significance of localized or bound state solutions. Since the sound channeling problem introduced here provides an ideal opportunity for examining the role played by bound-state solutions in eigenfunction expansions, we now take up a somewhat lengthy presentation of a specific problem in which the role played by the bound state solutions may be understood quite readily.

Let us consider in some detail the problem of a two-dimensional line source of strength $Q$ and frequency $\omega$ that is located at the depth of the velocity minimum in an infinite medium. The analysis is quite simple if $\mu^2(z)$ in (2.7.2) is given by $2\operatorname{sech}^2(\omega z/c)$. Writing $k_0$ for $\omega/c$, the governing equation is

$$\frac{\partial^2 p}{\partial x^2} + \frac{\partial^2 p}{\partial z^2} + k_0^2(1 + 2\operatorname{sech}^2 k_0 z)p = -Q\delta(x)\delta(z) \qquad (2.7.5)$$

with the radiation boundary condition that there be no incoming waves as $x$ and $z$ approach $\pm\infty$. The first and most obvious approach to solving this problem is to introduce a Fourier transform representation for the sound field by writing

$$p(x,z) = \int_{-\infty}^{\infty} \frac{dk_x}{2\pi} e^{ik_x x}\bar{p}(k_x,z) \qquad (2.7.6)$$

Also expressing $\delta(x)$ in the integral form $\delta(x) = \int_{-\infty}^{\infty}(dk_x/2\pi)e^{ik_x x}$, we find that the transform variable $\bar{p}(k_x,z)$ satisfies

$$\frac{d^2\bar{p}}{dz^2} + \left(k_0^2 - k_x^2 + 2k_0^2\operatorname{sech}^2 k_0 z\right)\bar{p} = -Q\delta(z) \qquad (2.7.7)$$

This inhomogeneous equation may be solved by integrating the equation across the singularity at $z=0$ as outlined in Ex. 4. Referring to the transform of the sound pressure in the upper and lower half-spaces as $\bar{p}_>$ and $\bar{p}_<$, respectively, we obtain

$$\left.\frac{d\bar{p}_>}{dz}\right|_0 - \left.\frac{d\bar{p}_<}{dz}\right|_0 = -Q \qquad (2.7.8)$$

For $z \neq 0$, $\bar{p}$ satisfies a homogeneous equation of the form of (2.6.14). Since $\bar{p}$ must contain only waves moving away from the source, the solutions $\bar{p}_>$ and $\bar{p}_<$ are seen from (2.6.15) to be

$$\bar{p}_> = A(k_x) e^{i\eta z} (i\eta - k_0 \tanh k_0 z)$$

$$\bar{p}_< = B(k_x) e^{-i\eta z} (i\eta + k_0 \tanh k_0 z) \qquad (2.7.9)$$

where $\eta = \sqrt{k_0^2 - k_x^2}$ . The sound field must be continuous at $z = 0$, so $\bar{p}_>(k_x,0) = \bar{p}_<(k_x,0)$ and therefore $A(k_x) = B(k_x)$.

From (2.7.8) we find that

$$A(k_x) = \frac{Q}{2(2k_0^2 - k_x^2)} \qquad (2.7.10)$$

Returning to (2.7.6), we see that the sound field is

$$p(x,z) = \frac{Q}{4\pi} \int_{-\infty}^{\infty} dk_x e^{ik_x x + i\eta|z|} \frac{i\eta - k_0 \tanh k_0 |z|}{2k_0^2 - k_x^2} \qquad (2.7.11)$$

We thus obtain a Fourier integral representation for the sound field. The solution is merely an integral over the complete set of functions $e^{ik_x x}$.

Instead of expressing the sound field in terms of an integral over the wave function for the $x$ coordinate as in (2.7.11), the solution may be recast in terms of an integral over the solutions of (2.6.14), the wave functions for the $z$ coordinate. In addition to an integral over the propagating solutions, we shall find a contribution that is proportional to the bound-state solution $\operatorname{sech} k_0 z$. Only when this bound-state contribution is included do we have a complete set of functions. To derive this result from the solution given above, let us replace (2.7.11) by a contour integral. For $x$ and $z$ positive, we can close the contour in the upper half of a complex $k_x$ plane provided that $\operatorname{Im}\eta > 0$. Because of the term $\exp(i\eta z)$, the contribution from the arc at infinity will then vanish exponentially. Since $\eta = \sqrt{k_0^2 - k_x^2}$ , the integrand is multiple-valued and we must choose the branch cut so as to guarantee $\operatorname{Im}\eta > 0$ everywhere. The appropriate choice of contour is easily determined if we first consider $k_0$ to be slightly complex so that $k_0 \rightarrow k_0 + i\epsilon$. [This would result either from the introduction of a small amount of dissipation into the medium or from the assumption that the time dependence of the problem is switched on gradually (adiabatically) by replacing $e^{-i\omega t}$ by $e^{-i(\omega+i\epsilon)t}$.] Setting $k_x = \sigma + i\tau$, we have $\eta = \eta_r + i\eta_i = \sqrt{(k_0 + i\epsilon)^2 - (\sigma + i\tau)^2}$. Squaring and separating real and imaginary parts, we obtain

$$\eta_r^2 - \eta_i^2 = k_0^2 - \epsilon^2 - \sigma^2 + \tau^2$$
$$\eta_r \eta_i = k_0 \epsilon - \sigma \tau \qquad (2.7.12)$$

Since we require $\eta_i > 0$, we first determine the locus of points for which $\eta_i = 0$. From the second of (2.7.12) this locus is seen to be the equilateral hyperbola $\sigma\tau = k_0\epsilon$. From the first of (2.7.12) we see that on this hyperbola $\eta_r$ will be zero when $\sigma = k_0$ and $\tau = \epsilon$. Since $\eta_r^2$ is positive, we must confine $\sigma$ and $\tau$ to the portion of the hyperbola for which $\sigma < k_0$ and $\tau > \epsilon$. The branch cut $\Gamma$ in the first quadrant thus proceeds from $(\sigma, \tau) = (k_0, \epsilon)$ to $(\sigma, \tau) = (0, \infty)$ as shown in Figure 2.3. The function $\eta$ will thus be real along $\Gamma$ but discontinuous across $\Gamma$. To determine the nature of the discontinuity, note that if we move off of $\Gamma$ to a point above $\Gamma$ by either increasing $\sigma$ at constant $\tau$ or increasing $\tau$ at constant $\sigma$, then the right-hand side of (2.7.12) becomes negative. The left-hand side must also be negative and since $\eta_i > 0$, we see that $\eta_r$ must be negative above $\Gamma$. Similarly, $\eta_r < 0$ below $\Gamma$.

The integrand also contains poles at $k_x = \pm\sqrt{2}\,k_0$. Again by replacing $k_0$ by $k_0 + i\epsilon$, the poles may be moved into the first and third quadrants. The contour for evaluating the integral in (2.7.11) is shown in Figure 2.3. The Fourier representation of the sound field may therefore be rewritten as

$$p(x,z) = 2\pi i \left(\text{residue of pole at } k_x = \sqrt{2}\,k_0\right)$$
$$-\frac{Q}{4\pi}\int_\Gamma dk_x e^{ik_x x + i\eta z}\frac{i\eta - \tanh k_0 z}{2k_0^2 - k_x^2} \qquad (2.7.13)$$

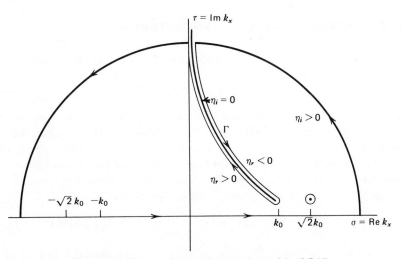

**Figure 2.3** Contour for transforming integral in (2.7.11).

The contribution from the pole is $(iQ/4\sqrt{2})e^{i\sqrt{2}\,k_0 x}\operatorname{sech} k_0 z$, while the integral around the branch cut is an integral with $\eta$ varying from $\eta = -\infty$ to $\eta = 0$ along the upper side of the cut plus an integral from $\eta = 0$ to $\eta = +\infty$ along the lower side of the cut. Converting to an integral over $\eta$ by writing $k_x = \sqrt{k_0^2 - \eta^2}$ and $dk_x = -\eta\,d\eta/k_x$, we finally obtain

$$p(x,z) = \frac{Q}{4\sqrt{2}}\, e^{i\sqrt{2}\,k_x x + i\pi/2}\operatorname{sech} k_0 z$$

$$+ \frac{Q}{4\pi}\int_{-\infty}^{\infty} d\eta\, e^{i\eta z}(i\eta - k_0\tanh k_0 z)F(\eta) \qquad (2.7.14)$$

where

$$F(\eta) = \frac{\eta e^{ix\sqrt{k_0^2 - \eta^2}}}{\sqrt{k_0^2 - \eta^2}\,(k_0^2 + \eta^2)} \qquad (2.7.15)$$

As mentioned above, the functions $e^{i\eta z}(i\eta - k_0\tanh k_0 z)$ with $\eta$ real do not of themselves constitute a complete set of functions. We have seen that the bound-state contribution $\operatorname{sech} k_0 z$ must also be included. Note that it may be obtained by evaluating the function $e^{i\eta z}(i\eta - k_0\tanh k_0 z)$ at $\eta = ik_0$.

*Exercise 13*

Show that the bound-state solution is orthogonal to the propagating solutions, that is,

$$\int_{-\infty}^{\infty} dz\operatorname{sech} k_0 z(i\eta - k_0\tanh k_0 z)e^{i\eta z} = 0$$

## 2.8   A GENERAL APPROACH TO SCATTERING

For localized potentials $u(x)$, all solutions of the Schrödinger equation

$$y'' + [k^2 - u(x)]y = 0 \qquad (2.8.1)$$

will reduce to a linear combination of the functions $e^{\pm ikx}$ as $x$ approaches $\pm\infty$. For scattering problems on the infinite interval it has become customary (in this section we follow the development of Faddeev, 1958, 1967) to express all solutions of the Schrödinger equation as linear combinations of a solution $f_1(x,k)$ that reduces to $e^{ikx}$ as $x$ approaches $+\infty$ and another solution $f_2(x,k)$ that reduces to $e^{-ikx}$ as $x$ approaches $-\infty$, that is,

$$\lim_{x\to+\infty}\left[e^{-ikx}f_1(x,k)\right] = 1; \qquad \lim_{x\to-\infty}\left[e^{ikx}f_2(x,k)\right] = 1 \qquad (2.8.2)$$

These solutions are frequently referred to as the fundamental solutions of the Schrödinger equation.

## Fundamental Solutions

Although the fundamental solutions cannot be determined explicitly for an arbitrary potential, many of their properties can be ascertained by examining certain integral equations that they satisfy. These integral equations may be obtained by treating the term $u(x)y$ in (2.8.1) as an inhomogeneous term on the right-hand side of that equation and then using the method of variation of parameters. Considering the two cases that have the limiting forms listed in (2.8.2), we readily find (see Ex. 14) that

$$f_1(x,k) = e^{ikx} - \frac{1}{k}\int_x^\infty dx' \sin k(x-x')u(x')f_1(x',k) \qquad (2.8.3)$$

$$f_2(x,k) = e^{-ikx} + \frac{1}{k}\int_{-\infty}^x dx' \sin k(x-x')u(x')f_2(x',k) \qquad (2.8.4)$$

When both $u(x)$ and $k$ are real, we have the equalities $f_1(x, -k) = f_1^*(x,k)$ and $f_2(x, -k) = f_2^*(x,k)$, where the asterisk indicates complex conjugate. These equalities follow from the fact that the functions satisfy the same integral equation and have the same asymptotic form.

If we were to begin to solve these integral equations by an iteration procedure, for example by substituting $e^{ikx}$ for $f_1(x,k)$ in (2.8.3) and so on, we would find that the resulting integrals would converge for $\text{Im}\,k > 0$. Furthermore, it is known that for integral equations of Volterra type such as those in (2.8.3) and (2.8.4), the resulting series expansion is always convergent (Lovitt, 1924, p. 13). Hence $f_1(x,k)$ is an analytic function in the upper half of the complex $k$ plane. The function $f_2(x,k)$ is similarly found to be analytic in the upper half of the $k$ plane. For $\text{Im}\,k > 0$, the functions $f_1$ and $f_2$ thus satisfy (2.8.2) and (2.8.3), respectively.

### Exercise 14

Obtain the fundamental solutions given in (2.8.3) and (2.8.4) by the variation-of-parameters method. More specifically, write the solution of (2.8.1) in the form $y = A(x)e^{ikx} + B(x)e^{-ikx}$ and show that one may obtain $A(x) = (2ik)^{-1}\int_0^x dx' uy e^{-ikx'} + C_1$, $B(x) = (2ik)^{-1}\int_0^x dx' uy e^{ikx'} + C_2$. For $f_1(x,k)$ obtain $C_1 = 1 - (2ik)^{-1}\int_0^\infty dx' uy e^{-ikx'}$ and $C_2 = (2ik)^{-1}\int_0^\infty dx' uy e^{ikx'}$. For $f_2(x,k)$ obtain $C_1 = (2ik)^{-1}\int_{-\infty}^0 dx' uy e^{-ikx'}$ and $C_2 = 1 - (2ik)^{-1}\int_{-\infty}^0 dx' uy e^{ikx'}$. The corresponding functions $A(x)$ and $B(x)$ now yield the results given in (2.8.3) and (2.8.4).

### Wronskian Relations

The Wronskian of any two solutions of the Schrödinger equation (2.8.1) will be defined as

$$W(y_1; y_2) \equiv y_1' y_2 - y_1 y_2' \qquad (2.8.5)$$

(which, for later convenience, differs by a minus sign from the usual definition). The Schrödinger equation (2.8.1) contains no first derivative term. Therefore, the Wronskian of any two linearly independent solutions is a constant that may depend upon $k$. Since the fundamental solutions $f_1(x,k)$ and $f_2(x,k)$ take on simple limiting forms as indicated in (2.8.2), certain Wronskian relations are readily evaluated. In particular, evaluating $f_1$ as $x \to + \infty$ and $f_2$ as $x \to - \infty$, we obtain

$$W\big[ f_1(x,k); f_1(x, -k) \big] = 2ik, \qquad W\big[ f_2(x,k); f_2(x, -k) \big] = -2ik$$

$$(2.8.6)$$

Since any third solution can be expressed as a linear combination of two linearly independent solutions, we may write

$$f_2(x,k) = c_{11}(k) f_1(x,k) + c_{12}(k) f_1(x, -k) \qquad (2.8.7a)$$

$$f_1(x,k) = c_{21}(k) f_2(x, -k) + c_{22}(k) f_2(x,k) \qquad (2.8.7b)$$

From the limiting forms of the fundamental solutions, (2.8.7a) represents a solution to the Schrödinger equation that reduces to $e^{-ikx}$ as $x$ approaches $- \infty$ and to the linear combination $c_{11} e^{ikx} + c_{12} e^{-ikx}$ as $x$ approaches $+ \infty$. This solution therefore corresponds to a scattering problem in which a wave of amplitude $c_{12}$ is incident from $x = + \infty$ upon the scattering potential $u(x)$. The wave is reflected with an amplitude $c_{11}$ and transmitted to $- \infty$ with an amplitude of unity as indicated in Figure 2.4$a$. Similarly, (2.8.7b) represents a scattering solution in which the incident wave comes from $- \infty$, as shown in Figure 2.4$b$. [The relation between the various solutions is analogous to that encountered in (2.5.10) for the hypergeometric functions.]

In terms of the customary reflection and transmission coefficients for an incident wave of unit amplitude, we have

$$R_R(k) = \frac{c_{11}(k)}{c_{12}(k)}, \qquad T_R(k) = \frac{1}{c_{12}(k)} \qquad (2.8.8)$$

where the subscript $R$ refers to a wave incident from the right. Similarly, the ratios

$$R_L(k) = \frac{c_{22}(k)}{c_{21}(k)}, \qquad T_L(k) = \frac{1}{c_{21}(k)} \qquad (2.8.9)$$

refer to a wave incident from the left with unit amplitude.

There are a number of relations among the coefficients $c_{ij}(k)$ that will now be summarized. Substituting $f_1(x,k)$ from (2.8.7b) into (2.8.7a) and equating

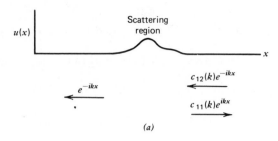

(a)

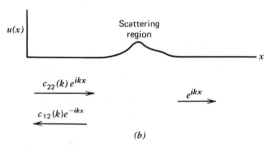

(b)

**Figure 2.4** Scattering of waves (a) incident from the right (b) incident from the left.

coefficients of $f_2(x, k)$ and $f_2(x, -k)$, we find that consistency of these equations requires that

$$1 = c_{11}(k)c_{22}(k) + c_{12}(k)c_{21}(-k)$$
$$0 = c_{11}(k)c_{21}(k) + c_{12}(k)c_{22}(-k) \qquad (2.8.10)$$

A similar substitution of (2.8.7a) into (2.8.7b) yields

$$1 = c_{21}(k)c_{12}(-k) + c_{22}(k)c_{11}(k)$$
$$0 = c_{21}(k)c_{11}(-k) + c_{22}(k)c_{12}(k) \qquad (2.8.11)$$

Furthermore, substitution of (2.8.6) and (2.8.7) into the various possible Wronskian relations and use of the fact that $W[f_1(x, k); f_1(x, k)] = W[f_2(x, k); f_2(x, k)] = 0$ yields

$$c_{11}(k) = \frac{1}{2ik} W[f_2(x, k); f_1(x, -k)]$$

$$c_{22}(k) = \frac{1}{2ik} W[f_2(x, -k); f_1(x, k)] \qquad (2.8.12)$$

$$c_{12}(k) = c_{21}(k) = \frac{1}{2ik} W[f_1(x, k); f_2(x, k)]$$

From the last of these relations it follows that $T_L(k) = T_R(k)$. The relations $f_i^*(x, -k) = f_i(x,k)$ for $i = 1$ and 2 can be used to show that for $u$ and $k$ real,

$$c_{12}(-k) = c_{12}^*(k), \qquad c_{11}(k) = -c_{22}^*(k) = -c_{22}(-k) \qquad (2.8.13)$$

The consistency equations (2.8.10) and (2.8.11) then yield

$$|c_{12}(k)|^2 = 1 + |c_{11}(k)|^2 = 1 + |c_{22}(k)|^2 \qquad (2.8.14)$$

which, according to (2.8.8) and (2.8.9), may be written

$$1 = |T(k)|^2 + |R_R(k)|^2 = |T(k)|^2 + |R_L(k)|^2 \qquad (2.8.15)$$

From (2.8.10) and (2.8.11), we obtain

$$R_R(k)T(-k) + R_L(-k)T(k) = 0 \qquad (2.8.16)$$

This latter result provides a useful relation between $R_L$ and $R_R$. Finally, we have $R_L^*(-k) = R_L(k)$ and similarly for $R_R(k)$.

*Exercise 15*

Compare the asymptotic forms of (2.8.4) and (2.8.7a) as $x \to +\infty$ and show that

$$c_{11}(k) = \frac{1}{2ik} \int_{-\infty}^{\infty} dx\, e^{-ikx} u(x) f_2(x,k)$$

$$c_{12}(k) = 1 - \frac{1}{2ik} \int_{-\infty}^{\infty} dx\, e^{ikx} u(x) f_2(x,k)$$

Thus show that as long as $\int_{-\infty}^{\infty} dx'\, u(x') f_2(x,0) \neq \infty$,

$$R_R(0) = \frac{c_{11}(0)}{c_{12}(0)} = -1$$

A similar result holds for $R_L(0)$.

## Poles of the Transmission Coefficient

We have already seen some specific examples in which the locations of the poles of the transmission and reflection coefficients in the upper half-plane were used to yield information about the localized or bound-state solutions. We now consider this topic in a more general way. From (2.8.8) and (2.8.9.) it is clear that we are interested in the zeros of $c_{12}(k)$. First, it is evident from (2.8.14), which holds only for real $k$, that $c_{12}$ can never be zero when $\text{Im}\,k = 0$. Thus all zeros of $c_{12}$ are off of the real axis. We can also show that for real potentials any poles in the upper half-plane must be on the imaginary axis. To see this, assume that $k_0$ is the location of one of the poles and write

$$y'' + k_0^2 y = u(x)y$$

$$y*'' + k_0^{*2} y* = u(x)y*$$

(2.8.17)

where $y$ may be either of the two fundamental solutions and $y*$ is the complex conjugate of that solution. Multiplying the first of these equations by $y*$, the second by $y$, subtracting and integrating over all $x$ yields

$$(y*y' - yy*')|_{-\infty}^{\infty} = -(k_0^2 - k_0^{*2}) \int_{-\infty}^{\infty} dx |y|^2$$

(2.8.18)

The left-hand side vanishes since it is either of the Wronskians given in (2.8.6) and thus has the same value at both limits. Writing $k_0 = k_{0r} + ik_{0i}$, we find that the imaginary part of (2.8.18) is

$$k_{0r} k_{0i} \int_{-\infty}^{\infty} dx |y|^2 = 0$$

(2.8.19)

Thus $k_{0r} = 0$ since $k_{0i} \neq 0$ as noted above and the integral is a finite positive quantity when $\text{Im} k > 0$.

Finally, when $c_{12}$ vanishes, $f_1(x, k_0)$ and $f_2(x, k_0)$ are linearly dependent since (2.8.7a) and (2.8.7b) then yield

$$f_2(x, k_0) = c_{11}(k_0) f_1(x, k_0)$$

(2.8.20)

and $c_{22}(k_0) = 1/c_{11}(k_0)$.

In later applications, it will be necessary to know the value of the residue at each pole of $1/c_{12}(k)$. Hence we must evaluate $\dot{c}_{12}(k_l) \equiv dc_{12}(k)/dk|_{k=k_l}$, where $k_l$ is the $l$th zero of $c_{12}$ on the imaginary axis in the upper half-plane. From the last of (2.8.12),

$$\dot{c}_{12}(k_l) = \frac{1}{2ik_l} \{ W[\dot{f}_1; f_2] + W[f_1; \dot{f}_2] \}$$

$$= \frac{1}{2ik_l} \{ c_{11}(k_l) W[\dot{f}_1; f_1] + c_{22}(k_l) W[f_2; \dot{f}_2] \}$$

(2.8.21)

The Wronskians in the last expression may be evaluated as follows. From the Schrödinger equation we can write

$$f''_1(x, k) + k^2 f_1(x, k) = u(x) f_1(x, k)$$

$$f''_1(x, k_l) + k_l^2 f_1(x, k_l) = u(x) f_1(x, k_l)$$

(2.8.22)

Multiplying the first of these by $f(x,k_l)$, the second by $f(x,k)$ and subtracting, we find that

$$\frac{d}{dx}\left[ f_1'(x,k_l)f_1(x,k) - f_1'(x,k)f_1(x,k_l) \right] - \left( k^2 - k_l^2 \right)f_1(x,k)f_1(x,k_l) = 0$$

(2.8.23)

Differentiating this result with resect to $k$ and then setting $k = k_l$, we obtain

$$\frac{d}{dx} W\left[ f_1(x,k_l); \dot{f}_1(x,k_l) \right] = 2k_l\left[ f_1(x,k_l) \right]^2$$

(2.8.24)

Integration from $x$ to $+\infty$ gives

$$W\left[ f_1(x,k_l); \dot{f}_1(x,k_l) \right] = -2k_l\int_x^\infty dx'\left[ f_1(x',k_l) \right]^2$$

(2.8.25)

since the Wronskian evaluated at the upper limit will vanish for $\mathrm{Im}\,k > 0$. A similar calculation shows that

$$W\left[ f_2(x,k_l); \dot{f}_2(x,k_l) \right] = 2k_l\int_{-\infty}^x dx'\left[ f_2(x',k_l) \right]^2$$

(2.8.26)

and (2.8.21) for $c_{12}(k_l)$ then yields the formula

$$i\gamma_l \equiv \mathrm{Res}\left[ \frac{1}{\dot{c}_{12}(k_l)} \right] = \frac{1}{\dot{c}_{12}(k_l)} = \frac{i}{\int_{-\infty}^\infty dx f_1(x,k_l)f_2(x,k_l)}$$

(2.8.27)

Equation 2.8.20 may be used to rewrite this result as

$$\int_{-\infty}^\infty dx c_{11}(k_l)\gamma_l\left[ f_1(x,k_l) \right]^2 = 1$$

(2.8.28a)

and

$$\int_{-\infty}^\infty dx c_{22}(k_l)\gamma_l\left[ f_2(x,k_l) \right]^2 = 1$$

(2.8.28b)

Thus the quantities $[\gamma_l c_{11}(k_l)]^{1/2}$ and $[\gamma_l c_{22}(k_l)]^{1/2}$ are the normalization constants for the bound-state wave functions $f_1(x,k_l)$ and $f_2(x,k_l)$, respectively.

Use of (2.8.20) also yields

$$\dot{c}_{12}(k_l) = -ic_{11}(k_l)\int_{-\infty}^\infty dx\left[ f_1(x,k_l) \right]^2$$

(2.8.29)

This result may be employed to show that $\dot{c}_{12}(k_0)$ can never vanish as long as the potential is real. Hence all zeros of $c_{12}$, that is, all poles of the transmission coefficient, must be simple. The nonvanishing of $\dot{c}_{12}(k_0)$ follows both from the normalization condition (2.8.2), which requires that $c_{11}$ be nonzero at a zero of $c_{12}$, and from (2.8.3), which shows that for a real potential, $f_1(x,k)$ is real at the zeros of $c_{12}$ where $k$ is purely imaginary; thus the integrand in (2.8.29) is positive and the integral is nonzero.

Since $f_1$ and $f_2$ are real at the zeros of $c_{12}$, we also see from (2.8.27) that $\gamma_l$ is real. Finally, using (2.8.27) we may write the normalization constants as

$$m_{Rl} \equiv \gamma_l c_{11}(k_l) = -i\frac{c_{11}(k_l)}{\dot{c}_{12}(k_l)} = \left[\int_{-\infty}^{\infty} dx f_2^2(x,k_l)\right]^{-1} \qquad (2.8.30a)$$

$$m_{Ll} \equiv \gamma_l c_{22}(k_l) = -i\frac{c_{22}(k_l)}{\dot{c}_{12}(k_l)} = \left[\int_{-\infty}^{\infty} dx f_1^2(x,k_l)\right]^{-1} \qquad (2.8.30b)$$

Since $f_1$ and $f_2$ are real for imaginary $k$, it follows that $m_{Rl}$ and $m_{Ll}$ are also real and positive.

As noted in Section 1.5, we shall encounter linear eigenvalue problems that are equivalent to a Schrödinger equation with a complex potential. In those cases the poles need no longer be simple, nor must they be confined to the imaginary axis.

In later applications of these results it will be useful to have expressions for the Fourier transforms of the reflection and transmission coefficients that were introduced in (2.8.8) and (2.8.9). Since the transmission coefficient approaches unity for large $k$, it is appropriate to define the transform

$$\Gamma(z) = \int_{-\infty}^{\infty} \frac{dk}{2\pi} e^{-ikz}\left[T(k) - 1\right] \qquad (2.8.31)$$

as well as

$$r_R(z) = \int_{-\infty}^{\infty} \frac{dk}{2\pi} e^{ikz}R_R(k)$$

$$r_L(z) = \int_{-\infty}^{\infty} \frac{dk}{2\pi} e^{-ikz}R_L(k) \qquad (2.8.32)$$

As noted previously, when the potential is real, $R_L^*(-k) = R_L(k)$ and similarly for $R_R(k)$. Therefore, $r_R(z)$ and $r_L(z)$ are real for real potentials. The integral for $\Gamma(z)$ is easily evaluated when $z < 0$. If there are no poles in the upper half-plane then $\Gamma(z)$ vanishes. For bound-state problems in which there are a number of poles on the imaginary axis in the upper half-plane at points $k = i\kappa_l$, we have

$$\Gamma(z) = -\sum_{l=1}^{n} \gamma_l e^{\kappa_l z}, \qquad z < 0 \qquad (2.8.33)$$

where $\gamma_l$ is the expression in (2.8.27) evaluated at $k_l = i\kappa_l$.

*Exercise 16*

The fundamental solutions for the potential $u(x) = -2\,\mathrm{sech}^2 x$ are seen from (2.6.14) and (2.6.15) to be $f_1(x,k) = e^{ikx}(ik - \tanh x)/(ik - 1)$ and $f_2(x,k) = e^{-ikx}(ik + \tanh x)/(ik - 1)$. Show that $c_{12}(k) = (k - i)/(k + i)$ and $c_{11}(k) = c_{22}(k) = 0$. Thus $R_L = R_R = 0$, and the potential is reflectionless. Use $f_1$ for $\psi$ in the expression for particle flux (2.4.3) and show that the flux is $\hbar k/m$, as it would be for the plane wave $e^{ikx}$.

*Exercise 17*

Use the result of Ex. 11 to show that for the potential $u(x) = -6\,\mathrm{sech}^2 x$, the fundamental solutions are

$$f_1(x,k) = e^{ikx}\frac{1 + k^2 + 3ik\tanh x - 3\tanh^2 x}{(k+i)(k+2i)}$$

$$f_2(x,k) = e^{-ikx}\frac{1 + k^2 - 3ik\tanh x - 3\tanh^2 x}{(k+i)(k+2i)}$$

and that

$$c_{12}(k) = \frac{(k-i)(k-2i)}{(k+i)(k+2i)}$$

*Exercise 18*

By differentiating the differential equations satisfied by $f_1(x,k)$ and $f_2(x,k)$ with respect to $k$, show that

$$W(\dot{f}_1; \dot{f}_2)\Big|_{x=-\infty}^{x=\infty} = 2k\int_{-\infty}^{\infty} dx\,(f_2\dot{f}_1 - f_1\dot{f}_2)$$

*Exercise 19*

For the delta function $u(x) = 2b\,\delta(x)$, the Schrödinger equation (2.8.1) becomes $y'' + [k^2 - 2b\delta(x)]y = 0$. Show that the fundamental solutions are

$$f_1(x,k) = \begin{cases} e^{ikx}, & x \geqslant 0 \\ e^{ikx} - \dfrac{2b}{k}\sin kx, & x \leqslant 0 \end{cases}$$

$$f_2(x,k) = \begin{cases} e^{-ikx} + \dfrac{2b}{k}\sin kx, & x \geqslant 0 \\ e^{-ikx}, & x \leqslant 0 \end{cases}$$

Show also that the coefficients $c_{ij}(k)$ are $c_{11}(k) = -c_{22}(-k) = -ib/k$ and $c_{12}(k) = 1 + ib/k$, so that $R_L(k) = -ib/(k + ib) = R_R(k)$. These results hold for both positive and negative values of $b$.

## Relation between Transmission and Reflection Coefficients

The transmission and reflection coefficients are related by the energy conservation relation (2.8.15) and by (2.8.16). We might consider whether it is possible to express the transmission coefficient in terms of a single reflection coefficient. We now show that if the location of the poles and zeros of $T(k)$ are known in the upper half-plane, then the transmission coefficient may be expressed in terms of either $R_L(k)$ or $R_R(k)$.

We first consider the case in which $T(k)$ possesses neither poles nor zeros in the upper half-plane. The logarithm of the transmission coefficient is then also free of singularities in this half-plane and by Cauchy's theorem

$$\ln T(k) = \oint_C \frac{d\zeta}{2\pi i} \frac{\ln T(\zeta)}{\zeta - k} \qquad (2.8.34)$$

Since $T(k) \to 1$ as $k \to \infty$, we have $\ln T \to 0$ in this limit. The contour $C$ may be chosen to be along the real axis from $-\infty$ to $+\infty$ and then closed by a semicircular arc at $\infty$ in the upper half-plane. There is no contribution from the arc at infinity, since $\ln T(\zeta) \to 0$ for $|\zeta| \to \infty$ and we may write

$$\ln T(k) = \int_{-\infty}^{\infty} \frac{d\zeta}{2\pi i} \frac{\ln T(\zeta)}{\zeta - k} \qquad (2.8.35)$$

Since $k^*$ is in the lower half-plane and therefore not enclosed by the contour $C$, we may also write

$$0 = \oint_C \frac{d\zeta}{2\pi i} \frac{\ln T(\zeta)}{\zeta - k^*}$$

$$= \int_{-\infty}^{\infty} \frac{d\zeta}{2\pi i} \frac{\ln T(\zeta)}{\zeta - k^*} \qquad (2.8.36)$$

where the contour at infinity in the upper half-plane has again been ignored in writing the second equality. Taking the complex conjugate of this last equation, subtracting the result from (2.8.35) and using (2.8.15), we have

$$\ln T(k) = \int_{-\infty}^{\infty} \frac{d\zeta}{2\pi i} \frac{\ln\left[1 - |R(\zeta)|^2\right]}{\zeta - k} \qquad (2.8.37)$$

where $R(\zeta)$ may be either $R_R(\zeta)$ or $R_L(\zeta)$. We thus have a prescription for obtaining $T(k)$ from $R(k)$ when $\ln T(k)$ is free from singularities in the upper half-plane. For computational purposes it is convenient to rewrite the result in the form

$$\frac{d}{dk} \ln T(k) = -\int_{-\infty}^{\infty} \frac{d\zeta}{2\pi i} \frac{\frac{d}{d\zeta}|R(\zeta)|^2}{(\zeta - k)\left[1 - |R(\zeta)|^2\right]} \qquad (2.8.38)$$

where an integration by parts has been performed. The integral in (2.8.38) is particularly easy to evaluate whenever $R(\zeta)$ is a ratio of polynomials in $\zeta$ times a phase factor $e^{ia\zeta}$. In obtaining $T(k)$ from (2.8.38), a final constant of integration is determined by requiring that $T(k) \rightarrow 1$ as $k \rightarrow \infty$.

If the integrand in (2.8.38) contains a first-order pole on the real axis, the original contour in (2.8.34) would have to be indented with a semicircle in the upper half-plane. The integral in (2.8.38) would then be interpreted as a principal value.

As an example, let us consider the repulsive delta function potential. From (2.4.4) we have $R_L(k) = -i\beta/(k+i\beta), \beta > 0$. Equation 2.8.38 then becomes

$$\frac{d}{dk}\ln T(k) = -2\beta^2 P \int_{-\infty}^{\infty} \frac{d\zeta}{2\pi i} \frac{1}{\zeta(\zeta-k)(\zeta^2+\beta^2)}$$

$$= \frac{1}{k} - \frac{1}{k+i\beta} \tag{2.8.39}$$

Requiring $T(k)$ to approach unity for large $k$, we thus arrive at the companion expression for the transmission coefficient. It is also given in (2.4.4). We have

$$T(k) = \frac{k}{k+i\beta}, \qquad \beta > 0 \tag{2.8.40}$$

If $T(k)$ contains first-order zeros or poles in the upper half-plane at points $\alpha_l$ and $\beta_l$, respectively, then a calculation similar to that given above may be carried out for the function

$$\tilde{T}(k) = T(k) \prod_l \left( \frac{k-\alpha_l^*}{k-\alpha_l} \right) \left( \frac{k-\beta_l}{k-\beta_l^*} \right) \tag{2.8.41}$$

which now has the properties required of $T(k)$ in the previous calculation. Noting that for real $k$,

$$|\tilde{T}(k)|^2 = |T(k)|^2 = 1 - |R(k)|^2 \tag{2.8.42}$$

we may proceed along the same lines as in the previous calculation and obtain the result

$$T(k) = \prod_l \left( \frac{k-\alpha_l}{k-\alpha_l^*} \right) \left( \frac{k-\beta_l^*}{k-\beta_l} \right) \exp \left\{ \int_{-\infty}^{\infty} \frac{d\zeta}{2\pi i} \frac{\ln[1-|R(\zeta)|^2]}{\zeta-k} \right\} \tag{2.8.43}$$

As an example, consider the attractive delta function potential for which $R_L = -i\beta/(k+i\beta), \beta < 0$. The transmission coefficient also has a pole at $k = i|\beta|$. We thus set

$$\tilde{T}(k) = T(k) \frac{k-i|\beta|}{k+i|\beta|} = T(k) \frac{k+i\beta}{k-i\beta} \tag{2.8.44}$$

and then obtain the same integral as that obtained in (2.8.39) except that now $\beta < 0$. Integration yields

$$\frac{d}{dk} \ln \tilde{T}(k) = \frac{1}{k} - \frac{1}{k - i\beta} \qquad (2.8.45)$$

Thus $\tilde{T}(k) = k/(k - i\beta)$ and from (2.8.44) we again obtain $T(k) = k/(k + i\beta)$, where now $\beta < 0$.

## Asymptotic Solution

Although the solution of a second-order differential equation with variable coefficients cannot in general be expressed in terms of quadratures (i.e., explicit integrals), it is possible to obtain an asymptotic solution of the Schrödinger equation for large $k$ in terms of the potential and its derivatives (Verde, 1955; Calogero and Degasperis, 1968). In addition to being a useful form of the solution for large $k$, the asymptotic expansion provides an infinite sequence of conserved quantities that will play an important role when the relationship of scattering theory to nonlinear partial differential equations is considered.

The asymptotic expansion for the fundamental solution $f_2$ may be obtained by writing it in the form

$$f_2(x,k) = e^{-ikx + h(x,k)} \qquad (2.8.46)$$

According to (2.8.2), we must require that $h(-\infty, k) = 0$. In terms of the function $h(x,k)$, the Schrödinger equation (2.8.1) becomes

$$h'' - 2ikh' + (h')^2 = u \qquad (2.8.47)$$

which is a Riccati equation in $h'$. This equation may be solved by introducing the expansion

$$h'(x,k) = \sum_{n=0}^{\infty} \frac{g_n(x)}{(2ik)^{n+1}} \qquad (2.8.48)$$

The series may be shown to be asymptotic (Verde, 1955) in the sense that (Jeffreys and Jeffreys, 1956, Chapter 17)

$$\lim_{k \to \infty} \left\{ h'(x,k) - \sum_{n=0}^{N} \frac{g_n(x)}{(2ik)^n} \right\} k^N = 0 \qquad (2.8.49)$$

When (2.8.48) is substituted into the Riccati equation (2.8.47) and terms of equal powers of $k$ are equated, we obtain $g_0 = -u, g_1 = g_0' = -u'$, and

$$g_{n+1} = g_n' + \sum_{m=0}^{n-1} g_m g_{n-m-1}, \qquad n \geq 1 \qquad (2.8.50)$$

The first few coefficients are

$$g_0 = -u, \qquad g_1 = -u', \qquad g_2 = -u'' + u^2, \qquad g_3 = -u''' + 4uu',$$
$$g_4 = -u^{(iv)} + 5(u')^2 + 6uu'' - 2u^3, \qquad \text{etc.}$$
$$(2.8.51)$$

From (2.8.7a) we see that the solution $f_2(x,k)$ has the limiting forms

$$f_2(x,k) \sim \begin{cases} e^{-ikx}, & x \to -\infty \\ c_{11}(k)e^{ikx} + c_{12}(k)e^{-ikx}, & x \to +\infty \end{cases} \qquad (2.8.52)$$

Hence, as $x \to +\infty$, $f_2(x,k)e^{ikx} \to c_{12}(k) + c_{11}(k)e^{2ikx}$ and for $\mathrm{Im}\,k > 0$,

$$\lim_{x \to \infty} f_2(x,k)e^{ikx} \to c_{12}(k) \qquad (2.8.53)$$

Combining (2.8.46) and (2.8.52) we have $h(\infty,k) = \ln c_{12}(k)$. Since $h(-\infty) = 0$, the integration of (2.8.48) over all $x$ yields

$$h(\infty,k) = \ln c_{12}(k) = \sum_{n=0}^{\infty} \frac{1}{(2ik)^{n+1}} \int_{-\infty}^{\infty} dx\, g_n(x). \qquad (2.8.54)$$

Now, as $k$ becomes large, so that $k^2 \gg u$, the potential in the Schrödinger equation becomes negligible and all wave energy is transmitted beyond the scatterer. Thus $\lim k = \infty\, c_{12}(k) \to 1$. Accordingly, $\ln c_{12}(k)$ approaches zero in this limit. If we introduce the asymptotic expansion $\ln c_{12}(k) = \sum_{n=1}^{\infty} c_n / k^n$ and compare coefficients with (2.8.54), we find that

$$c_n = \frac{1}{(2i)^n} \int_{-\infty}^{\infty} dx\, g_{n-1}(x), \qquad n = 1,2,3,\dots \qquad (2.8.55)$$

This result will be extremely useful when we consider the use of these asymptotic expansions in solving evolution equations.

Some insight into the significance of the functions $g_n$ in (2.8.51) may be obtained by considering certain standard limiting cases. If the potential varies slowly in a wavelength, a WKB type of solution of the wave equation is usually appropriate (Schiff, 1949, Sec. 28). It may be recovered here by neglecting all the derivatives in the $g_n(x)$ and using the approximate relations $g_0 = -u, g_2 = u^2, g_4 = -2u^3, \dots$ and $g_{2k+1} = 0$ for $k = 0,1,2,\dots$. Similarly, if the potential is weak so that $u \ll k^2$, an approximate solution known as the Born approximation is frequently used (Schiff, 1949, Sec. 26). It may be recovered here by retaining only terms that are linear in $u$ and its derivatives. This yields $g_0 = -u, g_1 = -u', g_2 = -u'', g_3 = -u''', g_4 = -u^{(iv)}, \dots$.

*Exercise 20*

If $u(x)$ varies slowly in the distance $k^{-1}$, we might expect that the function $f_2(x,k)$ would be of the form $f_2 = \exp\{-i\int_{-\infty}^{x} dx'(\sqrt{k^2-u} - k) - ikx\}$. Show that when this expression is introduced into (2.8.46), the resulting values of $g_n$ agree with those quoted above as the WKB limit.

*Exercise 21*

The Born approximation for $f_2(x,k)$ is obtained by setting $f_2(x,k)$ equal to $e^{-ikx}$ in the integrand of (2.8.4). Treat the integral as a small correction to $e^{-ikx}$, and show by repeated partial integration of the expression for $f_2'(x,k)/f_2(x,k)$ that the coefficients of $(2ik)^{-n}$ agree with those given above for the Born approximation limit.

## Number of Poles of the Transmission Coefficient

The number of poles of the transmission coefficient in the upper half-plane will not, in general, be evident from inspection of the function $c_{12}(k)$. Especially when the poles are being located by a numerial search procedure, it is useful to have a method for determining the total number of poles that must be found. This may be done by noting that the number of (simple) poles, $N$, is given by

$$N = \oint_C \frac{dk}{2\pi i} \frac{\dot{c}_{12}(k)}{c_{12}(k)} \qquad (2.8.56)$$

where the dot refers to differentiation with respect to $k$ and where $C$ is a contour in the upper half-plane that encloses all zeros of $c_{12}(k)$. This result is obtained by first noting that $c_{12}(k)$ is constructed according to (2.8.12) from fundamental solutions $f_1$ and $f_2$ that are analytic in the upper half-plane. Therefore, $c_{12}(k)$, and also $\dot{c}_{12}(k)$, have no poles within the contour $C$. Only zeros of $c_{12}(k)$ can contribute poles within this contour. If we take the contour $C$ to be along the whole real axis and closed by a semicircular arc at infinity in the upper half-plane, then the contribution from the arc at infinity will vanish. This follows from the asymptotic form of $c_{12}(k)$, which is $c_{12}(k) - 1 \sim k^{-1}$ as $|k| \to \infty$. Thus $\dot{c}_{12}/c_{12} \sim k^{-2}$ and the integral vanishes on the semicircular arc. If we write $c_{12}(k) = |c_{12}| \exp[i\varphi(k)]$ and note that $|c_{12}| \to 1$ as $|k| \to \infty$, we may integrate (2.8.56) and obtain

$$N = \frac{1}{2\pi i} \ln c_{12}(k) \Big|_{-\infty}^{\infty} = \frac{1}{2\pi}[\varphi(+\infty) - \varphi(-\infty)] \qquad (2.8.57)$$

which relates the number of poles to the total change in phase of $c_{12}$ along the real axis. In quantum scattering theory, this result is referred to as Levinson's Theorem (Newton, 1960, Sec. 5).

As a simple example, recall from Ex. 15 that for the potential $u(x) = -2\,\text{sech}^2 x$ we obtained $c_{12}(k) = (k-i)/(k+i)$. The phase of $c_{12}$ is given by

$\varphi = \tan^{-1}[2k/(1-k^2)]$. As $k$ proceeds from $-\infty$ to $-1$, the phase increases from 0 to $\pi/2$. As $k$ increases from $-1$ to 0, the phase increases from $\pi/2$ to $\pi$ and similarly increases by $\pi$ again as $k$ goes to $+\infty$. The total change in phase is thus $2\pi$ and (2.9.10) yields $N=1$ as expected.


## 2.9  TRUNCATED POTENTIALS*

The potential shapes that lead to simple analytical results can be extended by employing certain general results on truncated potentials. As shown in Figure 2.5, we consider a potential $u(x)$ and also the truncated potential

$$u^T(x) = u(x)\theta(x-x_1) \tag{2.9.1}$$

where $\theta(x)$ is the unit step function. If $f_1(x,k)$ and $f_2(x,k)$ refer to the fundamental solutions for the potential $u(x)$ while $f_1^T(x,k)$ and $f_2^T(x,k)$ refer to those for the truncated potential, then for $x > x_1$ we readily see from (2.8.3) and (2.8.4) that

$$f_{1>}^T(x,k) = f_1(x,k)$$

$$f_{2>}^T(x,k) = e^{-ikx} + \frac{1}{k}\int_{x_1}^x dx' \sin k(x-x')u(x')f_2^T(x',k) \tag{2.9.2}$$

while for $x < x_1$,

$$f_{1<}^T(x,k) = e^{ikx} - \frac{1}{k}\int_{x_1}^{\infty} dx' \sin k(x-x')u(x')f_1(x',k)$$

$$f_{2<}^T(x,k) = e^{-ikx} \tag{2.9.3}$$

Both $f_1^T(x,k)$ and $f_2^T(x,k)$ and their first derivatives with respect to $x$ are continuous at $x = x_1$.

Defining

$$c_{12}^T(k) = \frac{1}{2ik} W\left[f_1^T(x,k); f_2^T(x,k)\right] \tag{2.9.4}$$

and evaluating the Wronskian at $x = x_1$, we obtain

$$c_{12}^T(k) = \frac{e^{-ikx_1}}{2ik}\left[f_1'(x_1,k) + ikf_1(x_1,k)\right] \tag{2.9.5}$$

where $f_1'(x_1,k) = \partial f_1(x,k)/\partial x|_{x=x_1}$.

Similarly, we find that

$$c_{11}^T(k) = -\frac{e^{-ikx_1}}{2ik}\left[f_1'(x_1,-k) + ikf_1(x_1,-k)\right]$$

$$c_{22}^T(k) = -\frac{e^{ikx_1}}{2ik}\left[f_1'(x_1,k) - ikf_1(x_1,k)\right] \tag{2.9.6}$$

---

*I am indebted to J. R. Cox for acquainting me with the approach used here.

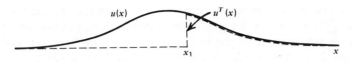

**Figure 2.5** Arbitrary potential $u(x)$ and associated potential $u^T(x)$ that is truncated on the left at $x = x_1$.

As $x_1$ approaches $+\infty$, we obtain $c_{12}^T(k) \to 1, c_{11}^T(k) \to 0$, and $c_{22}^T \to 0$, which are expected results in the limit that the potential vanishes. Also, as $x_1 \to -\infty$, we may use (2.8.7b) to express $f_1$ in terms of $f_2$ and again obtain the expected results $c_{ij}^T(k) \to c_{ij}(k)$. This limit will be discussed in more detail after we consider the following example.

As noted in Ex. 16, the potential $u(x) = -2\,\mathrm{sech}^2 x$ has the fundamental solution $f_1 = e^{ikx}(ik - \tanh x)/(ik - 1)$. Substitution into the foregoing results shows that for the truncated potential $u(x) = -2\theta(x - x_1)\,\mathrm{sech}^2 x$ the coefficients $c_{ij}^T$ are

$$c_{12}^T = \frac{k^2 + 1 + (k + i\tanh x_1)^2}{2k(k + i)} \tag{2.9.7}$$

and

$$c_{11}^T(k) = -c_{22}^T(-k) = \frac{e^{-2ikx_1}\,\mathrm{sech}^2 x_1}{2k(k - i)} \tag{2.9.8}$$

These expressions are quite helpful in providing an understanding of the analytic structure of the coefficients $c_{ij}(k)$. In particular, we have seen that the reflectionless potential has coefficients $c_{11}(k) = c_{22}(k) = 0$ and $c_{12}(k) = (k - i)/(k + i)$ for real $k$ while at the zero of $c_{12}$ we have $c_{11}(i) = f_2(x, i)/f_1(x, i) = 1 = 1/c_{22}(i)$. Since an analytic function is zero everywhere if it is zero on even a finite segment of a line (Churchill, 1948, p. 188), the foregoing results might seem paradoxical. However, by considering the expressions for the truncated coefficients given above, we can see that all the results for the reflectionless potential are obtained in the limit $x_1 \to -\infty$. In particular, the zero of $c_{12}^T(k)$ in the upper half-plane is found to be at $k_0 = \frac{1}{2}i(-\tanh x_1 + \sqrt{1 + \mathrm{sech}^2 x_1}\,)$. We also have

$$c_{22}^T(k_0) = -\frac{e^{2ik_0x_1}\,\mathrm{sech}^2 x_1}{2k_0(k_0 + i)} \tag{2.9.9}$$

as $x_1$ approaches $-\infty, k_0$ approaches $i$ and we obtain $c_{22}^T(i) \to 1$. For real $k$, on the other hand, we recover $c_{22}^T(k) \to 0$ as $x_1 \to -\infty$. A similar result is obtained from $c_{11}^T(k)$, but in this case an indeterminate form must be evaluated since both numerator and denominator vanish as $x_1 \to -\infty$ and $k_0 \to i$.

The poles of the transmission coefficient are determined by the roots of the equation $c_{12}^T(k) = 0$. From the root of this equation given above, and referred to as $k_0$, we see that one of the two roots is always in the upper half-plane. Thus there is always a bound state no matter how small the truncated attractive potential is. This is a particular example of the general result that, for one-dimensional problems, all real attractive potentials that vanish at infinity have at least one bound state (Morse and Feshbach, 1953, p. 1654).

*Exercise 22*

Show that when the potential is truncated on the right so that $u^T(x) = u(x)\theta(x_1 - x)$, the various coefficients are

$$c_{12}^T(k) = \frac{e^{ikx_1}[-f_2'(x_1,k) + ikf_2(x_1,k)]}{2ik}$$

$$c_{11}^T(k) = \frac{e^{-ikx_1}[f_2'(x_1,k) + ikf_2(x_1,k)]}{2ik}$$

$$= -c_{22}^T(-k)$$

*Exercise 23*

Another potential related to $u(x)$ for which the coefficients $c_{ij}(k)$ may readily be determined is the potential $\tilde{u}(x) = u(x)\theta(x - x_1) + a\delta(x - x_1)$, where $a$ is a constant. The fundamental solutions now have a discontinuity in the first derivative at $x = x_1$ with $f_>'(x_1,k) - f_<'(x_1,k) = -af(x_1,k)$. If the coefficients related to the potential $\tilde{u}(x)$ are denoted by $\tilde{c}_{ij}(k)$, show that

$$\tilde{c}_{12}(k) = e^{-ikx_1}\frac{f_1'(x_1,k) + (ik - a)f_1(x_1,k)}{2ik}$$

$$\tilde{c}_{11}(k) = -e^{-ikx_1}\frac{f_1'(x_1, -k) + (ik - a)f_1(x_1, -k)}{2ik}$$

$$\tilde{c}_{22}(k) = -e^{ikx_1}\frac{f_1'(x_1,k) - (ik + a)f_1(x_1,k)}{2ik}$$

## 2.10  SCATTERING OF PULSES—MARCHENKO EQUATIONS

In previous considerations of scattering, the incident wave was a steady wave at a single frequency. Although the scattering of an incident pulse could be constructed from the results of such steady-state calculations by Fourier synthesis as in (2.2.9), there are some advantages to considering the scattering of pulses directly in the time domain. In particular, we shall be led quite readily to a new representation for the fundamental solutions $f_1(x,k)$ and $f_2(x,k)$. The simplest pulse to consider is, of course, the delta function. Pulses $\delta(t \pm x/c)$, since they are functions of $t \pm x/c$, are exact solutions of the simple wave equation $y_{tt} - c^2 y_{xx} = 0$. However, for the elastically braced string that is described by the equation

$$\frac{\partial^2 y}{\partial x^2} - \frac{1}{c^2}\frac{\partial^2 y}{\partial t^2} - \mu^2(x)y = 0 \qquad (2.10.1)$$

Such functions are no longer solutions. If $\mu^2(x)$ is localized, we would expect that there could be an exact solution that would reduce to a delta function in a part of the string that is remote from the elastic region. Because of the elastic region, we would expect the pulse to be partly scattered and thus leave some sort of wake behind a sharp leading edge. Hence, it is reasonable to write a solution to (2.10.1) in the form (Balanis, 1972)

$$y_1(x,t) = \delta\left(t - \frac{x}{c}\right) + c\theta\left(t - \frac{x}{c}\right)A_R(x,ct) \qquad (2.10.2)$$

where $\theta(x)$ is again the unit step function and $A_R(x,ct)$ is the function that describes the scattering or wake. The factors of $c$ have been introduced for later convenience. Taking the Fourier transform of the solution above and then setting $ct = x'$ and $k = \omega/c$, we find that

$$Y_1(x,\omega) \equiv \int_{-\infty}^{\infty} dt\, e^{i\omega t} y_, (x,t)$$
$$= e^{ikx} + \int_{x}^{\infty} dx' A_R(x,x')e^{ikx'} \qquad (2.10.3)$$

This solution has the property that $\lim_{x\to\infty}[e^{-ikx}Y_1(x,\omega)] = 1$. Now, with $\mu^2(x) = u(x)$ the function $Y_1(x,\omega)$ is a solution of the equation

$$\frac{d^2 Y}{dx^2} + \left[\left(\frac{\omega}{c}\right)^2 - u(x)\right]Y = 0 \qquad (2.10.4)$$

for which we have introduced in Section 2.8 a fundamental solution with the property $\lim_{x\to\infty}[e^{-ikx}f_1(x,k) = 1$. Since this solution is unique, we conclude that $Y_1(x,\omega)$ must be another representation for the fundamental solution $f_1(x,k)$. Hence we may write

$$f_1(x,k) = e^{ikx} + \int_{x}^{\infty} dx' A_R(x,x')e^{ikx'} \qquad (2.10.5)$$

This will be a convenient representation in our later use of the fundamental solutions.*

*The solution $y_1(x,t)$ in (2.10.2) represents a scattering process in which an appropriate set of incident waves conspire to produce an outgoing delta function receding to plus infinity. This is due to the appearance of $e^{i\omega t}$ as the time dependence in (2.10.3) and the choice of signs for the fundamental solutions given in (2.10.5). The more natural scattering solution describing the scattering of an incident delta function pulse would require a change in one of these sign conventions, which does not seem to be warranted.

Similarly, a pulse traveling in the negative $x$ direction may be written

$$y_2(x,t) = \delta\left(t + \frac{x}{c}\right) + c\theta\left(t + \frac{x}{c}\right) A_L(x, -ct) \qquad (2.10.6)$$

The Fourier transform of this expression leads to the result

$$Y_2(x,w) = f_2(x,k) = e^{-ikx} + \int_{-\infty}^{x} dx' A_L(x,x') e^{-ikx'} \qquad (2.10.7)$$

Certain analytic properties of the fundamental solutions may be recovered from the foregoing results. The inverse Fourier transform of (2.10.5) is

$$A_R(x,x') = \int_{-\infty}^{\infty} \frac{dk}{2\pi} e^{ik(x-x')} \left[ f_1(x,k) e^{-ikx} - 1 \right] \qquad (2.10.8)$$

with $A_R(x,x')$ equal to zero for $x' < x$. Since the integral in (2.10.8) is closed in the upper half-plane for $x' < x$, we conclude that $f_1(x,k)e^{-ikx} - 1$ must be analytic in the upper half-plane so that the integral will yield $A_R(x,x') = 0$ for $x' < x$. Similarly, (2.10.7) leads to the conclusion that $f_2(x,k)e^{ikx} - 1$ is also analytic in the upper half-plane. The analyticity of these expressions was recognized previously in our consideration of (2.8.3) and (2.8.4) and could have been made the basis for deriving the representations given here in (2.10.5) and (2.10.7).

The Fourier transform of the solution $f_1(x, -k) = Y_1(x, -\omega)$ yields

$$\int_{-\infty}^{\infty} \frac{d\omega}{2\pi} e^{-i\omega t} f_1(x, -k) = y_1(x, -t)$$

$$= \delta\left(t + \frac{x}{c}\right) + c\theta(-ct - x) A_R(x, -ct) \qquad (2.10.9)$$

This is a time-reversed solution representing a pulse that moves in the negative $x$ direction with a disturbance ahead of it. Similarly, we find that $f_2(x, -k)$ is associated with a solution moving in the positive $x$ direction with a disturbance ahead of it. The solution in the time domain is

$$y_2(x, -t) = \delta\left(t - \frac{x}{c}\right) + c\theta(x - ct) A_L(x, ct) \qquad (2.10.10)$$

The significance of these four solutions, especially those involving the time-reversed solutions, are readily understood by considering scattering by a delta function potential in the time domain. The functions $f_1(x,k)$ and $f_2(x,k)$ are given in Ex. 19. The corresponding solutions $y_1(x,t)$ and $y_2(x,t)$ are obtained by a Fourier transform of the fundamental solutions given in Ex. 19.

The calculation is readily carried out by first considering $\partial y_1 / \partial t$. For $x < 0$ we find

$$\frac{\partial y_1(x,t)}{\partial t} = \delta'\left(t - \frac{x}{c}\right) + \frac{bc}{2} \int_{-\infty}^{\infty} \frac{d\omega}{2\pi} e^{-i\omega t}(e^{i\omega x/c} - e^{-i\omega x/c})$$

$$= \delta'\left(t - \frac{x}{c}\right) + \frac{bc}{2}\left[\delta\left(t - \frac{x}{c}\right) - \delta\left(t + \frac{x}{c}\right)\right] \qquad (2.10.11)$$

where $\delta'$ refers to the derivative of the delta function with respect to its argument. Hence

$$y_1(x,t) = \begin{cases} \delta\left(t - \dfrac{x}{c}\right) \\ \delta\left(t - \dfrac{x}{c}\right) + \dfrac{bc}{2}\left[\theta\left(t - \dfrac{x}{c}\right) - \theta\left(t + \dfrac{x}{c}\right)\right] \end{cases} \qquad (2.10.12)$$

Since $\theta(t - x/c) - \theta(t + x/c) = \theta(t - x/c)\theta(-t - x/c)$ for $x < 0$, we also find from (2.10.2) that $A_R(x, ct) = (b/2)\theta(-ct - x)$ in this example. The time-reversed solution is

$$y_1(x, -t) = \begin{cases} \delta\left(t + \dfrac{x}{c}\right) \\ \delta\left(t + \dfrac{x}{c}\right) + \dfrac{bc}{2}\left[\theta(-ct - x) - \theta(-ct + x)\right] \end{cases} \qquad (2.10.13)$$

Examination of (2.10.12) shows that $y_1(x, t)$ represents a scattering process in which a delta function pulse followed by a positive step of amplitude $bc/2$ is incident from $x = -\infty$. After interaction with the delta function potential at the origin, the delta function pulse proceeds in the positive $x$ direction while a reflected negative step of amplitude $-bc/2$ returns to $x = -\infty$ and annihilates the positive amplitude left by the incident step. Examination of the solution $y_1(x, -t)$ shows that it represents a scattering process in which there is both a delta function pulse incident from $x = +\infty$ and a step function of amplitude $bc/2$ incident from $x = -\infty$. After scattering from the delta function potential at the origin, the delta function pulse proceeds in the negative $x$ direction along with a negative step of amplitude $-bc/2$ that annihilates the amplitude of the incident step. Similar results obtain for the solution $y_2(x, t)$.

According to (2.8.7a) the fundamental solutions are related by

$$T(k)f_2(x, k) = R_R(k)f_1(x, k) + f_1(x, -k) \qquad (2.10.14)$$

where the reflection and transmission coefficients have been introduced according to (2.8.8). We now derive the corresponding relation in the time domain. This is done by taking the Fourier transform of (2.10.14). Using

(2.8.31) and (2.8.32), the Fourier transform of the right-hand side of (2.10.14) is

$$
\begin{aligned}
R.H.S. &= \int_{-\infty}^{\infty} \frac{d\omega}{2\pi} e^{-i\omega t} \int_{-\infty}^{\infty} dz\, e^{-i\omega z/c} r_R(z) \int_{-\infty}^{\infty} dt'\, e^{i\omega t'} y_1(x,t') \\
&\quad + y_1(x,-t) \\
&= \int_{-\infty}^{\infty} dz \int_{-\infty}^{\infty} dt'\, r_R(z) y_1(x,t') \delta\!\left(t'-t-\frac{z}{c}\right) + y_1(x,-t) \\
&= c \int_{-\infty}^{\infty} dt'\, r_R\big[c(t'-t)\big] y_1(x,t') + y_1(x,-t) \\
&= c r_R(x-ct) + c^2 \int_{x/c}^{\infty} dt'\, r_R\big[c(t'-t)\big] A_R(x,ct') \\
&\quad + \delta\!\left(t+\frac{x}{c}\right) + c\theta(-ct-x) A_R(x,-ct)
\end{aligned}
\tag{2.10.15}
$$

A similar transformation of the left-hand side yields

$$
\begin{aligned}
L.H.S. &= \delta\!\left(t+\frac{x}{c}\right) + c\theta\!\left(t+\frac{x}{c}\right) A_L(x,-ct) + c\Gamma(ct+x) \\
&\quad + c^2 \int_{-x/c}^{\infty} dt'\, \Gamma\big[c(t-t')\big] A_L(x,-ct')
\end{aligned}
\tag{2.10.16}
$$

Let us now consider the case $t+x/c<0$. The integral in (2.8.31) that defines $\Gamma(t+x/c)$ will then be closed in the upper half-plane. If there are no poles in the upper half of the $k$ plane, then the function $\Gamma(t+x/c)$ will be zero. Also, the integral in (2.10.16) is only over the region in which $\Gamma[c(t-t')]$ is zero. Hence this integral also vanishes. Hence for $t+x/c<0$, all terms in (2.10.16) vanish. On setting $ct'=x'$ and $-ct=y$, the Fourier transform of (2.10.14) is of the form

$$
0 = r_R(x+y) + A_R(x,y) + \int_{x}^{\infty} dx'\, r_R(x'+y) A_R(x,x'), \qquad x<y
\tag{2.10.17}
$$

A similar procedure applied to (2.8.7b) yields

$$
0 = r_L(x+y) + A_L(x,y) + \int_{-\infty}^{x} dx'\, r_L(x'+y) A_L(x,x'), \qquad x>y
\tag{2.10.18}
$$

Equations 2.10.17 and 2.10.18 are frequently referred to as Marchenko equations (Agranovich and Marchenko, 1964). In Chapter 3 they will be the central equations in our consideration of the inverse scattering problem. There they will be used to determine $A_R(x,y)$ or $A_L(x,y)$ when one of the reflection coefficients $r_L(x)$ or $r_R(x)$ is specified. At that time we will also consider the generalization required when there are poles of $T(k)$ in the upper half-plane.

The Marchenko equations could also be used to solve a direct scattering problem, that is, determine the reflection coefficients when the potential and hence the fundamental solutions and the functions $A_R$ or $A_L$ are known. However, once the fundamental solutions are known, it is much simpler to determine the coefficients $c_{ij}(k)$ and thus obtain the Fourier transform of the reflection coefficient directly. For the inverse scattering problem no such direct route is available and the integral equation must be considered.


## 2.11   TWO-COMPONENT SCATTERING

Thus far we have been examining scattering problems associated with the Schrödinger equation in which the potential is real. The motivation for these considerations was briefly outlined in Chapter 1, where the relation between the Schrödinger equation and the Korteweg–deVries equation was indicated. As mentioned in Section 1.5, a number of other nonlinear partial differential equations that exhibit soliton behavior are associated in a similar way with a different scattering problem. The ordinary differential equation in this case is most conveniently written as a pair of first-order equations. The simplest form of these equations is

$$\frac{dn_1}{dx} + ikn_1 = u(x)n_2$$
$$\frac{dn_2}{dx} - ikn_2 = -u(x)n_1$$

$$(2.11.1)$$

where $k$ is again the eigenvalue parameter. In order to introduce the procedure in its simplest form, the function $u(x)$ will be assumed to be real. As noted in Section 1.5, (2.11.1) is equivalent to a Schrödinger equation with a complex potential. More general pairs of equations that contain two different potential functions which may be complex will be considered in Section 5.4.


### A Time-Dependent Problem

In order to develop a simple intuitive understanding of the linear equations given above, let us introduce a time-dependent problem, as was done for the Schrödinger equation in Section 2.10. Consider two types of particles that may be endowed with positive and negative charge. They pass through each other in opposite directions with a velocity $c$ as shown in Figure 2.6. They are separated by a barrier from a reservoir containing additional positive and negative particles. A segment of the barrier contains a semipermeable membrane that allows both positive and negative particles to pass through from the reservoir to the upper region, where the positive ones then proceed to the

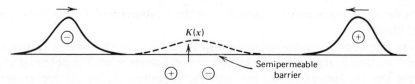

**Figure 2.6**    Interaction of particles via semipermeable membrane.

left with velocity $c$ while the negative ones proceed to the right at the same speed. The rate at which positive (negative) particles pass upward through the barrier at any point $x$ along the barrier is assumed to be proportional to the excess in the number of negative (positive) particles over an average value that are located above the barrier at that point. The proportionality factor is allowed to depend upon $x$. Representing the excess density of positive and negative particles at $x$ at time $t$ by $n_+(x,t)$ and $n_-(x,t)$, respectively, and denoting the proportionality factor in the permeability by $K(x)$, we have the following equations for the change in $n_+$ and $n_-$:

$$\frac{\partial n_+}{\partial t} - c\frac{\partial n_+}{\partial x} = K(x)n_-$$

$$\frac{\partial n_-}{\partial t} + c\frac{\partial n_-}{\partial x} = K(x)n_+$$

(2.11.2)

Only one space dimension will be considered. Spatial inhomogeneities associated with height above the barrier are ignored. Let us now introduce Fourier transforms by the relations

$$n_1(x,k) = \int_{-\infty}^{\infty} dt\, e^{ikct} n_+(x,t)$$

$$n_2(x,k) = \int_{-\infty}^{\infty} dt\, e^{ikct} n_-(x,t)$$

(2.11.3)

and assume that $n_+$ and $n_-$ vanish as $t$ approaches $\pm\infty$. The transform of (2.11.2) takes the form of (2.11.1) when we set $u(x) = -K(x)/c$.

Before considering the transformed equations, let us first examine a possible form of the solution in the time domain. By analogy with the solutions introduced in Section 2.10 for the second-order equation, let us consider a situation in which there is only a delta function of positive particles arriving at $x = -\infty$. [The comment in the footnote after (2.10.5) is again applicable.] This solution could be expected to be of the form

$$\varphi_+(x,t) = \delta\left(t + \frac{x}{c}\right) + c\theta\left(t + \frac{x}{c}\right)A_1(x, -ct)$$

$$\varphi_-(x,t) = c\theta\left(t + \frac{x}{c}\right)A_2(x, -ct)$$

(2.11.4)

where $\theta(x)$ is again the unit step function. As in Section 2.10, the functions $A_1$ and $A_2$ represent the wake as well as the reflected particles that could be expected to occur for $K(x) \neq 0$. Similarly, a delta function of negative particles arriving at $x = +\infty$ would be representable as

$$\psi_+(x,t) = c\theta\left(t - \frac{x}{c}\right)B_1(x,ct)$$

$$\psi_-(x,t) = \delta\left(t - \frac{x}{c}\right) + c\theta\left(t - \frac{x}{c}\right)B_2(x,ct)$$
(2.11.5)

After minor changes of variable, the Fourier transform of these solutions yields

$$\varphi_1(x,k) = e^{-ikx} + \int_{-\infty}^{x} dx' A_1(x,x')e^{-ikx'}$$

$$\varphi_2(x,k) = \int_{-\infty}^{x} dx' A_2(x,x')e^{-ikx'}$$
(2.11.6)

and

$$\psi_1(x,k) = \int_{x}^{\infty} dx' B_1(x,x')e^{ikx'}$$

$$\psi_2(x,k) = e^{ikx} + \int_{x}^{\infty} dx' B_2(x,x')e^{ikx'}$$
(2.11.7)

where the connection between subscripts $+, -$ and $1, 2$ is as in (2.11.3). As in the case of the Schrödinger equation, these forms of the solution will play a central role when the inverse scattering problem for (2.11.1) is considered.

## Fundamental Solutions

The four functions $\varphi_1$, $\varphi_2$, $\psi_1$, and $\psi_2$ can be considered as the fundamental solutions of the transformed problem (2.11.1). We now derive the integral equations satisfied by these solutions. They correspond to those given in (2.8.3) and (2.8.4) for the Schrödinger equation. It is convenient to first convert (2.11.7) into a vector equation for the two-component vector $n(x) = (n_1, n_2)^T$, where $T$ indicates the transpose operation, that is, $(n_1, n_2)^T = \begin{pmatrix} n_1 \\ n_2 \end{pmatrix}$. Introducing the matrix

$$A = ik\begin{pmatrix} 1 & 0 \\ 0 & -1 \end{pmatrix}$$
(2.11.8)

and the vector $w(x) = (un_2, -un_1)^T$, we may write (2.11.1) in the matrix form

$$n_x + An = w$$
(2.11.9)

As in obtaining (2.8.3) and (2.8.4) for the Schrödinger equation, let us convert (2.11.9) into an integral equation for the vector $n(x)$ by using a variation-of-parameters technique.

Since the homogeneous version of (2.11.9) has the two linearly independent solutions

$$n_1 = \begin{pmatrix} e^{-ikx} \\ 0 \end{pmatrix} \quad \text{and} \quad n_2 = \begin{pmatrix} 0 \\ e^{ikx} \end{pmatrix}$$

we may write a solution of (2.11.9) as the linear combination

$$n(x) = \alpha(x)n_1 + \beta(x)n_2 \tag{2.11.10}$$

where the amplitudes $\alpha$ and $\beta$ are to be determined. Substitution into (2.11.9) yields

$$\alpha_x n_1 + \beta_x n_2 = w \tag{2.11.11}$$

Separating into component form and integrating, we have

$$\alpha = \alpha_0 + \int_{x_0}^{x} dx' \, un_2 e^{ikx'}$$

$$\beta = \beta_0 - \int_{x_0}^{x} dx' \, un_1 e^{-ikx'} \tag{2.11.12}$$

where the constants $\alpha_0$, $\beta_0$, and $x_0$ are determined by initial conditions imposed upon the solution. If we consider the solution $n = \varphi = \begin{pmatrix} \varphi_1 \\ \varphi_2 \end{pmatrix}$ given in (2.11.6) which satisfies

$$\lim_{x \to -\infty} n(x) = \begin{pmatrix} 1 \\ 0 \end{pmatrix} e^{-ikx'} \tag{2.11.13}$$

then from (2.11.10) we see that

$$\alpha_0 + \int_{x_0}^{-\infty} dx' \, un_2 e^{ikx'} = 1$$

$$\beta_0 - \int_{x_0}^{-\infty} dx' \, un_1 e^{-ikx'} = 0 \tag{2.11.14}$$

and therefore, setting $n_1 = \varphi_1, n_2 = \varphi_2$,

$$\varphi_1(x,k) = e^{-ikx} + \int_{-\infty}^{x} dx' \, e^{ik(x'-x)} u(x')\varphi_2(x',k)$$

$$\varphi_2(x,k) = -\int_{-\infty}^{x} dx' \, e^{-ik(x'-x)} u(x')\varphi_1(x',k) \tag{2.11.15}$$

In a similar way the solution having the limiting form

$$\lim_{x \to \infty} n(x) = \begin{pmatrix} 0 \\ 1 \end{pmatrix} e^{ikx} \tag{2.11.16}$$

may be identified with the solution $\psi$ given in (2.11.7). We obtain

$$\psi_1(x,k) = -\int_x^\infty dx' \, e^{-ik(x'-x)} u(x') \psi_2(x',k)$$

$$\psi_2(x,k) = e^{ikx} + \int_x^\infty dx' \, e^{ik(x'-x)} u(x') \psi_1(x',k) \tag{2.11.17}$$

Since $u(x)$ is real, we have for real $k$ that $\varphi^*(x, -k) = \varphi(x,k)$ and $\psi^*(x, -k) = \psi(x,k)$.

## Wronskian Relations

If $\theta = (\theta_1, \theta_2)^T$ and $\chi = (\chi_1, \chi_2)^T$ are any two solutions of (2.11.1) for the same value of $k$, their linear independence is guaranteed by the nonvanishing of the expression

$$W(\theta; \chi) \equiv \theta_1 \chi_2 - \theta_2 \chi_1 \tag{2.11.18}$$

When (2.11.1) are written out for $\theta$ and $\chi$, they may be combined to show that $dW/dx = 0$ or $W = $ constant.

We now investigate the linear independence of various solutions of the transformed equations (2.11.1). First, it should be noted that if $n(x,k) = (n_1(x,k), n_2(x,k))^T$ represents a solution of (2.11.1) for some value of $k$, then by examining the equations obtained by replacing $k$ by $-k$ we see that the solution $\bar{n}(x,k) \equiv (n_2(x, -k), -n_1(x, -k))^T$ is also a solution of (2.11.1). It should be noted that $\bar{\bar{n}}(x,k) = -n(x,k)$. Also, the character of the solution of the linear equations (2.11.1) does not depend critically upon the sign of $u$ as it did in the case of the Schrödinger equation. More precisely, if $n(x,k) = (n_1(x, k), n_2(x, k))^T$ is a solution of (2.11.1), then $n(x, k) = (n_2(x, -k), n_1(x, -k))^T$ is a solution of these equations when $u$ is replaced by $-u$. The solutions $\varphi(x, k) = (\varphi_1(x, k), \varphi_2(x, k))^T$ and $\bar{\varphi}(x, k) = (\varphi_2(x, -k), -\varphi_1(x, -k))^T$ are readily shown to be linearly independent since evaluation of the Wronskian $W(\varphi, \bar{\varphi})$ at $x = -\infty$ yields $W(\bar{\varphi}, \varphi) = 1$. Similarly, we find that $W(\bar{\psi}, \psi) = 1$.

Since any third solution may be expressed as a linear combination of two linearly independent solutions, we may write

$$\varphi(x,k) = c_{11}(k) \psi(x,k) + c_{12}(k) \bar{\psi}(x,k) \tag{2.11.19a}$$

$$\psi(x,k) = c_{21}(k) \bar{\varphi}(x,k) + c_{22}(k) \varphi(x,k) \tag{2.11.19b}$$

and

$$\bar{\varphi}(x,k) = c_{11}(-k)\bar{\psi}(x,k) - c_{12}(-k)\psi(x,k) \qquad (2.11.20a)$$

$$\bar{\psi}(x,k) = -c_{21}(-k)\varphi(x,k) + c_{22}(-k)\bar{\varphi}(x,k) \qquad (2.11.20b)$$

where the $c_{ij}(k)$ are scalars.

As was the case in the treatment of the Schrödinger equation, there are a number of relations among the various coefficients $c_{ij}(k)$. These coefficients have the following definitions in terms of Wronskians:*

$$c_{11}(k) = W(\bar{\psi},\varphi) = c_{22}(-k)$$
$$c_{12}(k) = W(\varphi,\psi) = -c_{21}(k)$$
$$c_{22}(k) = W(\bar{\varphi},\psi) = c_{11}(-k) \qquad (2.11.21)$$
$$c_{12}(-k) = W(\bar{\varphi},\bar{\psi}) = -c_{21}(-k)$$

Since $\varphi^*(x,k) = \varphi(x,-k)$ and similarly for $\psi$, we obtain

$$c_{11}^*(k) = c_{22}(k), \qquad c_{12}^*(k) = c_{12}(-k) \qquad (2.11.22)$$

Compatibility of both (2.11.19) and (2.11.20) requires that

$$1 = c_{11}(k)c_{22}(k) - c_{12}(k)c_{21}(-k)$$
$$0 = c_{11}(k)c_{21}(k) + c_{12}(k)c_{22}(-k) \qquad (2.11.23)$$
$$0 = c_{11}(k)c_{21}(-k) + c_{12}(-k)c_{22}(-k)$$

The linear combinations in (2.11.19) and (2.11.20) may be interpreted as scattering problems, as was the case for the Schrödinger equation. Let us consider (2.11.19a) in detail. From the definition of $\varphi$ and $\psi$ we see that (2.11.19a) is a solution that reduces to $\binom{1}{0}e^{-ikx}$ as $x$ approaches $-\infty$ and, using the definition of $\bar{\psi}$, reduces to $c_{11}(k)\binom{0}{1}e^{ikx} + c_{12}(k)\binom{1}{0}e^{-ikx}$ as $x$ approaches $+\infty$. In the time domain the term $\binom{1}{0}e^{-ikx}$ would represent a delta function of positive particles receding to $-\infty$. The term $c_{11}(k)\binom{0}{1}e^{ikx}$ represents a flux of negative particles receding to $+\infty$ while $c_{12}(k)\binom{1}{0}e^{-ikx}$ represents an incident flux of positive particles. A unit flux (i.e., a delta function) of positive particles from $+\infty$ is obtained by dividing (2.11.19a) by $c_{12}$. We then have reflection and transmission coefficients as for the Schrödinger equation

$$R_R(k) = \frac{c_{11}(k)}{c_{12}(k)}; \qquad T(k) = \frac{1}{c_{12}(k)} \qquad (2.11.24)$$

*The $c_{ij}(k)$ are frequently written in terms of the two coefficients $a(k) = c_{12}(k)$ and $b(k) = c_{11}(k)$.

Equation 2.11.19b reduces to $\begin{pmatrix} 0 \\ 1 \end{pmatrix} e^{ikx}$ as $x$ approaches $+\infty$ and reduces to $-c_{21}(k)\begin{pmatrix} 0 \\ 1 \end{pmatrix} e^{ikx} + c_{22}(k)\begin{pmatrix} 1 \\ 0 \end{pmatrix} e^{-ikx}$ as $x$ approaches $-\infty$. We may therefore define

$$R_L(k) = -\frac{c_{22}(k)}{c_{21}(k)}; \qquad T(k) = -\frac{1}{c_{21}(k)} \tag{2.11.25}$$

as the reflection and transmission coefficients for a unit flux of negative-type particles incident from $-\infty$. The latter two of the equations in (2.11.23) are satisfied identically by virtue of (2.11.21). They may be rewritten as

$$R_R(k)T(-k) - R_L(-k)T(k) = 0 \tag{2.11.26}$$

Equations 2.11.22 and 2.11.23 yield

$$1 = |c_{11}(k)|^2 + |c_{12}(k)|^2 \tag{2.11.27}$$

or

$$|T(k)|^2 = 1 + |R_L(k)|^2 = 1 + |R_R(k)|^2 \tag{2.11.28}$$

These results should be compared with (2.8.14) and (2.8.15) for the Schrödinger equation. In the model described at the beginning of this section, it was seen that the number of particles participating in the process was not conserved. This has resulted in a relation among the coefficients that differs from the energy conservation law expressed by (2.8.14).

## Poles of the Transmission Coefficient

It is again the poles of the transmission coefficient or the zeros of $c_{12}(k) = -c_{21}(k)$ that are important in determining the bound states. In scattering problems governed by the Schrödinger equation with a real potential, it was determined from (2.8.19) that all poles of the transmission coefficient that are in the upper half-plane must be on the imaginary axis. This is not necessarily the case for the present problem since, as noted in the introduction, the pair of linear equations (2.11.1) are equivalent to a Schrödinger equation with a complex potential. However, as long as the function $u(x)$ in (2.11.1) is real, a restriction on the location of the poles can be inferred from the relation

$$c_{12}(k) = c_{12}^*(-k) \tag{2.11.29}$$

given in (2.11.22). To obtain this restriction we first recall that the fundamental solutions $f_1$ and $f_2$ were analytic in the upper half-plane. The fundamental solutions $\varphi$ and $\psi$ in the relation $c_{12} = W(\varphi, \psi)$ may also be extended into the upper half-plane. Hence in the upper half-plane a corresponding extension (analytic continuation) of (2.11.29) for real $k$ may be introduced. The extension of (2.11.29) into the upper half-plane that reduces to (2.11.29) on the real axis is

$$c_{12}(\zeta) = c_{12}^*(-\zeta) \qquad (2.11.30)$$

where $\zeta = k + i\eta$ and $\eta > 0$. Writing $c_{12}(\zeta) = u(k,\eta) + iv(k,\eta)$, we see that (2.11.30) implies that $u(k,\eta) = u(-k,\eta)$ and $v(k,\eta) = -v(-k,\eta)$. A zero of $c_{12}(\zeta)$ thus implies that $u$ and $v$ must vanish for both sets of arguments. This can be accomplished either by having $k = 0$ as was the case for the Schrödinger equation with a real potential, or by having the zeros come in pairs at the points $\zeta_l = \pm k_l + i\eta_l$, that is, symmetrically placed about the imaginary axis.

To find the residues at the points $\zeta_l$ we must again evaluate $\dot{c}_{12}(\zeta_l) = dc_{12}(\zeta)/d\zeta|_{\zeta=\zeta_l}$. From the Wronskian definition of $c_{12}$ and the fact that $\varphi = c_{11}\psi$ and $\psi = c_{22}\varphi$ at the zeros of $c_{12}$, we find that

$$\dot{c}_{12} = c_{22}(\dot{\varphi}_1\varphi_2 - \dot{\varphi}_2\varphi_1) + c_{11}(\dot{\psi}_2\psi_1 - \dot{\psi}_1\psi_2) \qquad (2.11.31)$$

This result may be rewritten with the help of a general relation that exists among the solutions of the linear equations (2.11.1). If $n = (n_1, n_2)^T$ and $m = (m_1, m_2)^T$, refer to solutions of (2.11.1) for $k = \zeta_1$ and $\zeta_2$, respectively, then

$$\frac{d}{dx}(m_2 n_1 - m_1 n_2) + i(\zeta_1 - \zeta_2)(m_2 n_1 + m_1 n_2) = 0 \qquad (2.11.32)$$

Setting $n = \varphi(x, \zeta)$ and $m = \varphi(x, \zeta_l)$ in (2.11.32), then differentiating the resulting equation with respect to $\zeta$ and finally setting $\zeta = \zeta_l$, we obtain

$$\frac{d}{dx}(\varphi_2\dot{\varphi}_1 - \varphi_1\dot{\varphi}_2) = -2i\varphi_1\varphi_2 \qquad (2.11.33)$$

Integration now yields

$$\varphi_2\dot{\varphi}_1 - \varphi_1\dot{\varphi}_2 = -2i\int_{-\infty}^{x} dx'\,\varphi_1\varphi_2 \qquad (2.11.34)$$

Similarly, choosing $n = \psi(x, \zeta)$ and $m = \psi(x, \zeta_l)$, we also obtain

$$\psi_2\dot{\psi}_1 - \psi_1\dot{\psi}_2 = 2i\int_{x}^{\infty} dx'\,\psi_1\psi_2 \qquad (2.11.35)$$

With these results we may finally obtain $\dot{c}_{12}$ from (2.11.31). The result is

$$\dot{c}_{12}(\zeta_l) = -2i\int_{-\infty}^{\infty} dx\,\psi_1(x,\zeta_l)\varphi_2(x,\zeta_l) \qquad (2.11.36)$$

Since $c_{12}(\zeta_l) = 0$, we may use (2.11.19) to derive the relations

$$m_R(\zeta_l) \equiv -i\frac{c_{11}(\zeta_l)}{\dot{c}_{12}(\zeta_l)} = \frac{1}{2}\left[\int_{-\infty}^{\infty} dx\, \psi_1(x,\zeta_l)\psi_2(x,\zeta_l)\right]^{-1} \quad (2.11.37a)$$

$$m_L(\zeta_l) \equiv -i\frac{c_{22}(\zeta_l)}{\dot{c}_{12}(\zeta_l)} = \frac{1}{2}\left[\int_{-\infty}^{\infty} dx\, \varphi_1(x,\zeta_l)\varphi_2(x,\zeta_l)\right]^{-1} \quad (2.11.37b)$$

Since the four solutions $\psi_1$, $\psi_2$, $\varphi_1$, and $\varphi_2$ are all purely real when $\zeta_l$ is purely imaginary, the constants $m_R(\zeta_l)$ and $m_L(\zeta_l)$ are also purely real in this case. For two zeros $\zeta_1$ and $\zeta_2$ that are located symmetrically about the imaginary axis so that $\zeta_2 = -\zeta_1^*$, the constants $m_R(\zeta_1)$ and $m_R(\zeta_2)$ are related by

$$m_R(\zeta_2) = [m_R(\zeta_1)]^* \quad (2.11.38)$$

This follows from the elimination of $\psi_1$ in (2.11.17), which then shows that $\psi_2(x,\zeta_1)$ satisfies

$$\psi_2(x,\zeta_1) = e^{i\zeta_1 x} - \int_x^{\infty} dx'\, e^{i\zeta_1(x'-x)}u(x')\int_{x'}^{\infty} dx''\, e^{-i\zeta_1(x''-x')}u(x'')\psi_2(x'',\zeta_1)$$

$$(2.11.39)$$

As long as $u(x)$ is real, this equation is unchanged when $\zeta_1$ is replaced by $-\zeta_1^*$ and the complex conjugate of the resulting equation is taken. Therefore, $\psi_2(x_1,\zeta_1) = [\psi_2(x_1, -\zeta_1)]^*$. From the first of (2.11.17), we see that $\psi_1(x,\zeta_1)$ also shares this property. The relation between $m_R(\zeta_2)$ and $m_R(\zeta_1)$ now follows from (2.11.37a).

### An Example

Many potentials are known for which the Schrödinger equation may be solved analytically (see, for example, Brekhovskikh, 1960, p. 447). A list of functions for which the linear equations in (2.11.1) may be solved analytically is not as readily available. An important class of functions for which these linear equations can be solved will be obtained later by inverse scattering techniques. As a simple preliminary example of these results, we may proceed along the lines developed by Bargmann for the Schrödinger equation and outlined in Section 1.3. We look for a solution of (2.11.1) in the form

$$n_1 = e^{-ikx}[4ik + a(x)]$$

$$n_2 = e^{-ikx}b(x)$$

$$(2.11.40)$$

Since no term proportional to $e^{ikx}$ has been included, we expect that the potential will be found to be reflectionless. Equations 2.11.1 impose the relations

$$a' = ub, \qquad b = 2u, \qquad b' = -ua \qquad (2.11.41)$$

Eliminating $u$ between the first and third of these and then integrating, we obtain $a^2 + b^2 = 4\mu^2$, where $\mu$ is a constant of integration. From the first and second we also have $b^2 = 2a'$. We thus find that the function $a(x)$ satisfies (1.3.5), the same first-order nonlinear equation as that satisfied by the function $a(x)$ that was introduced in solving the Schrödinger equation. The solution may again be written

$$a = \frac{2w'}{w} \qquad (2.11.42)$$

where

$$w = \alpha e^{\mu x} + \beta e^{-\mu x}$$

$$= 2e^{\phi}\cosh(\mu x - \phi) \qquad (2.11.43)$$

and again $\phi = \frac{1}{2}\ln(\beta/\alpha)$. Then

$$u^2 = \frac{1}{2}a' = \mu^2 \operatorname{sech}^2(\mu x - \phi) \qquad (2.11.44)$$

We thus obtain

$$u = \pm \mu \operatorname{sech}(\mu x - \phi) \qquad (2.11.45)$$

It will be sufficient to consider $\mu > 0$. The functions $a(x)$ and $b(x)$ are also easily determined. A fundamental solution of (2.11.1) is found from (2.11.40) to be

$$\varphi_1(x,k) = e^{-ikx}\frac{2ik + \mu\tanh(\mu x - \phi)}{2ik - \mu}$$

$$(2.11.46)$$

$$\varphi_2(x,k) = \pm \mu e^{-ikx}\frac{\operatorname{sech}(\mu x - \phi)}{2ik - \mu}$$

Repeating the foregoing calculation with

$$n_1 = e^{ikx}q(x)$$

$$(2.11.47)$$

$$n_2 = e^{ikx}[4ik + p(y)]$$

we find that $q = b$ and $p = -a$. The fundamental solution is now

$$\psi_1(x,k) = \pm \mu e^{ikx} \frac{\operatorname{sech}(\mu x - \phi)}{2ik - \mu}$$

$$\psi_2(x,k) = e^{ikx} \frac{2ik - \mu \tanh(\mu x - \phi)}{2ik - \mu} \qquad (2.11.48)$$

From the Wronskian relation for $c_{12}$, we find that

$$c_{12}(k) = \frac{k - i\mu/2}{k + i\mu/2} \qquad (2.11.49)$$

The zero of $c_{12}$ is at $k = i\mu/2$ and $\operatorname{Res}[1/c_{12}(i\mu/2)] = i\mu$. We also find that $c_{11}(k) = c_{22}(k) = 0$ so that the potential is reflectionless, as expected. The localized solutions are

$$\varphi_1\left(x, \frac{i\mu}{2}\right) = \tfrac{1}{2} e^{-\mu x/2} \operatorname{sech}\mu x$$

$$\varphi_2\left(x, \frac{i\mu}{2}\right) = \mp \tfrac{1}{2} e^{\mu x/2} \operatorname{sech}\mu x$$

$$\psi_1\left(x, \frac{i\mu}{2}\right) = \mp \tfrac{1}{2} e^{-\mu x/2} \operatorname{sech}\mu x \qquad (2.11.50)$$

$$\psi_2\left(x, \frac{i\mu}{2}\right) = \tfrac{1}{2} e^{\mu x/2} \operatorname{sech}\mu x$$

From (2.11.37) we see that the normalization constants are

$$m_R\left(\frac{i\mu}{2}\right) = m_L\left(\frac{i\mu}{2}\right) = \mp \mu \qquad (2.11.51)$$

where the upper and lower signs are associated with the upper and lower signs, respectively, in $u(x)$ as given in (2.11.45).

*Exercise 24*

Another soluble profile is the rectangular barrier $u(x) = u_0$ for $0 < x < a$ and $u(x) = 0$ elsewhere. Use (2.11.19a) to match solutions of (2.11.1) in the three regions and show that

$$R_R(k) = -u_0 e^{-2ika} \frac{\sin \alpha a}{D}$$

$$T(k) = \frac{\alpha e^{-ika}}{D}$$

where $\alpha = (k^2 + u_0^2)^{1/2}$ and $D = \alpha \cos \alpha a - ik \sin \alpha a$.

**Asymptotic Solution**

An asymptotic solution analogous to that developed for the Schrödinger equation in Section 2.8 may be obtained for the two-component system (2.11.1). When the equations for the two-component system are combined to yield a second-order equation for $n_1$, we find that

$$n_1'' - \left(\frac{u'}{u}\right)n_1' + \left(k^2 + u^2 - \frac{iku'}{u}\right)n_1 = 0 \qquad (2.11.52)$$

If we consider $n_1$ to be the solution $\varphi_1$, given in (2.11.15), we may write

$$n_1 = e^{-ikx + h(x,k)} \qquad (2.11.53)$$

where $h(-\infty, k) = 0$. The equation satisfied by $h$ is found from (2.11.52) to be

$$2ikh' = u^2 + (h')^2 + u\frac{d}{dx}\left(\frac{h'}{u}\right) \qquad (2.11.54)$$

If we substitute the expansion

$$h' = \sum_{n=0}^{\infty} \frac{g_n(x)}{(-2ik)^{n+1}} \qquad (2.11.55)$$

into (2.11.54), we find that $g_0 = -u^2, g_1 = uu'$, and

$$g_{n+1} = -u\frac{d}{dx}\left(\frac{g_n}{u}\right) - \sum_{m=0}^{n-1} g_m g_{n-m-1}, \qquad n \geqslant 1 \qquad (2.11.56)$$

The first few coefficients are

$$g_0 = -u^2, \qquad g_1 = uu', \qquad g_2 = -(uu'' + u^4), \qquad g_3 = uu''' + 5u^3u',$$

$$g_4 = -uu^{(iv)} - 11u^2(u')^2 - 7u^3u'' - 2u^6 \qquad (2.11.57)$$

$$= -uu^{(iv)} + 10u^2(u')^2 - 2u^6 - 7(u^3u')'$$

**Truncated Potentials**

We may extend the range of potentials for which the two-component scattering problem may be solved analytically by again considering truncated potentials. The calculation parallels that of Section 2.9 for the Schrödinger equation. We consider in detail the case in which the potential has been truncated on the left. Then, $u^T(x) = u(x)\theta(x - x_1)$ and we find from (2.11.15) and (2.11.17) that in the region $x < x_1$ the fundamental solutions are

$$\psi_{1<}^T = -\int_{x_1}^{\infty} dx' e^{-ik(x' - x)} u(x') \psi_2(x', k)$$

$$\psi_{2<}^T = e^{ikx} + \int_{x_1}^{\infty} dx' e^{ik(x' - x)} u(x') \psi_1(x', k) \qquad (2.11.58)$$

$$\varphi_{1<}^T = e^{-ikx}$$

$$\varphi_{2<}^T = 0$$

For $x > x_1$ the fundamental solutions are

$$\psi_{1>}^T = \psi_1(x, k)$$

$$\psi_{2>}^T = \psi_2(x, k)$$

$$\varphi_{1>}^T = e^{-ikx} + \int_{x_1}^{x} dx' e^{ik(x' - x)} u(x') \varphi_{2>}^T(x', k) \qquad (2.11.59)$$

$$\varphi_{2>}^T = \int_{x_1}^{x} dx' e^{ik(x' - x)} u(x') \varphi_{1>}^T(x', k)$$

These solutions must be continuous across $x = x_1$.

The coefficients $c_{ij}^T$ associated with this truncated potential now follow from (2.11.21). We find that

$$c_{11}^T(k) = e^{-ikx_1} \psi_1(x_1, -k)$$

$$c_{12}^T(k) = -c_{21}^T(k) = e^{-ikx_1} \psi_2(x_1, k) \qquad (2.11.60)$$

$$c_{22}^T(k) = e^{ikx_1} \psi_1(x_1, k)$$

The fundamental solutions for the potential $u(x) = \mu \operatorname{sech} \mu x$ were given in (2.11.46) and (2.11.48). These previous results may now be used to determine the coefficients $c_{ij}^T$ for the truncated potential $u^T(x) = \pm \mu \theta(x - x_1) \operatorname{sech} \mu x$. We obtain

$$c_{11}^T(k) = \mp e^{-2ikx_1} \mu \frac{\operatorname{sech} \mu x_1}{2ik + \mu}$$

$$c_{12}^T(k) = -c_{21}^T(k) = \frac{2ik - \mu \tanh \mu x_1}{2ik - \mu} \qquad (2.11.61)$$

$$c_{22}^T(k) = \pm e^{2ikx_1} \mu \frac{\operatorname{sech} \mu x_1}{2ik - \mu}$$

The pole of the transmission coefficient is located at $k = -\frac{1}{2} i\mu \tanh \mu x_1$. For $\mu > 0$ it is in the upper half-plane for $x_1 < 0$ and is on the real $k$ axis for $x_1 = 0$.

Note that this result differs from that for the Schrödinger equation with a real potential that was obtained in (2.9.7). There it was found that the pole could never be on the real $k$ axis and there was always a pole in the upper half-plane, no matter how small the negative potential is.

## 2.12 RELATION BETWEEN ONE- AND TWO-COMPONENT EQUATIONS—RICCATI EQUATIONS

In Section 1.5 we saw that the linear equations

$$y_x + iuy = \zeta z$$
$$z_x - iuz = -\zeta y \qquad (2.12.1)$$

are equivalent to second-order equations of Schrödinger type in which the potential is complex. More specifically, we found that

$$y_{xx} + (\zeta^2 + u^2 + iux)y = 0 \qquad (2.12.2)$$

Similarly, we can obtain

$$z_{xx} + (\zeta^2 + u^2 - iu_x)z = 0 \qquad (2.12.3)$$

We now recover this result by an alternative method which proceeds by way of a nonlinear first-order differential equation known as a Riccati equation. Riccati equations will occur again in later developments of soliton theory and for later convenience we now summarize some of the properties of this equation (Davis, 1960).

If the first and second of equations (2.12.1) are multiplied by $z$ and $y$, respectively, and subtracted, the resulting equation, after being divided by $z^2$, is the nonlinear equation

$$\varphi_x + 2iu\varphi = \zeta(1 + \varphi^2) \qquad (2.12.4)$$

where $\varphi \equiv y/z$. This first-order nonlinear equation is an example of a Riccati equation. For the present we merely note that the substitution $\varphi = -\theta_x/\zeta\theta$ leads to the second-order linear equation

$$\theta_{xx} + 2iu\theta_x + \zeta^2\theta = 0 \qquad (2.12.5)$$

When the first derivative term in this equation is eliminated by the substitution $\theta = \psi \exp(-i\int u \, dx)$, we obtain

$$\psi_{xx} + (\zeta^2 + u^2 - iu_x)\psi = 0 \qquad (2.12.6)$$

which is of the same form as the equation for $z$ in (2.12.3). Thus $\psi$ will be equal to $z$ if both functions are chosen to have the same boundary conditions.

To return from the Riccati equation (2.12.4) to the linear system (2.12.1) is somewhat more subtle, since the substitution $\varphi = y/z$ converts the Riccati equation to $z(y_x + iuy - \zeta z) - y(z_x - iuz + \zeta y) = 0$, which can be satisfied by (2.12.1), although not uniquely. However, if we note that

$$\varphi = -\frac{\theta x}{\zeta \theta} = -\frac{\psi_x - iu\psi}{\zeta \psi} \qquad (2.12.7)$$

we may set $\psi_x - iu\psi = \lambda y$ and $\zeta \psi = \lambda z$, where $\lambda$ is to be determined. Eliminating $\psi_x$, we see that the second of (2.12.1) is recovered if we choose $\lambda$ to be a constant.

For future reference we now briefly consider certain general properties of Riccati equations. A standard form for this equation is

$$\varphi_x + A(x)\varphi + B(x)\varphi^2 = C(x) \qquad (2.12.8)$$

The substitution

$$\varphi = \frac{\theta_x}{B\theta} \qquad (2.12.9)$$

transforms the Riccati equation to the second-order linear equation

$$\theta_{xx} + \left(A - \frac{B_x}{B}\right)\theta_x - BC\theta = 0 \qquad (2.12.10)$$

The general structure of the solution of the Riccati equation may be inferred from the general solution of this linear equation. If $\theta_1$ and $\theta_2$ are two linearly independent solutions of (2.12.10), then the general solution is of the form $\theta = a\theta_1 + b\theta_2$, where $a$ and $b$ are constants. From (2.12.8) we see that the Riccati equation has the general solution $\varphi = [(a/b)\theta_{1x} + \theta_{2x}]/[(a/b)\theta_1 + \theta_2]$, which is of the form

$$\varphi = \frac{kP + Q}{kR + S} \qquad (2.12.11)$$

where $k(=a/b)$ is the constant of integration and $P$, $Q$, $R$, and $S$ are functions of $x$.

Although a Riccati equation may always be transformed to a linear equation, as shown above, the general solution of this associated linear equation may be difficult to obtain. The situation is much simpler, however, whenever a particular solution of the Riccati equation is available. If a particular solution is denoted by $\varphi_1$, the function $f_1$ defined by $1/f_1 \equiv \varphi - \varphi_1$ is readily shown to satisfy the linear equation

$$f_{1x} - (A + 2B\varphi_1)f_1 - B = 0 \qquad (2.12.12)$$

The general solution of this equation, and hence that of the original Riccati equation can now be obtained.

If an additional particular solution $\varphi = \varphi_2$ is known, then we may similarly introduce a function $f_2$ by $1/f_2 \equiv \varphi - \varphi_2$ and find that

$$f_{2x} - (A + 2B\varphi_2)f_2 - B = 0 \qquad (2.12.13)$$

If (2.12.11) and (2.12.12) are divided by $f_1$ and $f_2$ respectively, we obtain

$$\left(\frac{f_1}{f_2}\right)_x = B(\varphi_1 - \varphi_2)\left(\frac{f_1}{f_2}\right) \qquad (2.12.14)$$

which has the solution

$$\frac{f_1}{f_2} = ae^{\int dx B(\varphi_1 - \varphi_2)} \qquad (2.12.15)$$

where $a$ is the constant of integration. From the definition of $f_1$ and $f_2$ given above, the general solution of the Riccati equation is found to be

$$\frac{\varphi - \varphi_2}{\varphi - \varphi_1} = ae^{\int B(\varphi_1 - \varphi_2)dx} \qquad (2.12.16)$$

Finally, we list the general form of the transformation used above for eliminating the first derivatives in the linear second-order equation (2.12.5). In the differential equation

$$\theta_{xx} + a(x)\theta_x + b(x)\theta = 0 \qquad (2.12.17)$$

the substitution $\theta = \psi \exp[-\frac{1}{2}\int a(x)dx]$ transforms the equation to

$$\psi_{xx} + \left[b - \frac{1}{4}(a^2 + 2a_x)\right]\psi = 0 \qquad (2.12.18)$$

*Exercise 25*

Show that the linear system (1.5.3) is equivalent to the Riccati equation $\varphi_x + 2i\zeta\varphi = u(1 + \varphi^2)$, where $\varphi = n_1/n_2$. Convert this Riccati equation to the Schrödinger equation $\psi_{xx} + [\zeta^2 + u^2 + i\zeta(u_x/u) - \frac{1}{4}(u_x/u)^2 + \frac{1}{2}(u_x/u)_x]\psi = 0$ by the transformations $\varphi = -\theta_x/u\theta$ and $\theta = \psi u^{1/2}e^{-i\zeta x}$. By using the method outlined in the text, return to the linear equations from the Riccati equation.

*Exercise 26*

Consider the Riccati equation that occurs when we set $u = \text{sech}\, x$ in (2.12.4). If we set $\varphi = i\bar{\varphi}$, the equation for $\bar{\varphi}$ is in a form that will occur in Chapter 8. Show that the resulting linear equation may be transformed to $\psi_{yy} + (4\zeta^2 + 2\,\text{sech}^2 y)\psi = 0$, where

$=\frac{1}{2}x+\frac{1}{4}i\pi$. Use (2.6.14) to show that a particular solution of the Riccati equation for $\bar{\varphi}$ is $\bar{\varphi}_1 = -[2\zeta + i\exp(-i\sigma/2)]/[2\zeta + i\exp(i\sigma/2)]$, where $\sigma = 4\tan^{-1}e^y$.

Use $\varphi_2 = -1/\varphi_1^*$ as the second solution and employ (2.12.15) to show that the general solution of the Riccati equation

$$\bar{\varphi}_x + 2i(\operatorname{sech} x)\bar{\varphi} + i\zeta(1-\varphi^2) = 0$$

is of the form of (2.12.11) with

$$P = e^{-i\nu x}(\nu + ie^{-i\sigma/2})$$
$$Q = \nu - ie^{-i\sigma/2}$$
$$R = e^{-i\nu x}(\nu + ie^{i\sigma/2})$$
$$S = -(\nu - ie^{i\sigma/2})$$

where $\nu = 2\zeta$ and $\sigma = 4\tan^{-1}e^x$.

Show that the general solution of the Riccati equation with $u$ and $\zeta$ interchanged, that is, $\chi_x + 2i\zeta\chi + i(\operatorname{sech} x)(1-\chi^2) = 0$, is

$$\chi = \frac{\bar{\varphi}+1}{\bar{\varphi}-1}$$

### Exercise 27

When (2.12.16) is solved for $\varphi$, we obtain $\varphi = (\varphi_2 - af\varphi_1)/(1-af)$, where $f = \exp[\int B(\varphi_2 - \varphi_1)dx]$. Four different values of the constant $a$ yield four different solutions of the Riccati equation (2.12.8). Show that these four solutions satisfy the cross-ratio theorem

$$\frac{(\varphi_1 - \varphi_3)(\varphi_2 - \varphi_4)}{(\varphi_1 - \varphi_4)(\varphi_2 - \varphi_3)} = \frac{(a_1 - a_3)(a_2 - a_4)}{(a_1 - a_4)(a_2 - a_3)} = \text{constant}$$

From this result it follows that when three solutions of a Riccati equation are known, the general solution may be obtained without any integration.

# CHAPTER 3

# Inverse Scattering in
# One Dimension

In Chapter 2 we considered examples of what could be called the direct scattering problem, that is, the determination of scattered waves when the shape of the scattering potential or refractive index profile has been specified. The use of scattering theory in the study of solitons requires consideration of a problem that is the inverse of this, namely the determination of the potential when information concerning the scattered wave (and bound states if present) is specified. In succeeding chapters, inverse scattering techniques will be used to analyze initial-value problems associated with various nonlinear evolution equations. In the present chapter we develop the topics in inverse scattering theory that will be needed in our consideration of these evolution equations.

## 3.1 RELATION BETWEEN THE POTENTIAL AND THE FUNCTIONS $A_R(x,y)$ AND $A_L(x,y)$

In Section 2.10 we saw that the Marchenko equation (2.10.17) provides a relation (an integral equation) between $r_R(x)$, the Fourier transform of the reflection coefficient $R_R(k)$, and the function $A_R(x,y)$ that appears in the definition of the fundamental solution $f_1(x,k)$ given in (2.10.5). We now consider how a knowledge of the function $A_R(x,y)$ enables us to determine the potential $U(x)$. We will then have a procedure for going from the reflection coefficient $R_R(k)$ to the potential $U(x)$ via the functions $r_R(x)$ and $A_R(x,y)$. A similar result obtains for the reflection coefficient $R_L(x)$ and the function $A_L(x,y)$.

It is clear from the discussion of the propagation of delta function pulses in Section 2.10 that the function $A_R(x,y)$ should be directly related to the potential. For if $U(x)=\mu^2(x)=0$ in the wave equation (2.10.1), the delta function $\delta(t-x/c)$ itself is an exact solution of the wave equation. The function $A_R(x,ct)$ in (2.10.2), which represents the wake behind the delta function pulse or, in general, the additional part of the solution that can be attributed to the function $U(x)$, is identically zero. To find the relation between $A_R(x,y)$ and $U(x)$ when the potential is not zero, we merely substitute the expression for $y_1(x,t)$ given in (2.10.2) into the wave equation

(2.10.1) and obtain

$$\left(1-\frac{x}{c}\right)\left[2\frac{\partial A_R(x,ct)}{\partial x}+\frac{2}{c}\frac{\partial A_R(x,ct)}{\partial t}+U(x)\right]$$

$$-c\theta\left(1-\frac{x}{c}\right)\left[\frac{\partial^2 A_R(x,ct)}{\partial x^2}-\frac{1}{c^2}\frac{\partial^2 A_R(x,ct)}{\partial t^2}-U(x)A_R(x,ct)\right]=0 \quad (3.1.1)$$

The coefficient of the step function shows, as expected, that the equation satisfied by $A_R(x,y)$ is no simpler than that for the function $y_1(x,t)$. The coefficient of the delta function, however, is the term of interest. Integrating (3.1.1) over time from $(x/c)-\epsilon$ to $(x/c)+\epsilon$, we obtain

$$2\frac{\partial A_R(x,ct)}{\partial x}\bigg|_{ct=x}+2\frac{\partial A_R(x,ct)}{\partial ct}\bigg|_{ct=x}+U(x)=0 \quad (3.1.2)$$

This provides the relation that we seek, namely

$$U(x)=-2\frac{d}{dx}A_R(x,x) \quad (3.1.3)$$

This is the prescription for obtaining the potential from the function $A_R(x,y)$. A similar calculation employing the function $y_2(x,t)$ given in (2.10.6) yields

$$U(x)=2\frac{d}{dx}A_L(x,x) \quad (3.1.4)$$

When the reflection coefficient is specified and either of the Marchenko equations (2.10.17) or (2.10.18) is solved for $A_R(x,y)$ or $A_L(x,y)$, equation (3.1.3) or (3.1.4) yields the potential that produced the given reflection coefficient.

Somewhat lengthy considerations lead to the criterion

$$\int_{-\infty}^{\infty}dx(1+|x|)|u(x)|<\infty \quad (3.1.5)$$

for the validity of the procedure (Faddeev, 1967). The method is thus inapplicable when the integral in (3.1.5) diverges. This happens if the potential is either too singular, such as $u(x)=\delta'(x)$ the derivative of the delta function, or too slowly converging for large values of $x$.

Unfortunately, solution of the Marchenko equation is no simple matter in most instances. It is an integral equation and, even though linear, can only be solved analytically in certain simple cases.

The derivation of the Marchenko equation given in Chapter 2 was restricted to situations in which the potential had no bound states. This restriction will be relaxed in the next section. Let us first indicate the method for reconstructing the potential from the reflection coefficient by considering a simple example in which no bound states are present. We consider a reflection coefficient that will lead to the repulsive delta function potential.

### Example—Repulsive Delta Function Potential

For waves incident from the left, the reflection coefficient for the potential $u(x) = 2b\,\delta(x)$ was obtained in Ex. 19 of Chapter 2. The result is

$$R_L(k) = \frac{-ib}{k+ib}, \qquad b > 0 \tag{3.1.6}$$

We now use this expression as our starting point for recovering the delta function potential. According to (2.8.32), the Fourier transform of (3.1.4) is

$$r_L(z) = -be^{-bz}\theta(z) \tag{3.1.7}$$

The Marchenko equation (2.10.18) then becomes

$$be^{-b(x+y)}\theta(x+y) + b\int_{-y}^{x} dx' e^{-b(x'+y)} A_L(x,x') - A_L(x,y) = 0 \tag{3.1.8}$$

For $x < -y$ both the step function and the integral are zero, so $A_L(x,y) =$ for $x+y < 0$. For $x+y > 0$ the integral equation can be satisfied by setting $A_L(x,y)$ equal to a constant. A general procedure leading to this result is described in Ex. 1. Setting $A_L(x,y) = K$, we find by direct substitution into the integral equation that $K = b$. The solution is then

$$A_L(x,y) = b\theta(x+y) \tag{3.1.9}$$

and the formula for the potential given in (3.1.4) yields the expected result

$$U(x) = 2b\frac{d}{dx}\theta(2x) = 2b\,\delta(x) \tag{3.1.10}$$

The following is an alternative approach to the foregoing problem.

### Exercise 1

It will be shown in Section 3.4 that when the reflection coefficient is a ratio of polynomials in $k$, as is the case for the reflection coefficient given above in (3.1.6), we may expect the solution of the Marchenko equation to be a sum of products of the form $A_L(x,y) = \sum_{i=1}^{N} X_i(x) Y_i(y)$. Choose $N = 1$ and make the further factorization

$r(y)=h(y)e^{-by}$. Show by substitution into (3.1.8) that $dh(y)/dy=be^{2by}h(-y)$. Use he equation obtained by changing $y$ to $-y$ to obtain the second-order equation

$$\left(\frac{d^2}{dy^2} -2b\frac{d}{dy} +b^2\right)h(y)=0$$

Solve this equation and determine the constants of integration so that the Marchenko equation is satisfied. Obtain the potential in the form $U(x)=2b\,\delta(x)$.

## 3.2   THE PRESENCE OF BOUND STATES

We now return to the derivation of the Marchenko equation that was given in Section 2.10 and consider the extension of that derivation that is required when bound states are present. As noted in Section 2.8, the bound states correspond to poles of the transmission coefficient at points $i\kappa_l$, on the imaginary axis in an upper half-plane. The Fourier transform of the left-hand side of (2.10.14) is again given by (2.10.16). However, for $t+x/c<0$ this expression no longer vanishes. According to (2.8.33) we now have

$$\Gamma(z)= -\sum \gamma_l e^{\kappa_l t}, \qquad t<0 \tag{3.2.1}$$

For $t+x/c<0$, (2.10.16) becomes

$$\text{L.H.S.} = -c\sum \gamma_l e^{\kappa_l(ct+x)} -c\sum \gamma_l e^{\kappa_l ct}\int_{-\infty}^{x} dx'\, e^{\kappa_l x'}A_L(x,x')$$

$$= -c\sum \gamma_l e^{\kappa_l ct}f_2(x,i\kappa_l) \tag{3.2.2}$$

where the definition of the fundamental solution given in (2.10.7) has been used. At $k=i\kappa_l$, the two fundamental solutions are linearly dependent and, according to (2.8.20),

$$f_2(x,i\kappa_l)=c_{11}(i\kappa_l)f_1(x,i\kappa_l) \tag{3.2.3}$$

Using the definition of $f_1(x,k)$ from (2.10.5) and again setting $ct'=x'$ and $-ct=y$, we may write

$$\text{L.H.S.} = -c\sum \gamma_l c_{11}(i\kappa_l)e^{-\kappa_l(x+y)}$$

$$-c\sum \gamma_l c_{11}(i\kappa_l)\int_{x}^{\infty} dx'\, A_R(x,x')e^{-\kappa_l(x'+y)} \tag{3.2.4}$$

The Marchenko equation will now have exactly the same form as in the case without bound states if we define

$$\Omega_R(z)\equiv r_R(z)+ \sum_{l=1}^{N} \gamma_l c_{11}(i\kappa_l)e^{-\kappa_l z} \tag{3.2.5}$$

According to (2.8.8) and (2.8.30a), this may be written

$$\Omega_R(z) = \int_{-\infty}^{\infty} \frac{dk}{2\pi} \frac{c_{11}(k)}{\dot{c}_{12}(k)} e^{ikz} - i \sum_{l=1}^{N} \frac{c_{11}(i\kappa_l)}{\dot{c}_{12}(i\kappa_l)} e^{-\kappa_l z} \qquad (3.2.6)$$

Finally, the integral equation is

$$\Omega_R(x+y) + A_R(x,y) + \int_{x}^{\infty} dx' \, \Omega_R(x'+y) A_R(x,x') = 0, \qquad x < y \quad (3.2.7)$$

with $\Omega_R$ given by (3.2.6). We thus obtain exactly the same integral equation as that obtained for the case of no bound states except that the reflection coefficient $r_R$ is replaced by $\Omega_R$. To determine $\Omega_R$ we must know not only the reflection coefficient $r_R(z)$ but also the number of bound states, the points $i\kappa_l$, that determine their location in the complex plane, and the normalization constants $m_{Rl} = \gamma_l c_{11}(i\kappa_l)/c_{12}(i\kappa_l)$ for each of the bound-state wave functions.

A similar procedure applied to (2.8.7b) yields

$$\Omega_L(x+y) + A_L(x,y) + \int_{-\infty}^{x} dx' \, A_L(x,x')\Omega_L(x'+y) = 0, \qquad x > y \quad (3.2.8)$$

where

$$\Omega_L(z) = \int_{-\infty}^{\infty} \frac{dk}{2\pi} \frac{c_{22}(k)}{c_{12}(k)} e^{-ikz} - i \sum_{l=1}^{N} \frac{c_{22}(i\kappa_l)}{\dot{c}_{12}(i\kappa_l)} e^{\kappa_l z} \qquad (3.2.9)$$

As a simple example in which a bound state is present, let us consider the attractive delta function potential.

### Example—Attractive Delta Function Potential

The scattering problem for waves incident from the left will be considered. The reflection coefficient is $R_L(k) = c_{22}(k)/c_{12}(k)$. It was found in Ex. 19 of Chapter 2 that the scattering coefficients for the delta function potential $u(x) = 2b\,\delta(x)$ are $c_{12}(k) = 1 + ib/k$ and $c_{22}(k) = -ib/k$. For the attractive potential, $b$ is negative and the zero of $c_{12}$ is in the upper half-plane at $k_1 = i|b|$. Also, the reflection coefficient is $R_L(k) = -ib/(k+ib)$. From (2.8.32) the Fourier transform of this reflection coefficient is

$$r_L(z) = \begin{cases} -|b|e^{|b|z}, & z < 0 \\ 0, & z > 0 \end{cases} \qquad (3.2.10)$$

According to (3.2.9), we must construct the quantity

$$\Omega_L(z) = r_L(z) - i \frac{c_{22}(i\kappa_l)}{\dot{c}_{12}(i\kappa_l)} e^{\kappa_l z} \qquad (3.2.11)$$

where $\kappa_l = |b|$. With the foregoing values of $c_{12}$ and $c_{22}$, we find that

$$\Omega_L(z) = \begin{cases} 0, & z < 0 \\ |b|e^{|b|z}, & z > 0 \end{cases}$$

$$= -be^{-bz}\theta(z) \qquad\qquad (3.2.12)$$

where $\theta(z)$ is the unit step function. This is the same expression as that obtained for $r_L(z)$ in (3.1.5) for the repulsive delta function potential. The solution of the Marchenko equation proceeds as before and the same result is obtained for the potential except that now $b < 0$.

### Exercise 2

For the truncated potential $u(x) = -2\theta(x)\operatorname{sech}^2 x$, we obtain from (2.9.7) and (2.9.8),

$$R_R(k) = \frac{k+i}{(k-i)(2k^2+1)}, \qquad R_L(k) = \frac{-1}{2k^2+1}, \qquad T(k) = \frac{2k(k+i)}{2k^2+1}$$

For the single pole at $k = i/\sqrt{2}$, show that $\gamma_1 = (1+\sqrt{2})/2\sqrt{2}$ and obtain

$$\Omega_L(z) = \begin{cases} 0, & z < 0 \\ \dfrac{1}{\sqrt{2}}\sinh(z/2), & z > 0 \end{cases}$$

$$\Omega_R(z) = \begin{cases} \dfrac{1}{2\sqrt{2}}\left(\dfrac{1-\sqrt{2}}{1+\sqrt{2}}\right)e^{z/\sqrt{2}} - \dfrac{1}{2\sqrt{2}}\left(\dfrac{1+\sqrt{2}}{1-\sqrt{2}}\right)e^{-z/\sqrt{2}}, & z < 0 \\ 2e^{-z}, & z > 0 \end{cases}$$

To recover the potential, a simple procedure is to use $\Omega_L$ for $z < 0$ to show that $U(x) = 0$ for $x < 0$ and then use $\Omega_R$ for $z > 0$. Show that $A_R(x,y) = -e^y \operatorname{sech} x$ and obtain the potential $U(x) = -2\operatorname{sech}^2 x$, $x > 0$.

### Exercise 3

As shown in Ex. 16 of Chapter 2, the reflection coefficients are equal to zero for the potential $u(x) = -2\operatorname{sech}^2 x$. Also, $c_{12}(k) = (k-i)/(k+i)$ and, from (2.9.7) and (2.9.8) in the limit $x_1 \to -\infty$, we find that $c_{22}(i) = 1$. Use this information to determine $\Omega_L$ and obtain the Marchenko equation (3.2.7) for this case. Solve by setting $A_L(x,y) = f(x)e^y$ and recover the potential $U(x) = -2\operatorname{sech}^2 x$.

*Exercise 4*

Solve the Marchenko (3.2.7) for the reflectionless potential for which there is one pole in the upper half-plane at $\kappa_1 = \kappa$. Set $m_{R1} = 2k\exp(2\kappa\xi)$ and show that

$$A_1(x,y) = \frac{-2\kappa e^{\kappa(2\xi-x-y)}}{1+e^{2\kappa(\xi-x)}}$$

$$\Omega_R(x) = 2\kappa e^{\kappa(2\xi-x)}$$

$$u(x) = -2\kappa^2 \operatorname{sech}^2\kappa(x-\xi)$$

Show that the fundamental solutions are

$$f_1(x,k) = e^{ikx}\frac{k+i\kappa\tanh z}{k+i\kappa}$$

$$f_2(x,k) = e^{-ikx}\frac{k-i\kappa\tanh z}{k+i\kappa}$$

where $z = \kappa(x-\xi)$. This form of the solution will be useful in Chapter 9.


## 3.3  REFLECTIONLESS POTENTIALS

One class of potentials for which the Marchenko equation may be solved completely is that of reflectionless potentials. The simplest example of this class was considered above in Ex. 3. We now take up a generalization of the procedure used to solve that example and consider the situation in which the reflection coefficient, say $R_L(k)$, equals zero while the transmission coefficient has $n$ poles on the positive imaginary axis at the specified points $i\kappa_l$ ($l = 1, 2, \ldots N$). According to (3.2.9),

$$\Omega_L(z) = \sum_{l=1}^{N} m_{Ll}e^{\kappa_l z} \tag{3.3.1}$$

where the normalization constants $m_{Ll}$ are also assumed to be given. The Marchenko equation (3.2.8) may now be conveniently solved by introducing certain $N$ component vectors. Defining

$$\Psi(z) = (m_{L1}e^{\kappa_1 z}, \ldots, m_{LN}e^{\kappa_N z}) \tag{3.3.2}$$

$$\Phi(z) = (e^{\kappa_1 z}, \ldots, e^{\kappa_N z}) \tag{3.3.3}$$

we may write

$$\Omega_L(x+y) = \Psi(x)\cdot\Phi(y) \tag{3.3.4}$$

The Marchenko equation may now be solved by writing the unknown function $A_L(x,y)$ in the form

$$A_L(x,y) = \mathcal{C}(x)\cdot\Phi(y) \tag{3.3.5}$$

where $\mathcal{C}(x)$ is to be determined. The Marchenko equation then becomes

$$\mathcal{C}(x)\cdot\left[1+\int_{-\infty}^{x}dx'\,\Phi(x')\Psi(x')\right]\cdot\Phi(y)+\Psi(x)\cdot\Phi(y)=0 \qquad (3.3.6)$$

where I is the $N\times N$ unit dyadic (matrix) $I=\delta_{ij}$. Setting

$$V(x)=1+\int_{-\infty}^{x}dx'\,\Phi(x')\Psi(x') \qquad (3.3.7)$$

we arrive at

$$\left[\mathcal{C}(x)\cdot V(x)+\Psi(x)\right]\cdot\Phi(y)=0 \qquad (3.3.8)$$

Since $\Phi(y)$ is nonzero and its orientation may be changed by varying $y$, it follows that this equation must be satisfied by setting the first factor equal to zero. Thus

$$\mathcal{C}(x)\cdot V(x)=-\Psi(x) \qquad (3.3.9)$$

or

$$\mathcal{C}(x)=-\Psi(x)\cdot V^{-1}(x) \qquad (3.3.10)$$

The dyadic $V$ always has an inverse (Kay and Moses, 1956). The unknown $A_L(x,y)$ is thus given by

$$A_L(x,y)=-\Psi(x)\cdot V^{-1}(x)\cdot\Phi(y) \qquad (3.3.11)$$

To obtain the potential we require only $A_L(x,x)$. This may be written

$$A_L(x,x)=-\Psi(x)\cdot V^{-1}(x)\cdot\Phi(x) \qquad (3.3.12)$$

An equivalent expression is

$$A_L(x,x)=-\text{Tr}\left[\Phi(x)\Psi(x)\cdot V^{-1}(x)\right] \qquad (3.3.13)$$

where Tr signifies the trace of the dyadic in brackets. The equivalence of these two expressions for $A_L(x,x)$ is easily established by introducing a set of unit vectors $e_i$ and writing $\Psi=e_i\psi_i$, $\Phi=e_l\varphi_l$, and $V^{-1}=v_{jk}^{-1}e_je_k$ with summation over repeated indices. We immediately find that both expressions for $A_L(x,x)$ reduce to $\psi_i\varphi_k v_{ik}^{-1}$. From the definition of $V(x)$ in (3.3.7) we have

$$\frac{dV}{dx}=\Phi(x)\Psi(x) \qquad (3.3.14)$$

so (3.3.13) may be written

$$A_L(x,x)=-\text{Tr}\left[\frac{dV}{dx}\cdot V^{-1}\right]$$

$$=-\frac{d}{dx}\ln(\det V) \qquad (3.3.15)$$

where $\det V$ signifies the determinant of the dyadic $V$. The last equality in (3.2.15) employs the result from matrix calculus (Coddington and Levinson, 1955, p. 28), that if $V$ and $A$ are matrices such that $dV/dx = AV$, and hence $dV/dx \cdot V^{-1} = A$, then $d(\det V)/dx = (\mathrm{Tr}\,A)\det V$. Thus

$$\mathrm{Tr}\,A = \mathrm{Tr}\left(\frac{dx}{dx} \cdot V^{-1}\right) = \frac{d(\det V)/dx}{\det V} = \frac{d}{dx}\left[\ln(\det V)\right] \quad (3.3.16)$$

To determine the potential, we now employ (3.1.4) and obtain

$$U(x) = -2\frac{d^2}{dx^2}\ln(\det V) \quad (3.3.17)$$

As an example of the use of this result, let us show that the case $N=2$ leads to the quadratic Bargmann potential considered in Chapter 1. It should be noted that the case $N=1$, which was considered in Ex. 3, yielded the linear Bargmann potential.

### Example—The Case $N=2$

To recover the numerical result for the quadratic Bargmann potential that was given in (1.3.25) we specify the values $\kappa_1 = 1$ and $\kappa_2 = 2$. Then

$$\Psi(x) = \left(m_{L1}e^x, m_{L2}e^{2x}\right), \qquad \Phi(x) = \left(e^x, e^{2x}\right) \quad (3.3.18)$$

and from (3.3.7)

$$V(x) = \begin{bmatrix} 1 + \frac{1}{2}m_{L1}e^{2x} & \frac{1}{3}m_{L2}e^{3x} \\ \frac{1}{3}m_{L1}e^{3x} & 1 + \frac{1}{4}m_{L2}e^{4x} \end{bmatrix} \quad (3.3.19)$$

The determinant of $V$ is

$$\det V = 1 + \frac{1}{2}m_{L1}e^{2x} + \frac{1}{4}m_{L2}e^{4x} + \frac{1}{72}m_{L1}m_{L2}e^{6x} \quad (3.3.20)$$

The function $A_L(x,x)$ is now readily obtained from (3.3.13). Setting $\lambda = \frac{1}{2}\ln(m_{L2}/2m_{L1})$ and $\mu = \frac{1}{2}\ln(m_{L1}m_{L2}/72)$, we find that

$$A_L(x,x) = -\frac{3e^{-(x+\lambda)} + 6e^{x+\lambda} + 3e^{3x+\mu}}{3\cosh(x+\lambda) + \cosh(3x+\mu)} \quad (3.3.21)$$

A subsequent differentiation and use of (3.1.3) yields

$$u(x) = -12\frac{3 + \cosh(4x+\mu+\lambda) + 4\cosh(2x+\mu-\lambda)}{\left[3\cosh(x+\lambda) + \cosh(3x+\mu)\right]^2} \quad (3.3.22)$$

When $\lambda$ and $\mu$, which are related to $m_{L1}$ and $m_{L2}$ as shown above, are chosen to make this result also satisfy the Korteweg–deVries equation, the expression given in (1.3.25) is recovered. This example will be considered again in Section 4.4.

## 3.4   REFLECTION COEFFICIENT A RATIONAL FUNCTION OF $k$

When a reflection coefficient, say $R_L(k)$, is a rational function of $k$, the Fourier transform $r_L(z)$ will again be a sum of exponential terms and will thus be of a form similar to that obtained for the reflectionless potentials in the previous section. The most general form for such a reflection coefficient is

$$R_L(k) = c \frac{\prod\limits^{m}(k - \alpha_i)}{\prod\limits_{n}(k - \beta_i)} \tag{3.4.1}$$

with $m \leqslant n - 1$. For $m = n - 1$, as in (2.4.4), the potential will contain a delta function. We shall only consider some specific examples. A more general discussion of the problem may be found in Kay (1960).

We first consider the case in which there are no poles of $R_L(k)$ in the upper half-plane. Then, according to the definition of $r_L(z)$ in (2.8.32), we have $r_L(z) = 0$ for $z < 0$. If, in addition, we require that $T(k)$ have no poles in the upper half-plane, then the Marchenko equation (2.10.18) may be used to determine the potential and we see that for $x + y < 0$, there is no contribution from the integral and $A_L(x,x) = 0$ for $x < 0$. Thus $U(x) = 0$ for $x < 0$.

For $x > 0$ the lower limit on the integral in the Marchenko equation is $-y$ rather than $-\infty$. As a result, any attempt to construct the solution of the Marchenko equation along the lines employed in the previous section for reflectionless potentials would be quite cumbersome. However, when $R_R(k)$ is obtained from a knowledge of $R_L(k)$ and $T(k)$ by means of (2.8.16) and (2.8.38), we find that $R_R(k)$ has singularities in both half-planes. Thus $R_R(z)$ does not vanish for $z > 0$. The upper limit in the Marchenko equation (2.10.17) remains $\infty$ and may then be solved for $A_R(x,y)$ by the techniques used in the previous section.

As an example, let us determine the potential that gives rise to the reflection coefficient

$$R_L(k) = \frac{\alpha\beta}{(k + i\alpha)(k + i\beta)} \tag{3.4.2}$$

where $\alpha$ and $\beta$ are real positive constants. From (2.8.32), we see that $r_L(z)$ is zero for $z < 0$. Thus $A_L(x,y)$ and $u(x)$ are also zero for $x < 0$. To obtain the potential for $x > 0$, we calculate $r_R(z)$. From (2.8.38), we then obtain

$$\frac{d}{dk}\ln T(k) = 2\alpha^2\beta^2 P \int_{-\infty}^{\infty} \frac{d\zeta}{2\pi i} \frac{2\zeta^2 + \gamma^2}{\zeta(\zeta - k)(\zeta^2 + \alpha^2)(\zeta^2 + \beta^2)(\zeta^2 + \gamma^2)} \tag{3.4.3}$$

where $\gamma^2 = \alpha^2 + \beta^2$ and $P$ signifies a principal value. Integration yields

$$T(k) = \frac{k(k + i\gamma)}{(k + i\alpha)(k + i\beta)} \tag{3.4.4}$$

From (2.8.16),

$$R_R(k) = -\frac{\alpha\beta(k + i\gamma)}{(k + i\alpha)(k + i\beta)(k - i\gamma)} \tag{3.4.5}$$

and for $z > 0$,

$$r_R(z) = m_0 e^{-\gamma z} \tag{3.4.6}$$

in which $m_0 = -2\alpha\beta/[(\gamma + \alpha)(\gamma + \beta)]$. The constant $m_0$ is thus negative. On setting $A_R(x,y) = \rho(x)e^{-\gamma y}$, we find that the Marchenko equation (2.10.17) reduces to the algebraic equation

$$m_0 e^{-\gamma x} + \rho(x) + \rho(x)\left(\frac{m_0}{2\gamma}\right)e^{-2\gamma x} = 0 \tag{3.4.7}$$

Solving for $\rho(x)$, we find that

$$A_1(x,x) = \gamma e^{-(\gamma x + \varphi)}\operatorname{csch}(\gamma x + \varphi) \tag{3.4.8}$$

where $e^\varphi = \sqrt{2\gamma/|m_0|}$. The potential is then

$$u(x) = 2\gamma^2 \theta(x)\operatorname{csch}^2(\gamma x + \varphi) \tag{3.4.9}$$

where $\theta(x)$ is the unit step function. Since $\gamma$ and $\varphi$ are both positive, the singularity in the expression for $u(x)$ is cut off by the step function. The wave function for this potential may be obtained by using the method of Darboux given in Section 2.6 with $A = \frac{1}{2}e^\varphi$, $B = -\frac{1}{2}e^{-\varphi}$ or by employing a general method described by Kay (1960). We obtain

$$y = Ae^{ikx}[ik - \gamma\coth(\gamma x + \varphi)] - Be^{-ikx}[ik + \coth(\gamma x + \varphi)] \tag{3.4.10}$$

which should be compared with (2.6.15)

When the reflection and transmission coefficients do have poles in the upper half-plane, we may proceed as in Section 2.8 by introducing the function $\tilde{T}(k)$ that is devoid of such singularities. The calculation will then be similar to that outlined above. As an example, consider the reflection coefficient

$$R_L(k) = \frac{-\alpha^2}{k^2 + \alpha^2} \tag{3.4.11}$$

with $\alpha$ real. For simplicity we choose the location of the pole of $T(k)$ to be also at $k = i\alpha$ in the upper half-plane. Then, introducing $\tilde{T}(k) = T(k)(k - i\alpha)/(k + i\alpha)$, we find that

$$\frac{d\ln\tilde{T}(k)}{dk} = 4\alpha^4 P \int_{-\infty}^{\infty} \frac{d\zeta}{2\pi i} \frac{1}{\zeta(\zeta - k)(\zeta^2 + \alpha^2)(\zeta^2 + 2\alpha^2)} \tag{3.4.12}$$

Integration leads to the transmission coefficient

$$T(k) = \frac{k(k + i\alpha\sqrt{2})}{k^2 + \alpha^2} \tag{3.4.13}$$

When $R_r(k)$ is obtained from (2.8.16), a Fourier inversion yields

$$r_R(z) = \begin{cases} 2\alpha\sqrt{2}\, e^{\alpha\sqrt{2}\,z}, & z < 0 \\ 2\alpha\sqrt{2}\, e^{-\alpha\sqrt{2}\,z} + \dfrac{\alpha}{2}\left(\dfrac{1 + \sqrt{2}}{1 - \sqrt{2}}\right)e^{-\alpha z}, & z > 0 \end{cases} \tag{3.4.14}$$

Since there is a pole of $T(k)$ in the upper half-plane, we must construct the function $\Omega_r(z)$ and consider the Marchenko equation in the form given in (2.10.17). Since there is only the one zero of $c_{12}(k)$ in the upper half-plane at $k = i\alpha$, $\Omega_R(z)$ is given by

$$\Omega_R(z) = r_R(z) + \gamma_0 c_{11}(i\alpha) e^{-\alpha z} \tag{3.4.15}$$

The value of $\gamma_0$ follows from (2.8.27). We obtain

$$i\gamma_0 = \text{Res}\left[\frac{1}{c_{12}(i\alpha)} - 1\right] = \frac{i\alpha}{2}(1 + \sqrt{2}) \tag{3.4.16}$$

Since $c_{11}(i\alpha) = R_r(i\alpha)/T(i\alpha) = (\sqrt{2} - 1)^{-1}$, we have

$$\gamma_0 c_{11}(i\alpha) = \frac{1}{2}\alpha\frac{\sqrt{2} + 1}{\sqrt{2} - 1}$$

Then for $z > 0$, (3.4.15) yields

$$\Omega_R(z) = 2\alpha\sqrt{2}\, e^{-\alpha\sqrt{2}\,z} \tag{3.4.17}$$

Introducing $A_R(x,y) = f(x)e^{-\alpha\sqrt{2}\,y}$, we find that the Marchenko equation (2.10.17) yields $f(x) = -2\gamma e^{-\gamma x}/(1 + e^{-\gamma x})$, where $\gamma = \alpha\sqrt{2}$. Finally, the potential is

$$U(x) = 2\frac{d}{dx}A_2(x,x) = -4\alpha^2\theta(x)\text{sech}^2\alpha\sqrt{2}\,x \tag{3.4.18}$$

For $\alpha^2 = \frac{1}{2}$, this is the truncated potential considered in Section 2.9 for the special case $x_1 = 0$.

## 3.5  BARGMANN POTENTIALS

In the two previous sections we have considered how an inverse scattering method can be devised to construct the potential from the reflection coefficient alone in the case of a repulsive potential or from a knowledge of the reflection coefficient plus information concerning the bound states (energy levels and normalization constants) in the case of attractive potentials. In the present section we consider a simpler technique for constructing certain potentials (those with zero reflection coefficient). Application of the method to the radial Schrödinger equation (Bargmann, 1949) predates the inverse scattering approach. It was results obtained by this method that provided impetus for development of the more complete inverse scattering techniques. The method has already been introduced in Chapter 1, where the linear and quadratic Bargmann potentials were related to the one- and two-soliton solution of the Korteweg–deVries equation. We now consider the method in more detail, especially in regard to the quadratic case. In that instance we will find that the energy levels associated with these solutions are dependent upon only two parameters, while the shape of the potential is characterized by four parameters. The shape of the potential can thus undergo continuous deformation while the eigenvalues remain fixed. That this can happen is the crucial fact that enables inverse scattering theory to be of use in the theory of nonsteady-state pulse propagation.

Bargmann's approach was to consider solutions of the radial Schrödinger equation in the form $e^{ikr}$ times a polynomial in $k$. In our use of the method we introduce fundamental solutions of the form

$$f_1(x,k) = e^{ikx} P_1(x,k)$$

$$f_2(x,k) = e^{-ikx} P_2(x,k) \tag{3.5.1}$$

where the functions $P_i(x,k)$ are polynomials in $k$. Only the linear and quadratic cases will be considered. Since the assumed forms of the solution contain only $e^{+ikx}$ or $e^{-ikx}$, we expect that the resulting potential will be reflectionless. From (2.8.7), which relates the fundamental solutions, we see that

$$P_2(x,k) = c_{12} P_1(x, -k),$$

$$P_1(x,k) = c_{21} P_2(x, -k) \tag{3.5.2}$$

Since $c_{12}(k) = c_{21}(k)$, we have

$$P_1(x,k) P_1(x, -k) = P_2(x,k) P_2(x, -k) \tag{3.5.3}$$

## Linear Case

The simplest example is the linear case and has already been considered in some detail in Chapter 1. The fundamental solutions are

$$f_1(x,k) = e^{ikx} \frac{2k + ia_1(x)}{2k + ia_1(\infty)}$$

$$f_2(x,k) = e^{-ikx} \frac{2k + ia_2(x)}{2k + ia_2(-\infty)} \qquad (3.5.4)$$

Substitution into the Schrödinger equation shows that $u(x) = a_2{}'(x) = -a_1{}'(x)$. From (3.5.2), we find that

$$c_{12}(k) = \frac{2k - ia_1(\infty)}{2k + ia_1(-\infty)} \qquad (3.5.5)$$

In Chapter 1 we obtained $a_1(x) = 2\mu \tanh(\mu x - \varphi)$, where $\mu$ and $\varphi$ are constants of integration. The zero of $c_{12}(k)$ is thus located at $\frac{1}{2}a_1(\infty) = i\mu$. The location of the pole thus depends only on the value of $\mu$ while the potential depends upon both $\mu$ and $\psi$. In this linear case the extra degree of freedom in the shape of the potential merely displaces the potential along the $x$ axis.

## Quadratic Case

In the quadratic case the potential is found to depend upon four parameters while only two of these parameters are associated with the location of the eigenvalues. The potential can then experience a continuous deformation as these two additional parameters are varied. As indicated in Chapter 1, this deformation leads to a solution that can be used to describe the interaction of two solitons. We now consider this quadratic solution in more detail.

Introducing the fundamental solutions

$$f_1(x,k) = e^{ikx} \frac{4k^2 + 2ika_1(x) + b_1(x)}{4k^2 + 2ika_1(\infty) + b_1(\infty)} \qquad (3.5.6)$$

and

$$f_2(x,k) = e^{-ikx} \frac{4k^2 + 2ika_2(x) + b_2(x)}{4k^2 + 2ika_2(-\infty) + b_2(-\infty)} \qquad (3.5.7)$$

we find from substitution into the Schrödinger equation that $a_2(x) = -a_1(x)$, $b_2(x) = b_1(x)$, and

$$a_1'(x) = -u(x) \qquad (3.5.8)$$

We also find that

$$L_1 \equiv b_1' + a_1'' - a_1 u = 0 \tag{3.5.9}$$

$$L_2 \equiv b_1'' - b_1 u = 0 \tag{3.5.10}$$

When (3.5.8) and (3.5.9) are combined and integrated, we obtain

$$b_1 + a_1' + \tfrac{1}{2} a_1^2 = 2c_1 \tag{3.5.11}$$

where $c_1$ is a constant of integration. Integration of $b_1 L_1 - a_1 L_2$ yields

$$\tfrac{1}{2} b_1^2 + b_1 a_1' - a_1 b_1' = 2c_2 \tag{3.5.12}$$

in which $c_2$ is a second constant of integration. Again introducing a function $w$ through the definition $a_1 = 2w'/w$ we obtain, from (3.5.11),

$$b_1 = 2\left(c_1 - \frac{w''}{w}\right) \tag{3.5.13}$$

Equation 3.5.12 now takes the form

$$2w'w''' - (w'')^2 - 2c_1 (w')^2 + w^2 (c_1^2 - c_2) = 0 \tag{3.5.14}$$

which, when differentiated, reduces to the much simpler linear equation,

$$w^{(iv)} - 2c_1 w'' + (c_1^2 - c_2) w = 0 \tag{3.5.15}$$

This equation has solutions of the form $e^{\omega x}$, with

$$\omega^2 = c_1 \pm \sqrt{c_2} \tag{3.5.16}$$

Indicating the roots by $\pm \rho = \pm \sqrt{c_1 + \sqrt{c_2}}$ and $\pm \sigma = \pm \sqrt{c_1 - \sqrt{c_2}}$, which implies that

$$c_1 = \tfrac{1}{2}(\rho^2 + \sigma^2), \qquad c_2 = \tfrac{1}{4}(\rho^2 - \sigma^2)^2 \tag{3.5.17}$$

we find that $w$ may be written

$$w = \gamma_1 e^{\rho x} + \gamma_2 e^{-\rho x} + \delta_1 e^{\sigma x} + \delta_2 e^{-\sigma x} \tag{3.5.18}$$

The requirement that $w$ must satisfy the third-order equation (3.5.14) imposes a constraint upon these four integration constants. Substituting into (3.5.14), we find that

$$\gamma_1 \gamma_2 \rho^2 = \delta_1 \delta_2 \sigma^2 \tag{3.5.19}$$

This condition may be satisfied by setting $\gamma_1 = \sigma\alpha$, $\gamma_2 = \sigma/\alpha$, $\delta_1 = \rho\beta$, and $\delta_2 = \rho/\beta$. With these definitions, (3.5.18) reduces to the result quoted in Chapter 1, namely

$$w = 2\sigma\cosh(\rho x - \phi) + 2\rho\cosh(\sigma x - \psi) \tag{3.5.20}$$

where $\alpha = e^{-\phi}$ and $\beta = e^{-\psi}$.

From (3.5.2), we have

$$c_{12}(k) = \frac{4k^2 - 2ika_1(\infty) + b_1(\infty)}{4k^2 - 2ika_1(-\infty) + b_1(-\infty)} \tag{3.5.21}$$

The zeros of $c_{12}(k)$ thus depend upon only the two parameters $a_1(\infty)$ and $b_1(\infty)$, while the potential depends upon four parameters. To see this in more detail, note that since we are considering $\rho > \sigma$, $w \sim e^{\rho x}$ as $x \to +\infty$. Also, $a_1 = 2w'/w \sim 2\rho$ in this limit. Similarly, $b_1 = 2c_1 - 2w''/w \sim \sigma^2 - \rho^2$ as $x \to +\infty$. Thus the location of the zeros of $c_{12}$ depends upon the two parameters $\rho$ and $\sigma$, while the potential depends upon $\varphi$ and $\psi$ as well. There can be continuous variation of the parameters $\varphi$ and $\psi$, and thus of the shape of the potential, without changing the two eigenvalues of the Schrödinger equation.

## 3.6  TWO-COMPONENT INVERSE METHOD FOR REAL POTENTIALS

It was shown in Section 2.11 that the pair of equations

$$\frac{dn_1}{dx} + ikn_1 = un_2$$

$$\frac{dn_2}{dx} - ikn_2 = -un_1 \tag{3.6.1}$$

can be treated in the same way that the Schrödinger equation was treated in Section 2.8; that is, fundamental solutions can be introduced and a rather general analysis of the scattering process carried out. These solutions were then related by

$$\varphi(x,k) = c_{11}(k)\psi(x,k) + c_{12}(k)\overline{\psi}(x,k) \tag{3.6.2a}$$

$$\psi(x,k) = c_{21}(k)\overline{\varphi}(x,k) + c_{22}(k)\varphi(x,k) \tag{3.6.2b}$$

where

$$\varphi(x,k) = \begin{pmatrix} \varphi_1(x,k) \\ \varphi_2(x,k) \end{pmatrix}, \qquad \overline{\varphi}(x,k) = \begin{pmatrix} \varphi_2(x,-k) \\ -\varphi_1(x,-k) \end{pmatrix} \tag{3.6.3}$$

and similarly for $\psi(x,k)$. By taking the Fourier transform of (3.6.2a), a pair of coupled integral equations of Marchenko type may be derived. A second pair of such integral equations also follows from (3.6.2b). A procedure for solving inverse scattering problems for the two-component equations can therefore be formulated. The analysis leading to the integral equations is identical with that used in Sections 2.10 and 3.2 for the Schrödinger equation, so the derivation will only be outlined.

The integral equations associated with the two components of (3.6.2b) are obtained by first writing them in the form

$$[T(k)-1+1]\psi_1(x,k)=-\varphi_1(x,-k)+R_L(k)\varphi_2(x,k) \qquad (3.6.4a)$$

$$[T(k)-1+1]\psi_2(x,k)=\varphi_2(x,-k)+R_L(k)\varphi_1(x,k) \qquad (3.6.4b)$$

where $T(k)=-1/c_{21}(k)$ and $R_L(k)=-c_{22}(k)/c_{21}(k)$, as in (2.11.25). Defining Fourier transforms of $T(k)-1$ and $R_L(k)$ as in (2.8.31) and (2.8.32) and then using (2.11.6) and (2.11.7) for the transforms of $\varphi$ and $\psi$, we obtain

$$c\int_{x/c}^{\infty}dt'\Gamma[c(t-t')]\psi_+(x,t')+\psi_+(x,t)$$
$$=-\varphi_-(x,-t)+c\int_{-x/c}^{\infty}dt'r_L[c(t-t')]\varphi_+(x,t') \qquad (3.6.5a)$$

and

$$c\int_{x/c}^{\infty}dt'\Gamma[c(t-t')]\psi_-(x,t')+\psi_-(x,t)$$
$$=\varphi_+(x,-t)+c\int_{-x/c}^{\infty}dt'r_L[c(t-t')]\varphi_-(x,t') \qquad (3.6.5b)$$

Since $\psi_+(x,t)$ and $\psi_-(x,t)$ both vanish for $t-x/c<0$, the second term on the left-hand side of each of these equations vanishes when this inequality is satisfied. Let us now concentrate on (3.6.5a) when the inequality $t-x/c<0$ is satisfied. With $\psi_+(x,t')$ given by (2.11.5) and $\Gamma(z)$ given by (2.8.32), although as noted in Section 2.11 the location of the poles $k_l$ need no longer be confined to the imaginary axis, the left-hand side of (2.6.5a) reduces to

$$\text{L.H.S.}=-c\sum_l\gamma_le^{-ik_lct}\psi_1(x,k_l) \qquad (3.6.6)$$

The definition of $\psi_1(x,k)$ given in (2.11.7) has been employed here. Since $\psi_1(x,k_l)=c_{22}(k_l)\varphi_1(x,k_l)$, we finally obtain

$$\text{L.H.S.}=-c\sum_l\gamma_lc_{22}(k_l)e^{-ik_l(x+ct)}$$
$$-c\int_{-\infty}^{x}dx'A_1(x,x')\sum_l\gamma_lc_{22}(k_l)e^{-ik_l(x'+ct)} \qquad (3.6.7)$$

Similarly, the right-hand side may be written

$$\text{R.H.S.} = -cA_2(x,ct) + cr_L(ct+x)$$

$$+ c^2 \int_{-x/c}^{\infty} dt' \, r_L[c(t-t')] A_1(x, -ct') \qquad (3.6.8)$$

Equating the two expressions and defining

$$\Omega_L(z) \equiv r_L(z) + \sum \gamma_l c_{22}(k_l) e^{-ik_l z}$$

$$= r_L(z) - i \sum \frac{c_{22}(k_l)}{\dot{c}_{12}(k_l)} e^{-ik_l z} \qquad (3.6.9)$$

we finally obtain the integral equation

$$-A_2(x,y) + \Omega_L(x+y) + \int_{-\infty}^{x} dx' A_1(x,x') \Omega_L(x'+y) = 0, \qquad x > y$$

$$(3.6.10a)$$

where $x' = ct'$. A similar reduction of (3.6.5b) leads to

$$A_1(x,y) + \int_{-\infty}^{x} dx' A_2(x,\dot{x}') \Omega_L(x'+y) = 0, \qquad x > y \qquad (3.6.10b)$$

The pair of equations (3.6.10) are the Marchenko integral equations for this two-component inverse scattering problem. A similar pair of equations follows from (3.6.2a). The result is

$$B_2(x,y) + \int_{x}^{\infty} dx' B_1(x,x') \Omega_R(x'+y) = 0$$

$$-B_1(x,y) + \Omega_R(x+y) + \int_{x}^{\infty} dx' B_2(x,x') \Omega_R(x'+y) = 0 \qquad x < y$$

$$(3.6.11)$$

where

$$\Omega_R(z) = r_R(z) + \sum \gamma_l c_{11}(k_l) e^{ik_l z}$$

$$= r_R(z) + i \sum \frac{c_{11}(k_l)}{\dot{c}_{21}(k_l)} e^{ik_l z} \qquad (3.6.12)$$

We must also determine the relation between the function $u(x)$ and the kernel functions $A_1$ and $A_2$ or $B_1$ and $B_2$. Again we follow closely the procedure used for the Schrödinger equation. Substituting the time-dependent solutions $\varphi_+(x,t)$ and $\varphi_-(x,t)$ into (2.11.2), we find that

$$c\theta\left(t+\frac{x}{c}\right)\left[\frac{\partial A_1}{\partial t}-c\frac{\partial A_1}{\partial x}-K(x)A_2\right]=0 \qquad (3.6.13)$$

and

$$c\delta\left(t+\frac{x}{c}\right)\left[2A_2-\frac{K(x)}{c}\right]$$

$$+c\theta\left(t+\frac{x}{c}\right)\left[\frac{\partial A_2}{\partial t}+c\frac{\partial A_2}{\partial x}-K(x)A_1\right]=0 \qquad (3.6.14)$$

For $t+x/c>0$, we obtain, on setting $y=-ct$ and $u=-K/c$,

$$\frac{\partial A_1(x,y)}{\partial y}+\frac{\partial A_1(x,y)}{\partial x}=u(x)A_2(x,y)$$

$$\frac{\partial A_2(x,y)}{\partial y}-\frac{\partial A_2(x,y)}{\partial x}=u(x)A_1(x,y) \qquad x>y \qquad (3.6.15)$$

Also, by integrating (3.6.14) with respect to time across the singularity in the delta function, we obtain

$$u(x)=-2A_2(x,x) \qquad (3.6.16)$$

On setting $x=y$ in the first of (3.6.15), there results

$$u^2(x)=-2\frac{dA_1(x,x)}{dx} \qquad (3.6.17)$$

A similar treatment based upon the solution $\psi_+(x,t)$ and $\psi_-(x,t)$ given in (2.11.5) yields

$$\frac{\partial B_1(x,y)}{\partial x}-\frac{\partial B_1(x,y)}{\partial y}=u(x)B_2(x,y) \qquad x<y \qquad (3.6.18)$$

$$\frac{\partial B_2(x,y)}{\partial x}+\frac{\partial B_2(x,y)}{\partial y}=-u(x)B_1(x,y)$$

$$u(x)=-2B_1(x,x) \qquad (3.6.19)$$

$$u^2(x)=2\frac{dB_2(x,x)}{dx} \qquad (3.6.20)$$

## 3.7 REFLECTIONLESS POTENTIALS FOR TWO-COMPONENT SYSTEMS

The coupled Marchenko equations obtained in the previous section may be solved exactly for reflectionless potentials. We follow the method used in Section 3.3 for the Schrödinger equation. Let us consider the solution of (3.6.10) with $r_L = 0$. We have

$$- A_2(x,y) + \Omega_L(x+y) + \int_{-\infty}^{x} dx' A_1(x,x')\Omega_L(x'+y) = 0$$

$$A_1(x,y) + \int_{-\infty}^{x} dx' A_2(x,x')\Omega_L(x'+y) = 0 \qquad x > y \,(3.7.1)$$

with

$$\Omega_L(z) = \sum_{1}^{N} m_{Ll} e^{-ik_l z} \tag{3.7.2}$$

where $m_{Ll} \equiv \gamma_l c_{22}(k_l)$. We again introduce the vectors

$$\Psi(z) = \left( m_{L1} e^{-ik_1 z}, \ldots, m_{LN} e^{-ik_N z} \right)$$

$$\Phi(z) = \left( e^{-ik_1 z}, \ldots, e^{-ik_N z} \right) \tag{3.7.3}$$

Setting

$$A_i(x,y) = \mathcal{C}_i(x) \cdot \Phi(y), \qquad i = 1,2$$

$$\Omega_L(x+y) = \Psi(x) \cdot \Phi(y) \tag{3.7.4}$$

we find that the integral equation may be reduced to algebraic equations as in Section 3.2. The result is

$$- \mathcal{C}_2(x) + \Psi(x) + \mathcal{C}_1(x) \cdot M(x) = 0$$

$$\mathcal{C}_1(x) + \mathcal{C}_2(x) \cdot M(x) = 0 \tag{3.7.5}$$

where $M(x)$ is the dyadic

$$M(x) \equiv \int_{-\infty}^{x} dx' \, \Phi(x')\Psi(x') \tag{3.7.6}$$

Eliminating $\mathcal{C}_1(x)$, we obtain

$$\mathcal{C}_2(x) = \Psi(x) \cdot D^{-1}(x) \tag{3.7.7}$$

where $D^{-1}(x)$ is the inverse of

$$D(x) \equiv 1 + M(x) \cdot M(x) \tag{3.7.8}$$

Following the procedure used in Section 3.3, we find that the potential may be expressed as

$$u(x) = -2A_2(x,x) = -2\Phi(x)\cdot \mathcal{C}_2(x)$$

$$= -2\Phi(x)\cdot \Psi(x)\cdot D^{-1}(x)$$

$$= -2\,\text{Tr}\left[\frac{dM}{dx}\cdot (1 + M\cdot M)^{-1}\right] \tag{3.7.9}$$

A formula for $u^2$ may also be obtained. We use (3.6.17) and note that

$$A_1(x,x) = \Phi(x)\cdot \mathcal{C}_1(x)$$

$$= -\Phi(x)\cdot \left[\Psi(x)\cdot D^{-1}(x)\right]\cdot M(x)$$

$$= -\text{Tr}\left[\Phi(x)\Psi(x)\cdot D^{-1}(x)\cdot M(x)\right]$$

$$= -\text{Tr}\left[M\cdot \frac{dM}{dx}\cdot D^{-1}(x)\right] \tag{3.7.10}$$

where (3.7.6) and the invariance of a trace under cyclic permutation have been employed. Employing (3.3.16) to write

$$\frac{d}{dx}(\ln \det D) = \text{Tr}\left(\frac{dD}{dx}\cdot D^{-1}\right)$$

$$= 2\,\text{Tr}\left(M\cdot \frac{dM}{dx}\cdot D^{-1}\right)$$

$$= -2A_1(x,x) \tag{3.7.11}$$

we can finally use (3.6.17) to obtain

$$u^2 = \frac{d^2}{dx^2}(\ln \det D) \tag{3.7.12}$$

A particularly simple result may be obtained by combining (3.6.16) and (3.6.17). Consider the expression

$$u^2 + iu_x = -2\frac{d}{dx}(A_1 + iA_2)$$

$$= 2\frac{d}{dx}\left[\text{Tr}\left(M\cdot \frac{dM}{dx}\cdot D^{-1} - i\frac{dM}{dx}\cdot D^{-1}\right)\right]$$

$$= -2i\frac{d}{dx}\left\{\text{Tr}\left[(1 + iM)\cdot \frac{dM}{dx}\cdot D^{-1}\right]\right\} \tag{3.7.13}$$

Again employing cyclic permutation of the trace and the fact that $D^{-1} = (I - iM)^{-1})(I + iM)^{-1}$, we obtain

$$u^2 + iu_x = -2i\frac{d}{dx}\left\{ \mathrm{Tr}\left[ \frac{dM}{dx}\cdot(I - iM)^{-1} \right] \right\}$$

$$= 2\frac{d^2}{dx^2}\left[ \ln\det(I - iM) \right] \tag{3.7.14}$$

where (3.7.11) has been used. Taking the imaginary part of this result and integrating, we find that

$$u = 2\frac{d}{dx}\left[ \mathrm{Im}\ln\det(I - iM) \right] \tag{3.7.15}$$

The final expression for $u(x,t)$ is thus

$$u = 2\frac{d}{dx}\tan^{-1}\left[ \frac{\mathrm{Im}\det(I - iM)}{\mathrm{Re}\det(I - iM)} \right] \tag{3.7.16}$$

From (3.7.3) and (3.7.6),

$$M_{ij} = \frac{im_{Lj}}{\kappa_i + \kappa_j}e^{-i(\kappa_i + \kappa_j)x} \tag{3.7.17}$$

It should be noted that this simple result is only obtained when $u$ is real. Some examples using these expressions to construct multisoliton solutions will be considered in Chapter 5.

## 3.8 REFLECTION COEFFICIENT FOR TWO-COMPONENT SYSTEM A RATIONAL FUNCTION OF $k$

When the reflection coefficient is a rational function of $k$, the coupled Marchenko equations (3.6.10) and (3.6.11) may be solved by using the technique developed in Section 3.4 for the Schrödinger equation. The only difference is that now the relation between transmission and reflection coefficients is $|T|^2 = 1 + |R|^2$, where $R$ may be either $R_L$ or $R_R$. For the case in which the transmission coefficient is free from poles in the upper half-plane, we now have

$$\frac{d\ln T(k)}{dk} = \int_{-\infty}^{\infty}\frac{d\zeta}{2\pi i}\frac{d|R(\zeta)2|/d\zeta}{(\zeta - k)[1 + |R(\zeta)|^2]} \tag{3.8.1}$$

When there are poles of $T(k)$ in the upper half-plane, we again introduce the function $\tilde{T}(k)$ and follow the method developed in Section 2.11. If there are poles on the real axis, (3.8.1) is interpreted as a principal value integral as in Section 2.8.

As a simple example, let us determine the potential that gives rise to the reflection coefficient $R_L(k) = i\beta/k$, where $\beta$ is a real positive constant. In this case, (3.8.2) becomes the principal value integral

$$\frac{d\ln T(k)}{dk} = -2\beta^2 P \int_{-\infty}^{\infty} \frac{d\zeta}{2\pi i} \frac{1}{\zeta(\zeta-k)(\zeta^2+\beta^2)} \tag{3.8.2}$$

which is the same integral as that obtained in (3.4.3). Hence

$$T(k) = \frac{k+i\beta}{k} \tag{3.8.3}$$

We shall first determine the potential for $x < 0$. Since there is a pole on the real axis at the origin, we first determine $\Omega_L(z) = r_L(z) + \frac{1}{2}\gamma_0(0)c_{22}(0)$. Taking the principal value in evaluating the integral, we find that $r_L(z) = \mathrm{sgn}(z)\frac{1}{2}\beta$. Also, from (2.8.27) we find that $\gamma_0 = -i\,\mathrm{Res}[T(0)] = \beta$ and $c_{22}(0) = R_L(0)/T(0) = 1$. Thus

$$\Omega_L(z) = \begin{cases} 0, & z < 0 \\ \beta, & z > 0 \end{cases} \tag{3.8.4}$$

Since $\Omega_L(z)$ vanishes for $z < 0$, we see from the Marchenko equation (3.6.10a) that $A_2(x,y)$ and hence the potential are zero for $x < 0$.

For $x > 0$ it is again easier to work with the integral equation involving $r_R(z)$. From (2.11.26),

$$R_R(k) = -\frac{i\beta}{k}\left(\frac{k+i\beta}{k-i\beta}\right) \tag{3.8.5}$$

The Fourier transform is

$$r_R(z) = \begin{cases} \beta\left(-\frac{1}{2}+2e^{-\beta z}\right), & z > 0 \\ -\frac{1}{2}\beta, & z < 0 \end{cases} \tag{3.8.6}$$

Also, since $\frac{1}{2}\gamma_0(0)c_{11}(0) = \frac{1}{2}\gamma_0(0)R_r(0)/T(0) = \frac{1}{2}\beta$, we obtain $\Omega_r(z) = 2\beta e^{-\beta z}$ for $z > 0$. Setting $B_i(x,y) = b_i(x)e^{-\beta y}$, $i = 1, 2$, we find that the coupled Marchenko equations (3.6.11) reduce to the algebraic equations

$$b_1 e^{-2\beta x} + b_2 = 0$$

$$b_1 - b_2 e^{-2\beta x} = 2\beta e^{-\beta x} \tag{3.8.7}$$

which have the solution $b_1 = 2\beta e^{-\beta x}/(1+e^{-4\beta x})$ and $b_2 = -2\beta e^{-3\beta x}/(1+e^{-4\beta x})$. From (3.6.19), we have $u(x) = -2B(x,x) = 2\beta \operatorname{sech} 2\beta x$ and the final result is

$$u(x) = -2\beta\theta(x)\operatorname{sech} 2\beta x \qquad (3.8.8)$$

When $\beta$ is replaced by $\mu/2$, this becomes the truncated potential considered in Section 2.11.

## 3.9 TWO-COMPONENT SYSTEM WITH A COMPLEX POTENTIAL

In some of the later applications of the two-component inverse method, complex potentials will arise. The linear equations will then be of the form

$$n_{1x} + ikn_1 = q(x)n_2$$
$$n_{2x} - ikn_2 = -q^*(x)n_1 \qquad (3.9.1)$$

where $q^*(x)$ is the complex conjugate of the complex potential $q(x)$. Following the procedure used in Section 2.11, we may view these equations as the Fourier transform of

$$\frac{\partial n_+}{\partial t} - c\frac{\partial n_+}{\partial x} = -cq(x)n_- \qquad (3.9.2a)$$

$$\frac{\partial n_-}{\partial t} + c\frac{\partial n_-}{\partial x} = -cq^*(x)n_+ \qquad (3.9.2b)$$

The fundamental solutions $\varphi$ and $\psi$ of the form used in Section 2.11 may be employed again. Substituting (2.11.4) into (3.9.2), we obtain

$$q^*(x) = -2A_2(x,x) \qquad (3.9.3)$$

$$|q(x)|^2 = -2\frac{d}{dx}A_1(x,x) \qquad (3.9.4)$$

Similarly, from the fundamental solutions given in (2.11.5), we find that

$$q(x) = 2B_1(x,x) \qquad (3.9.5)$$

$$|q(x)|^2 = 2\frac{d}{dx}B_2(x,x) \qquad (3.9.6)$$

In addition to the two fundamental solutions

$$\varphi(x,k) = \begin{pmatrix} \varphi_1(x,k) \\ \varphi_2(x,k) \end{pmatrix} \quad \text{and} \quad \psi(x,k) = \begin{pmatrix} \psi_1(x,k) \\ \psi_2(x,k) \end{pmatrix}$$

two other solutions may be obtained by examining the complex conjugate of the linear equations (3.9.1). If $k$ is complex, we then find that another solution of the system (3.9.1) is

$$\tilde{\varphi}(x,k) = \begin{pmatrix} \varphi_2^*(x,k^*) \\ -\varphi_1^*(x,k^*) \end{pmatrix} \qquad (3.9.7)$$

and similarly for $\tilde{\psi}$. The independence of these solutions may be shown by examining the various Wronskians as in Section 2.11. For real $k$ we may again introduce linear combinations that define various scattering problems and write

$$\varphi = c_{11}\psi + c_{12}\tilde{\psi} \qquad (3.9.8a)$$

$$\psi = c_{21}\tilde{\varphi} + c_{22}\varphi \qquad (3.9.8b)$$

and

$$\tilde{\varphi} = c_{11}^*\tilde{\psi} - c_{12}^*\psi \qquad (3.9.9a)$$

$$\tilde{\psi} = -c_{21}^*\varphi + c_{22}^*\tilde{\varphi} \qquad (3.9.9b)$$

These expressions are analogous to the relations introduced in (2.11.19) and (2.11.20). The various coefficients are related by

$$c_{11}(k) = w(\tilde{\psi},\varphi) = c_{22}^*(k)$$

$$c_{12}(k) = w(\varphi,\psi) = -c_{21}(k)$$

$$c_{22}(k) = w(\tilde{\varphi},\psi) = c_{11}^*(k) \qquad (3.9.10)$$

$$c_{12}^*(k) = w(\tilde{\varphi},\tilde{\psi}) = -c_{21}^*(k)$$

where the Wronskian is the linear combination of solutions defined in (2.11.18). The compatibility of (3.9.8) and (3.9.9) requires that

$$1 = c_{11}(k)c_{22}(k) - c_{12}(k)c_{21}^*(k)$$

$$0 = c_{11}(k)c_{21}(k) + c_{12}(k)c_{22}^*(k) \qquad (3.9.11)$$

$$0 = c_{11}^*(k)c_{21}(k) + c_{12}(k)c_{22}(k)$$

The reflection and transmission coefficients are the same as those obtained in Section 2.11 for real potentials and are related as in (2.11.16) and (2.11.28).

Since the integrals in the definition of $\varphi$ and $\psi$ in (2.11.15) and (2.11.17) still converge when $k$ is complex if $\text{Im}\,k > 0$, we may also use the definition of $c_{12}(k)$ given in (3.9.10) in the upper half of the $k$ plane. As in the case of real $q$ treated in Section 2.11, localized solutions are obtained from the zeros of $c_{12}$

in the upper half-plane. However, there is now no restriction that these zeros must be either on the imaginary axis or located symmetrically in pairs about the imaginary axis. We shall consider an example presently in which there is a single zero at an arbitrary point in the upper half-plane.

To obtain the coupled Marchenko equations, we take the Fourier transform of (3.9.8). The procedure is identical with that leading to (3.6.10) and (3.6.11) for real $q(x)$ except that the Fourier transform of the functions $\varphi^*(x,k)$ and $\psi^*(x,k)$ must be considered. To obtain these expressions, we use (2.11.3) and (2.11.4) to write

$$\varphi_2^*(x,k)=c\int_{-\infty}^{\infty} dt'\, e^{-ikct'}A_2^*(x,-ct')\theta\left(t'+\frac{x}{c}\right) \qquad (3.9.12)$$

Consequently, the Fourier transform is

$$\int_{-\infty}^{\infty}\frac{d\omega}{2\pi}e^{-i\omega t}\varphi_2^*(x,k)=cA_2^*(x,ct)\theta\left(-t+\frac{x}{c}\right) \qquad (3.9.13)$$

The final result of taking the Fourier transform of (3.9.8a) is

$$B_2^*(x,y)+\int_{x}^{\infty} dx'\, B_1(x,x')\Omega_R(x'+y)=0$$

$$-B_1^*(x,y)+\Omega_R(x+y)+\int_{x}^{\infty} dx'\, B_2(x,x')\Omega_R(x'+y)=0 \qquad x<y$$

$$(3.9.14)$$

where

$$\Omega_R(z)=r_R(z)+\sum m_{Rl}(k_l)e^{ik_l z} \qquad (3.9.15)$$

as in (3.6.12). From (3.9.8a), we obtain

$$A_1^*(x,y)+\int_{-\infty}^{x} dx'\, A_2(x,x')\Omega_L(x'+y)=0$$

$$-A_2^*(x,y)+\Omega_L(x+y)+\int_{-\infty}^{x} dx'\, A_1(x,x')\Omega_L(x'+y)=0 \qquad x>y$$

$$(3.9.16)$$

where

$$\Omega_L(z)=r_L(z)+\sum m_{Ll}(k_l)e^{-ik_l z} \qquad (3.9.17)$$

as in (3.6.9).

To obtain the purely reflectionless potentials, we follow the procedure used in Section 3.7. Let us consider the integral equations relating $B_1$ and $B_2$. We first define the $N$ component vectors

$$\Phi(z) = (e^{ik_1z}, \ldots, e^{ik_Nz})$$

$$\Psi(z) = (m_1 e^{ik_1z}, \ldots, m_N e^{ik_Nz}) \tag{3.9.18}$$

and then set $B_i(x,y) = \mathscr{B}_i(x) \cdot \Phi^*(y)$ for $i = 1, 2$. We then find $\mathscr{B}_1^* = \Psi \cdot (1 + M^* \cdot M)^{-1}$ and $\mathscr{B}_2 = -\mathscr{B}_1 \cdot M$, where

$$M(x) = \int_x^\infty dx' \, \Phi^*(x') \Psi(x') \tag{3.9.19}$$

The final expressions are obtained from (3.9.5) and (3.9.6). They are

$$u(x) = 2\Phi^* \cdot \Psi^* \cdot (1 + M \cdot M^*)^{-1} \tag{3.9.20}$$

and

$$|q(x)|^2 = \frac{d^2}{dx^2} \left\{ \ln\left[ \det(1 + M \cdot M^*) \right] \right\} \tag{3.9.21}$$

For $N = 1$ and $k_1 = \alpha + i\beta$, we find that $M(x) = (m_{R1}/2\beta)e^{-2\beta x}$. Setting $m_{R1}/2\beta = e^{\delta + i\theta}$, we obtain

$$q(x) = 2\beta \, \mathrm{sech}(2\beta x - \delta)e^{-i(2\alpha x + \theta)} \tag{3.9.22}$$

For $\alpha = \theta = 0$ the pole is on the imaginary axis, $u(x)$ is then real and the solution reduces to (2.11.45). These same results could, of course, be obtained by solving (3.9.16) for $A_1$ and $A_2$.

*Exercise 5*

The potential and a fundamental solution of (3.9.1) corresponding to one pole may be obtained by direct substitution into (3.9.1) if we assume that $n_1 = e^{ikx}q(x)$ and $n_2 = e^{ikx}[2ik + p(x)]$, where $p$ and $q$ are complex. With proper normalization this solution becomes the fundamental solution $\psi$. Show that the results are $u(x) = 2\beta \, \mathrm{sech}\,\zeta \exp[i(-2\alpha x + \theta)]$ and

$$\psi = \frac{e^{ikx}}{k - \alpha + i\beta} \begin{pmatrix} -i\beta e^{i(-2\alpha x + \theta)} \, \mathrm{sech}\,\zeta \\ k - \alpha + i\beta \tanh \zeta \end{pmatrix}$$

where $\zeta = 2\beta x + \zeta_0$ and $\alpha$, $\beta$, $\zeta_0$, and $\theta$ are constants of integration. In a similar way, obtain

$$\varphi = \frac{e^{-ikx}}{k - \alpha + i\beta} \begin{pmatrix} k - \alpha - i\beta \tanh \zeta \\ -i\beta e^{i(2\alpha x - \theta)} \, \mathrm{sech}\,\zeta \end{pmatrix}$$

These expressions will be useful in Chapter 9.

## Asymptotic Solution

In Section 2.11 an asymptotic solution of the two-component equations with a real potential was developed. The method used there is readily extended to the case of a complex potential (Zakharov and Shabat, 1972). When a second-order equation for $n_1$ is obtained from (3.9.1), we find that

$$n_{1xx} + (k^2 + |q|^2)n_1 - \frac{(n_{1x} + ikn_1)q_x}{q} = 0 \tag{3.9.23}$$

Identifying $n_1$ with the fundamental solution $\varphi_1$ given in (2.11.15) and writing

$$n_1 = e^{-ikx + h(x,k)} \tag{3.9.24}$$

with $\lim_{x \to -\infty} h(x,k) = 0$, so that $n_1$ will reduce to $\varphi_1$ in this limit, we find from (3.9.23) that $h(x,k)$ satisfies

$$2ikh_x = |q|^2 + (h_x)^2 + q\left(\frac{h_x}{q}\right)_x \tag{3.9.25}$$

Inserting the expansion

$$h_x = \sum_1^\infty \frac{g_n(x)}{(2ik)^n} \tag{3.9.26}$$

into (3.9.25) and equating coefficients of each power of $k$, we have

$$g_1 = |q|^2$$
$$g_{n+1} = q\left(\frac{g_n}{q}\right)_x + \sum_{k+l=n} g_k g_l, \quad n \geqslant 1 \tag{3.9.27}$$

The first few coefficients beyond $g_1$ are

$$g_2 = qq_x^*$$
$$g_3 = qq_{xx}^* + |q|^4 = |q|^4 - |q_x|^2 + (qq_x^*)_x$$
$$g_4 = qq_{xxx}^* + q^*|q|^2 q_x + 4q|q|^2 q_x^*$$
$$= qq_{xxx}^* + 3q|q|^2 q_x^* - \tfrac{1}{2}(|q|^4)_x \tag{3.9.28}$$
$$g_5 = q\left(\frac{g_4}{q}\right)_x + 2g_1 g_3 + g_2^2$$
$$= |q_{xx}|^2 + 2|q|^6 - 6|q|^2|q_x|^2 - \left[(|q|^2)_x\right]^2 + P_x$$

where $P = qq_{xxx}^* - q_x q_{xx}^* + q[(q^*)^2 q_x + 6|q|^2 q_x^*]$.

From (3.9.8a), we find $\lim_{x\to\infty} \varphi_1 e^{ikx} = c_{12}(k)$ for $\operatorname{Im} k > 0$. Therefore, $\lim_{x\to+\infty} h(x,k) = \ln c_{12}(k)$. Again using the fact that the transmission coefficient approaches unity as $k\to\infty$, we introduce the asymptotic expansion

$$\ln c_{12}(k) = \sum_{n=1}^{\infty} \frac{c_n}{k^n} \qquad (3.9.29)$$

Proceeding as in Section 2.8, we obtain the constants

$$(2i)^n c_n = \int_{-\infty}^{\infty} dx\, g_n(x) \qquad (3.9.30)$$

In using (3.9.30) we may, of course, discard the perfect derivatives indicated in (3.9.28) when $q(x)$ is localized.

# CHAPTER 4

# The Korteweg–deVries Equation

The mathematical techniques that will be used to solve nonlinear evolution equations by inverse scattering methods have been introduced in Chapters 2 and 3. In the present chapter we consider the use of these techniques for solving the Korteweg–deVries equation. This is the simplest example of the procedure since it requires only the inverse scattering methods used with the Schrödinger equation. Evolution equations solved by the inverse methods associated with the less familiar two-component systems are treated in Chapter 5. The properties of the Korteweg–deVries equation have been summarized many times in recent years. As an example, see Jeffrey and Kakutani (1972).

## 4.1  STEADY-STATE SOLUTION

Before approaching the general solution to the Korteweg–deVries equation, we shall first obtain the most general steady-state solution of this equation. In Chapter 1 we considered the Korteweg–deVries equation in the form

$$u_t - 6uu_x + u_{xxx} = 0 \qquad (4.1.1)$$

and found that it has the steady-state pulse solution $u = \frac{1}{2}c \, \text{sech}^2$ $[\frac{1}{2}\sqrt{c}\,(x - ct)]$. We now derive a more general steady-state solution. It has the form of an oscillation that can reduce to the previously considered single-pulse solution in the limit that the oscillation period tends to infinity. It will be of some convenience to examine the oscillatory solution that ultimately reduces to a pulse of positive amplitude. This and other algebraic conveniences are obtained by setting $u = -2f$ in (4.1.1). Writing the steady state solution in the form $f(x - ct)$, we have $f_t = -cf_x$, and the Korteweg–deVries equation reduces to an ordinary differential equation for which a first integral may be obtained by inspection. A second integration is then possible after the first integral is multiplied by $f_x$. We find that

$$(f_x)^2 = -4f^3 + cf^2 + af + b \qquad (4.1.2)$$

**113**

where $a$ and $b$ are constants of integration. Let us indicate the factored form of the cubic expression by

$$(f_x)^2 = -4(f - \alpha_1)(f - \alpha_2)(f - \alpha_3) = 4\varphi(f) \tag{4.1.3}$$

The graph of $\varphi(f)$ versus $f$ will be of the form shown in Figure 4.1 with three real roots $(\alpha_1 < \alpha_2 < \alpha_3)$. To see that all roots are real note that $\varphi(f)$ must be positive if $f_x$ is to be real. Thus for finite-amplitude oscillations, $f$ must be confined to the range $\alpha_2 \leqslant f \leqslant \alpha_3$. That is, there must be at least two real roots [and hence three since the coefficients of (4.1.2) are real]. Writing the only negative factor in (4.1.3) as $f - \alpha_3 = -g$, we find that the equation for $g$ is

$$(g_x)^2 = 4g(\alpha_3 - \alpha_2 - g)(\alpha_3 - \alpha_1 - g) \tag{4.1.4}$$

Setting $g = (\alpha_3 - \alpha_2)v^2$, we obtain

$$(v_x)^2 = (\alpha_3 - \alpha_1)(1 - v^2)(1 - k^2 v^2) \tag{4.1.5}$$

where $k^2 = (\alpha_3 - \alpha_2)/(\alpha_3 - \alpha_1)$.
    Now the equation

$$(w_z)^2 = (1 - w^2)(1 - k^2 w^2) \tag{4.1.6}$$

is the equation that defines the Jacobian elliptic function $sn(z, k)$ (Davis, 1959). The constant $k$ is referred to as the modulus of the elliptic function.

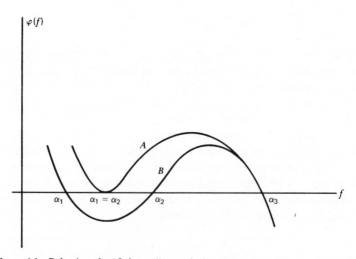

**Figure 4.1**  Behavior of $\varphi(f)$ for soliton solution (A) and oscillatory solution (B).

The function $\text{sn}(z,k)$ has the two limiting forms $\text{sn}(z,0)=\sin z$ and $\text{sn}(z,1)=\tanh z$. The solution of (4.1.5) is therefore $v(x)=\text{sn}\left(\sqrt{\alpha_3-\alpha_1}\ x,k\right)$ and thus

$$f(x-ct)=\alpha_3-(\alpha_3-\alpha_2)\text{sn}^2\left[\sqrt{\alpha_3-\alpha_1}\ (x-ct),k\right] \qquad (4.1.7)$$

Since the coefficient of the squared term in any cubic equation is equal to minus the sum of the roots, a comparison of (4.1.2) and (4.1.3) shows that $c=\frac{1}{4}(\alpha_1+\alpha_2+\alpha_3)$. The function $\text{sn}^2(z,k)$ oscillates between 0 and 1 with a period equal to $2K$ where $K=\int_0^1 dx[(1-x^2)(1-k^2x^2)]^{-1/2}$. Hence $f(x-ct)$ oscillates between $\alpha_2$ and $\alpha_3$ with a period $2K/(\alpha_3-\alpha_1)^{1/2}$. As $\alpha_2$ approaches $\alpha_1$, the modulus $k$ approaches unity and $K\rightarrow\infty$. We then obtain

$$f(x-ct)=\alpha_2+(\alpha_3-\alpha_2)\text{sech}^2\left[\sqrt{\alpha_3-\alpha_2}\ (x-ct)\right] \qquad (4.1.8)$$

which has the form of a pulse of half width $(\alpha_3-\alpha_2)^{-1/2}$ that rises to an amplitude $\alpha_3-\alpha_2$ above a reference level $\alpha_2$. If $\alpha_2=0$ (so that $\alpha_1=c/4$), we obtain

$$u(x-ct)=-2f(x-ct)=-\frac{1}{2}\text{sech}^2\left[\frac{1}{2}\sqrt{c}\ (x-ct)\right] \qquad (4.1.9)$$

which is the steady-state pulse obtained previously.

Since the Korteweg–deVries equation occurs in applications with various numerical coefficients, it is sometimes convenient to note that in the form $u_t+c_1uu_x+c_2u_{xxx}=0$, the equation has the single soliton solution $u=u_0\text{sech}^2[(x-Vt)/L]$, where $L=2\sqrt{3c_2/u_0c_1}$ and $V=u_0c_1/3$.

On the other hand, a steady-state solution in the form of a small-amplitude oscillation is obtained when $\alpha_3$ approaches $\alpha_2$ and $k$ becomes small. In that limit

$$f(x-ct)=\alpha_3-(\alpha_3-\alpha_2)\sin^2\left[\sqrt{\alpha_3-\alpha_1}\ (x-ct)\right] \qquad (4.1.10)$$

where $c=4(2\alpha_1+\alpha_3)$. Our future concern will be directed almost entirely toward pulse solutions, and these steady-state oscillatory solutions will not be considered further.

## 4.2  RESULTS OF NUMERICAL SOLUTIONS

The steady-state pulse solution considered in the previous section takes the form

$$u(x,t)=-2\,\text{sech}^2(x-4t) \qquad (4.2.1)$$

when we choose $c=4$. We may interpret this result by saying that this expression describes the propagation of the initial disturbance $u(x,0)=$

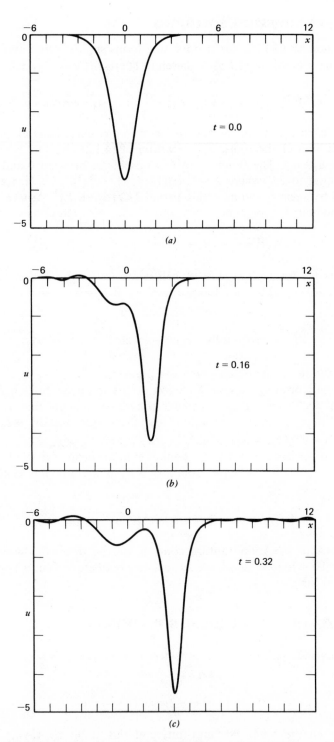

**Figure 4.2** Nonsoliton contribution to solution of Korteweg–deVries equation $u_t - 6uu_x + u_{xxx} = 0$ for initial profile $u(x,0) = \frac{15}{4} \operatorname{sech}^2 x$. Propagation is from left to right. Precursor is due to imposition of periodic boundary conditions. (Solution provided by W. E. Ferguson, Jr.)

$-2\operatorname{sech}^2 x$. Also, in Section 1.3 we saw that the initial profile $u(x,0) = -6\operatorname{sech}^2 x$ evolves into two solitons according to (1.3.25). As we shall see in Section 4.4, the initial profiles $u(x,0) = -n(n+1)\operatorname{sech}^2 x$, which correspond to the reflectionless potentials obtained in (3.5.15), will evolve into pure multisoliton solutions when $n$ is a positive integer. The two cases just mentioned correspond to $n=1$ and $n=2$.

The criterion for the occurrence of pure multisoliton solutions may be recast in various ways by rescaling the original equation. For instance, the coordinate transformations $\zeta = Lx$, $\tau = Tt$, and $v = -6(L/T)u$ yield $v_\tau + vv_\zeta + \beta v_{\zeta\zeta\zeta} = 0$, where $\beta = L^3/T$. The initial condition corresponding to $n$ solitons is then $v = 6n(n+1)(\beta/L^2)\operatorname{sech}^2(\zeta/L)$. If this amplitude is set equal to $u_0$ and the parameter $\sigma^2 = L^2 u_0/\beta$ is introduced, we see that the single-soliton solution corresponds to $\sigma = \sqrt{12}$. For larger values of $\sigma$, more than one soliton will emerge. For a discussion of the equation in this form, the paper by V. I. Karpman and V. P. Sokolov (1968) may be consulted.

When $u(x,0)$ is not one of these special profiles, the subsequent evolution of the pulse is found from numerical solution to include an oscillatory wave train in addition to the solitons. An example of the evolution of the initial pulse profile $u(x,0) = \frac{15}{4}\operatorname{sech}^2 x$, which corresponds to $n=\frac{3}{2}$ in the amplitude $n(n+1)$, is shown in Figure 4.2. The oscillatory tail is seen to be quite small. For larger nonintegral values of $n$ the ratios of energy in the tail to that in the solitons is even smaller. Both exact and approximate techniques for determining the amplitudes of the various solitons when an oscillatory tail is present

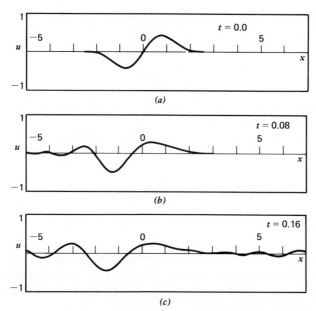

**Figure 4.3** Nonsoliton contribution to solution of Korteweg–deVries equation $u_t - 6uu_x + u_{xxx} = 0$ for initial profile $u(x,0) = xe^{-x^2}$. Propagation is from left to right. Precursor is due to imposition of periodic boundary conditions. (Solution provided by W. E. Ferguson, Jr.).

will be considered in subsequent sections of this chapter. For values of $n$ less than 1, a soliton will still emerge but a larger fraction of the energy will be associated with the oscillatory tail. Another type of initial profile that can be expected to enhance the nonsoliton portion of the solution is that in which $u(x,0)$ contains discontinuities or when the initial profile changes sign. An example of the latter is shown in Figure 4.3. The determination of the oscillatory tail may be made to depend upon the solution of a linear integral equation (the Marchenko equation), but the quantitative analysis of this equation, although linear, is quite laborious. Fortunately, when the initial profile varies smoothly and more than one soliton emerges, the oscillatory part of the solution is relatively small.

## 4.3 INVERSE SCATTERING AND THE KORTEWEG–deVRIES EQUATION

In Chapter 1 we saw that the Korteweg–deVries equation in the form

$$u_t - 6uu_x + u_{xxx} = 0 \tag{4.3.1}$$

could be viewed as a condition to be imposed upon the potential $u(x,t)$ in the equation

$$y_{xx}(x,t) + \left[ k^2 - u(x,t) \right] y(x,t) = 0 \tag{4.3.2}$$

so that the eigenvalue parameter $k^2$ would remain constant as the parameter $t$ is varied. In addition, the time dependence of the solution $u(x,t)$ was seen to be governed by

$$y_t = -4y_{xxx} + 6uy_x + 3u_x y \tag{4.3.3}$$

We now consider how these linear equations may be used to obtain solutions of the Korteweg–deVries equation.

As described in Chapter 2, two fundamental solutions of the Schrödinger equation (4.3.2) may be written

$$\begin{aligned} f_1(x,k;t) &= e^{ikx} + \int_x^\infty dx' A_R(x,x';t)e^{ikx'} \\ f_2(x,k;t) &= e^{-ikx} + \int_{-\infty}^x dx' A_L(x,x';t)e^{-ikx'} \end{aligned} \tag{4.3.4}$$

The parametric time dependence arising from the potential $u(x,t)$ has been indicated explicitly. As indicated in (2.8.6) and (2.8.7), the functions $f_1$ and $f_2$ may be written as linear combinations of each other. In particular, the functions $f_1$ and $f_2$ are related by

$$f_1(x,k;t) = c_{21}(k,t)f_2(x,-k;t) + c_{22}(k,t)f_2(x,k;t) \tag{4.3.5}$$

Also, according to (3.1.4), the potential in the Schrödinger equation is related to $A_L(x,x';t)$ by

$$u(x,t) = 2\frac{\partial}{\partial x} A_L(x,x;t) \tag{4.3.6}$$

The function $A_L(x,y;t)$ satisfies the integral equation given previously in (3.2.7), namely

$$\Omega_L(x+y;t) + A_L(x,y;t) + \int_{-\infty}^{x} dx' \Omega_L(x'+y;t) A_L(x,x';t) = 0 \qquad x > y \tag{4.3.7}$$

where

$$\Omega_L(z;t) = \int_{-\infty}^{\infty} \frac{dk}{2\pi} \frac{c_{22}(k,t)}{c_{21}(k,t)} e^{-ikz} + \sum_{1}^{N} m_{Ll}(i\kappa_l,t) e^{\kappa_l z} \tag{4.3.8}$$

We also have $m_{Ll}(i\kappa_l,t) = \gamma_l(t)c_{22}(i\kappa_l,t) = -ic_{22}(i\kappa_l,t)/\dot{c}_{21}(i\kappa_l,t)$ where the dot signifies a derivative with respect to $k$ rather than $t$. According to (4.3.6), the temporal variation of $u$ is known once $A_L(x,x;t)$ has been determined. The function $A_L(x,y;t)$ is related to $\Omega_L(x+y;t)$ through the integral equation (4.3.7). But the temporal variation of $\Omega_L(x+y;t)$ is unexpectedly simple! It is contained solely in the scattering coefficients $c_{21}$, $c_{22}$, and the $m_L$, which are related to the scattering coefficients. If we can determine the temporal variation of these terms, the temporal variation of $u(x,t)$, and thus the solution of the Korteweg–deVries equation will, at least in principle, be determined. We now show that the time dependence of the scattering coefficients is extremely simple and is readily provided by the linear equations (4.3.2) and (4.3.3). We need only consider these equations in the region of $x$ values for which $u(x,t) \to 0$, where (4.3.3) reduces to

$$y_t = -4y_{xxx} \tag{4.3.9}$$

This provides information on the time dependence of only the reflected and transmitted waves, of course, but the power of the inverse method is that information on reflection and transmission coefficients (and their analytic continuation into the upper half-plane) is sufficient for reconstructing the scattering potential itself.

Let us examine the solution $y(x,k,t)$ that is proportional to $f_1(x,k,t)$. Since the proportionality factor can be time-dependent, we write $y(x,k,t) = h(k,t)f_1(x,k;t)$. We first show that the function $h(k,t)$ can be determined by considering the solution as $x \to +\infty$. In that limit (4.3.9) shows that $h(k,t)$ satisfies $h_t = 4ik^3h$ so that $h(k,t) = h(k,0)\exp(4ik^3t)$. The time dependence of

$c_{21}$ and $c_{22}$ may now be obtained by considering $y(x,k,t)$ as $x$ approaches $-\infty$, where we can use (4.3.5) to write

$$y(x,k,t) \sim h(k,0)e^{4ik^3t}\left(c_{21}e^{ikx} + c_{22}e^{-ikx}\right) \qquad (4.3.10)$$

Again substituting into (4.3.9) and equating coefficients of $\exp(\pm ikx)$, we find that $c_{21_t} = 0$ and $c_{22_t} = -8ik^3c_{22}$. We thus obtain

$$c_{21}(k,t) = c_{21}(k,0) \qquad (4.3.11a)$$

$$c_{22}(k,t) = c_{22}(k,0)e^{-8ik^3t} \qquad (4.3.11b)$$

The normalization constant $m_{Ll}$ is found from (2.8.30b) to be

$$m_{Ll}(k_l,t) = -i\frac{c_{22}(k_l,t)}{\dot{c}_{21}(k_l,t)} = m_{Ll}(k_l,0)e^{-8ik_l^3t} \qquad (4.3.12)$$

As shown in Section 2.8, the points $k_l$ are located along the imaginary axis in the upper half-plane at points $k_l = i\kappa_l$.

The relations among the $c_{ij}$ given in (2.8.12) and (2.8.13) may be used to show that

$$c_{11}(k,t) = c_{11}(k,0)e^{8ik^3t} \qquad (4.3.13a)$$

$$c_{12}(k,t) = c_{12}(k,0) \qquad (4.3.13b)$$

and

$$m_{Rl}(k_l,t) = m_{Rl}(k_l,0)e^{8ik_l^3t} \qquad (4.3.14)$$

The reflection coefficient $R_L = c_{22}/c_{21}$ has the time dependence

$$R_L(k,t) = R_L(k,0)e^{-8ik^3t} \qquad (4.3.15)$$

The time dependence of $\Omega_L$ is thus completely determined and we have

$$\Omega_L(z;t) = \int_{-\infty}^{\infty}\frac{dk}{2\pi}R_L(k,0)e^{-i(kz+8k^3t)} + \sum_{1}^{N}m_{Ll}(0)e^{-8\kappa_l^3t+\kappa_l z} \qquad (4.3.16)$$

According to this result, once the reflection coefficient $R_L(k,0)$, the bound-state normalization constants $m_{Ll}(0)$, and pole locations $\kappa_l$ associated with the initial potential $u(x,0)$ have been obtained, the subsequent temporal evolution of $\Omega_L(z;t)$ is determined. The time dependence of $A_L(x,y;t)$ is then also determined by the Marchenko equation (4.3.7). The solution of the Korteweg–deVries equation then follows from

$$u(x,t) = 2\frac{\partial}{\partial x}A_L(x,x;t)$$

Unfortunately, the time-dependent phase term in the integral over the reflection coefficient $R_L(k,t)$ makes this integral difficult to evaluate analytically. However, this integration is avoided if we confine attention to reflectionless potentials, since then $R_L(k,0)=0$ and this integral does not occur. The Marchenko equation (4.3.7) is then readily solved and the solution of the Korteweg–deVries equation may be obtained in closed form. The solutions that we obtain are the pure multisoliton solutions.

A similar calculation for $\Omega_R(z;t)$ yields

$$\Omega_R(z;t)=\int_{-\infty}^{\infty}\frac{dk}{2\pi}R_R(k,0)e^{i(kz+8k^3t)}+\sum m_{Rl}(0)e^{8\kappa_l^3z-\kappa_l t}\qquad(4.3.17)$$

The solution to the Korteweg–deVries equation may again be obtained by using this expression for $\Omega_R(z,t)$ in the Marchenko equation (3.2.6). The solution of the Korteweg–deVries equation now follows from

$$u(x,t)=-2\frac{\partial}{\partial x}A_R(x,x;t)$$

when the integral equation (3.2.6) has been solved for $A_R$.

## 4.4  MULTISOLITON SOLUTIONS

The solution of the Marchenko equation for reflectionless potentials was carried out in Section 3.3. In application of inverse scattering theory to the solution of the Korteweg–deVries equation being discussed in this chapter, the time dependence enters only parametrically, so we again expect that exact solutions can be obtained when the initial condition of the Korteweg–deVries equation $u(x,0)$ corresponds to a reflectionless potential. The solutions obtained are the multisoliton solutions. The expression obtained for $A_L(x,x)$ in Section 3.3 is still valid; only a parametric time dependence need be incorporated. We therefore have

$$\Omega_L(z;t)=\sum_{1}^{N}m_{Ll}(t)e^{\kappa_l z}\qquad(4.4.1)$$

where $m_{Ll}(t)=m_{Ll}(0)\exp(-8\kappa_l^3 t)$ according to (4.3.12). From (3.3.16), the solution of the Marchenko equation is

$$u(x,t)=2\frac{\partial}{\partial x}A_L(x,x;t)=-2\frac{\partial^2}{\partial x^2}\ln[\det V(x,t)]\qquad(4.4.2)$$

where $V(x,t)$ is the $N\times N$ matrix with elements

$$V_{ij}(x,t)=\delta_{ij}+\frac{m_{Lj}(t)e^{(\kappa_i+\kappa_j)x}}{\kappa_i+\kappa_j}\qquad(4.4.3)$$

**Example—The Two-Soliton Solution ($N=2$)**

Let us consider the case in which there are two poles in the upper half-plane at points $i\kappa_1$ and $i\kappa_2$ ($\kappa_2 > \kappa_1$). Then

$$\det V = \begin{vmatrix} 1 + \dfrac{m_1(t)}{2\kappa_1} e^{2\kappa_1 x} & \dfrac{m_2(t)}{\kappa_1 + \kappa_2} e^{(\kappa_1 + \kappa_2)x} \\[2ex] \dfrac{m_1(t)}{\kappa_1 + \kappa_2} e^{(\kappa_1 + \kappa_2)x} & 1 + \dfrac{m_2(t)}{2\kappa_2} e^{2\kappa_2 x} \end{vmatrix} \tag{4.4.4}$$

Note that this determinant may be written in symmetric form if the $i$th column is divided by $\sqrt{m_i(t)/2\kappa_i} \, e^{\kappa_i x}$ and the $i$th row now is multiplied by this same factor. Such symmetrization may be carried out for arbitrary $N$ (Wadati and Toda, 1972).

We now differentiate $\ln(\det M)$ and multiply numerator and denominator of the resulting expression by

$$2\left(\frac{\kappa_2 + \kappa_1}{\kappa_2 - \kappa_1}\right)\left(\frac{\kappa_1 \kappa_2}{m_1 m_2}\right)\frac{1}{2} \exp\left[-(\kappa_1 + \kappa_2)x\right]$$

In the result, set $\exp(\lambda) = (m_2 \kappa_1 / m_1 \kappa_2)^{1/2}$ and

$$\exp(\mu) = \frac{1}{2}\left(\frac{m_1 m_2}{\kappa_1 \kappa_2}\right)^{1/2} \frac{\kappa_2 - \kappa_1}{\kappa_2 + \kappa_1}.$$

We find that

$$\frac{\partial}{\partial x}(\ln \det V) = \frac{\dfrac{\kappa_2 + \kappa_1}{\kappa_2 - \kappa_1}(\kappa_1 e^{-\varphi} + \kappa_2 e^{\varphi}) + (\kappa_1 + \kappa_2)e^{\psi}}{\cosh \varphi + \dfrac{\kappa_2 + \kappa_1}{\kappa_2 - \kappa_1}\cosh \psi} \tag{4.4.5}$$

where $\varphi = (\kappa_2 - \kappa_1)x + \lambda$ and $\psi = (\kappa_2 + \kappa_1)x + \mu$. A simpler form of the result is

$$\frac{\partial}{\partial x}(\ln \det V) = \kappa_1 + \kappa_2 + (\kappa_2^2 - \kappa_1^2)\frac{\sinh \varphi + \sinh \psi}{(\kappa_2 + \kappa_1)\cosh \varphi + (\kappa_2 - \kappa_1)\cosh \psi} \tag{4.4.6}$$

Writing

$$\sinh \varphi + \sinh \psi = 2\sinh\left(\frac{\varphi + \psi}{2}\right)\cosh\left(\frac{\varphi - \psi}{2}\right)$$

and using similar identities for $\cosh \varphi \pm \cosh \psi$, we obtain

$$\frac{\partial}{\partial x} (\ln \det V) = \kappa_1 + \kappa_2 + \frac{\kappa_2^2 - \kappa_1^2}{\kappa_2 \coth \gamma_2 - \kappa_1 \coth \gamma_1} \tag{4.4.7}$$

where $\gamma_2 = \frac{1}{2}(\psi + \varphi) = \kappa_2 x + \frac{1}{2}(\lambda + \mu)$ and $\gamma_1 = \frac{1}{2}(\psi - \varphi) = \kappa_1 x + \frac{1}{2}(\mu - \lambda)$. Since

$$e^{\mu + \lambda} = \frac{m_2(t)}{2\kappa_2} \left( \frac{\kappa_2 - \kappa_1}{\kappa_2 + \kappa_1} \right) = \frac{m_L(0)}{2\kappa_2} e^{-8\kappa_2^3 t} \left( \frac{\kappa_2 - \kappa_1}{\kappa_2 + \kappa_1} \right) \tag{4.4.8a}$$

and

$$e^{\mu - \lambda} = \frac{m_1(t)}{2\kappa_1} \left( \frac{\kappa_2 - \kappa_1}{\kappa_2 + \kappa_1} \right) = \frac{m_1(0)}{2\kappa_1} e^{-8\kappa_1^3 t} \left( \frac{\kappa_2 - \kappa_1}{\kappa_2 + \kappa_1} \right) \tag{4.4.8b}$$

we may write

$$\gamma_1 = \kappa_1 x - 4\kappa_1^3 t + \delta_1$$
$$\gamma_2 = \kappa_2 x - 4\kappa_2^3 t + \delta_2 \tag{4.4.9}$$

where

$$\delta_i = \frac{1}{2} \ln \left[ \frac{m_i(0)}{2\kappa_i} \left( \frac{\kappa_2 - \kappa_1}{\kappa_2 + \kappa_1} \right) \right]$$

When the second derivative of $\ln(\det V)$ is obtained from (4.4.7), we have

$$u(x, t) = -2(\kappa_2^2 - \kappa_1^2) \frac{\kappa_2^2 \operatorname{csch}^2 \gamma_2 + \kappa_1^2 \operatorname{csch}^2 \gamma_1}{(\kappa_2 \coth \gamma_2 - \kappa_1 \tanh \gamma_1)^2} \tag{4.4.10}$$

Except for the additional phase terms $\delta_1$ and $\delta_2$, which are the integration constants that were ignored integrating (1.3.24), this result is identical with that given in (1.3.19).

It was shown in Section 1.3 that, at times long before the two solitons interact, $u(x, t)$ reduces to

$$u(x, t) = -2\kappa_1^2 \operatorname{sech}^2 (\gamma_1 - \Delta) - 2\kappa_2^2 \operatorname{sech}^2 (\gamma_2 + \Delta) \tag{4.4.11}$$

where

$$\Delta = \tan^{-1} \left( \frac{\kappa_1}{\kappa_2} \right) = 2 \ln \left( \frac{\kappa_2 + \kappa_1}{\kappa - \kappa_1} \right)$$

At times long after interaction we obtain

$$u(x, t) = -2\kappa_1^2 \operatorname{sech}^2 (\gamma_1 + \Delta) - 2\kappa_2^2 \operatorname{sech}^2 (\gamma_2 - \Delta) \tag{4.4.12}$$

The location of the peak of the pulse associated with $\kappa_2$ is thus shifted forward by an amount $2\Delta/\kappa_2$ from $x_2 = 4\kappa_2^2 t - (\delta_2 - \Delta)/\kappa_2$ to $x_2 = 4\kappa_2^2 t - (\delta_2 + \Delta)/\kappa_2$. Similarly, the slower pulse is retarded by an amount $2\Delta/\kappa_1$. A detailed analysis of the interaction of an arbitrary number of solitons is available (Wadati and Toda, 1972).

The two-soliton interaction for $\kappa_1 = 1$ and $\kappa_2 = 2$ and $\delta_1 = \delta_2 = 0$ is shown in Figure 1.2. Both solitons merge into a single pulse of amplitude $-6$ at $t = 0$. [Note that $-u(x, t)$ is plotted in the figure.] In Figure 4.4 the merging of two solitons is shown for the case $\kappa_1 = 1$, $\kappa_2 = 1.5$, and $\delta_1 = \delta_2 = 0$. A single peak is not formed in this instance. To obtain a criterion for distinguishing the two cases, note that $u_{xx}(0,0) > 0$ for Figure 1.2 while $u_{xx}(0,0) < 0$ for Figure 4.4. An expression for $u_{xx}(0,0)$ may be obtained by using the Korteweg–deVries equation and (4.4.2). We find that

$$u_{xx}(x,t) = 3u^2(x,t) - 2\frac{\partial}{\partial t} A_L(x,x;t) \qquad (4.4.13)$$

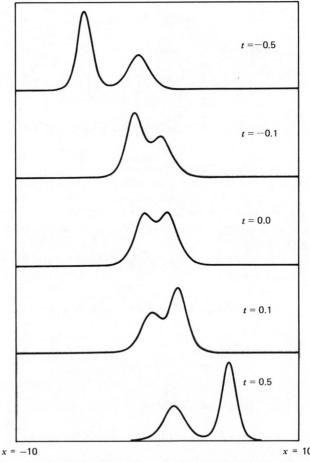

$t = -0.5$

$t = -0.1$

$t = 0.0$

$t = 0.1$

$t = 0.5$

$x = -10$                                              $x = 10$

**Figure 4.4**  Interaction of two solitons in which single peak is not formed.

Using (4.4.10) to obtain $u^2(0,0)$ and noting that $\frac{\partial}{\partial t} A_2(x,x;t)$ may be obtained from (4.4.7), we find that

$$u_{xx}(0,0) = 4(\kappa_2^2 - \kappa_1^2)(\kappa_2^2 - 3\kappa_1^2), \qquad \kappa_2 > \kappa_1 \qquad (4.4.14)$$

Therefore, the two solitons will merge into a single peak only when $\kappa_2/\kappa_1 > \sqrt{3}$ .

We have now seen that the evolution of an initial pulse profile is extremely simple when there is a vanishing reflection coefficient $R_L(k,0)$ for the scattering problem associated with $u(x,0)$. The $N$ poles of the transmission coefficient $T(k,0)$ lead to $N$ solitons with amplitudes $-2\kappa_n^2$. When the reflection coefficient is not zero, the solution still contains a soliton for each pole. However, there is also an oscillatory part as shown in Figure 4.2. The quantitative analytical description of the solution is now much more complicated and is currently an active area of research. However, the amplitude and velocity of each soliton can still be determined quite easily. The amplitudes are still equal to $-2\kappa_n^2$ and the velocity is still $4\kappa_n^2$, as in the reflectionless case. Hence, as long as the reflection coefficient can be determined for $u(x,0)$, the number of solitons and their amplitudes can be determined even when $R_L(k,0) \neq 0$. As an example, consider the initial pulse profile $u(x,0) = v \, \text{sech}^2 x$. The transmission coefficient for this potential was shown in Section 2.5 to be

$$T(k,0) = \frac{\Gamma(a)\Gamma(b)}{\Gamma(c)\Gamma(a+b-c)} \qquad (4.4.15)$$

where

$$a = \tfrac{1}{2} - ik + \sqrt{v + \tfrac{1}{4}}$$

$$b = \tfrac{1}{2} - ik - \sqrt{v + \tfrac{1}{4}} \qquad (4.4.16)$$

$$c = 1 - ik$$

For $v = \frac{15}{4}$, this becomes

$$T(k,0) = \frac{\Gamma(\tfrac{5}{2} - ik)\Gamma(-\tfrac{3}{2} - ik)}{\Gamma(1 - ik)\,\Gamma(-ik)} \qquad (4.4.17)$$

The factor $\Gamma(-\tfrac{3}{2} - ik)$ has poles at $-\tfrac{3}{2} - ik = -n, \, n = 0, 1, 2, \ldots$ . The poles are thus at $k = \kappa_n = \tfrac{3}{2}i, \tfrac{1}{2}i, -\tfrac{1}{2}i, \ldots$ . The two poles in the upper half-plane give rise to solitons. Since the soliton amplitudes are equal to $-2\kappa_n^2$, in this case the amplitudes are $-2(\tfrac{1}{2})^2 = -\tfrac{1}{2}$ and $-2(\tfrac{3}{2})^2 = -\tfrac{9}{2}$, which agree with the exact numerical solution shown in Figure 4.2.

## 4.5  CONSERVED QUANTITIES

In section 2.8 we saw that $c_{12}(k)$, the reciprocal of the transmission coefficient, has the asymptotic expansion

$$c_{12}(k) = \sum_1^\infty \frac{c_n}{k^n} \tag{4.5.1}$$

where

$$c_n = \frac{1}{(-2i)^{n+1}} \int_{-\infty}^\infty dx\, g_n(x) \tag{4.5.2}$$

The $g_n(x)$ are functions of the potential and its spatial derivatives. The first few $g_n$ are given in (2.8.51). Since $u$ depends upon time, we might expect that the $c_n$ and thus $c_{12}$ would also depend upon time. However, according to (4.3.13b), $c_{12}$ is independent of time for potentials that evolve in time according to the Korteweg–deVries equation. Hence the coefficients $c_n$ in the asymptotic expansion of $c_{12}$ must also be independent of time even though there is time dependence in the potential $u(x,t)$. The coefficients $c_n$ thus represent an infinite number of constants of the motion that may be associated with solutions of the Korteweg–deVries equation. From (2.8.50) we see that $g_1$ and $g_3$ are perfect spatial derivatives so that their integrals vanish for pulse solutions. This turns out to be true for all the higher odd values of $g_n$ (Miura et al., 1968). Hence only $g_0, g_2, g_4, \ldots$ are of interest. From (2.8.50) and (2.8.51) the first few conserved quantities are

$$c_1 = -\frac{i}{2} \int_{-\infty}^\infty dx\, g_0 = \frac{i}{2} \int_{-\infty}^\infty dx\, u(x,t)$$

$$c_3 = -\frac{i}{8} \int_{-\infty}^\infty dx\, g_2 = \frac{i}{8} \int_{-\infty}^\infty dx\, u^2(x,t) \tag{4.5.3}$$

$$c_5 = \frac{i}{2^5} \int_{-\infty}^\infty dx\, g_4 = \frac{i}{32} \int_{-\infty}^\infty dx \left[ 2u^3 - 6uu_{xx} - 5(u_x)^2 \right]$$

where terms in the $g_n$ that involve perfect derivatives have been ignored and an integration by parts has been performed in obtaining $c_5$.

The time independence of the $c_n$ may also be inferred directly from the Korteweg–deVries equation. Integrating this equation over all space yields

$$\frac{\partial}{\partial t} \int_{-\infty}^\infty dx\, u = \int_{-\infty}^\infty dx\, \frac{\partial}{\partial x} (3u^2 - u_{xx}) = 0 \tag{4.5.4}$$

which is equivalent to the first of (4.3.3). Multiplying the Korteweg–deVries equation by $u$ and then, integrating over all $x$, we obtain the second of (4.5.3) in the form

$$\frac{1}{2}\frac{\partial}{\partial t}\int_{-\infty}^{\infty} dx\, u^2 = \int_{-\infty}^{\infty} dx\, \frac{\partial}{\partial x}\left[2u^3 - uu_{xx} + \tfrac{1}{2}(u_x)^2\right] = 0 \qquad (4.5.5)$$

Similar verification of the higher conserved quantities may be carried out but increasingly elaborate manipulations are required. A list of the first 10 conserved quantities has been given by Miura et al. (1968).

The conserved quantities provide a simple scheme for obtaining approximate values of soliton amplitudes (Berezin and Karpman, 1967). The method has the disadvantage that it requires knowledge of the final number of pulses to be expected. Also, the method is inaccurate for rapidly varying initial pulse profiles and may not yield the amplitude of the smallest soliton very accurately. However, it has the advantage that it proceeds directly from the shape of the initial pulse profile and does not require a determination of the bound states associated with that profile. As an example, let us consider the initial pulse profile $u(x,0) = -v\,\mathrm{sech}^2 x$. For values of $v$ that lead to a decomposition into two solitons, we must have

$$u(x,t) \underset{t\to\infty}{\to} -2\kappa_1^2\,\mathrm{sech}^2(\kappa_1 x - v_1 t + \varphi_1) - 2\kappa_2^2\,\mathrm{sech}^2(x_2 x - v_2 t + \varphi_2)$$

$$(4.5.6)$$

where $v_i = 4\kappa_i^3$. As long as the phase terms $\varphi_i$ are not of interest, only the pulse amplitudes $\kappa_i$ need be determined to obtain the final form of the solution. Since the two quantities $F_1 = \int_{-\infty}^{\infty} dx\, u(x,t)$ and $F_2 = \int_{-\infty}^{\infty} dx\, u^2(x,t)$ are constant in time, we may evaluate them at $t = 0$ using the value of $u(x,0)$ given above and also as $t \to \infty$ using (4.3.5). We may then equate the results. In so doing, we obtain the pair of algebraic equations

$$\kappa_1 + \kappa_2 = \tfrac{1}{2}v$$

$$\kappa_1^3 + \kappa_2^3 = \tfrac{1}{4}v^2 \qquad (4.5.7)$$

where we have used the integrals* $\int_{-\infty}^{\infty} dx\,\mathrm{sech}^2 x = 2$ and $\int_{-\infty}^{\infty} dx\,\mathrm{sech}^4 x = \tfrac{4}{3}$. Equations 4.5.7 may be combined to yield the quadratic equation $\kappa_1^2 - \kappa_1\kappa_2 + \kappa_2^2 = \tfrac{1}{2}v$, which has the solutions $\kappa_{1,2} = \tfrac{1}{4}(v \pm \sqrt{v(8-v)/3}\,)$. For $v = \tfrac{15}{4}$, the value considered in the last section, we obtain $\kappa_1 = 1.51$ and $\kappa_2 = 0.363$. These values are to be compared with the exact values $\kappa_1 = 1.50$ and $\kappa_2 = 0.50$ that were obtained at the end of the last section. The exact solution, which contains a contribution from the nonsoliton part of the solution would, of course, satisfy all the higher conservation laws as well.

---

*The general result is $\int_{-\infty}^{\infty} dx\,\mathrm{sech}^n x = 2^{n-1}[\Gamma(n/2)]^2/\Gamma(n)$.

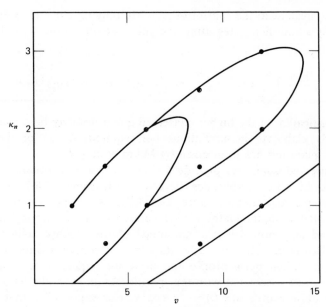

**Figure 4.5** Locus of roots for first three conservation laws for initial pulse profile $u(x,0) = v$ $\text{sech}^2 x$.

When three pulses emerge, their amplitudes may be estimated by employing the first three conservation laws. Again using the initial pulse profile $u(x,0) = v \, \text{sech}^2 x$ and the third conserved quantity from (4.5.3), we find that

$$\kappa_1 + \kappa_2 + \kappa_3 = \tfrac{1}{2} v$$

$$\kappa_1^3 + \kappa_2^3 + \kappa_3^3 = \tfrac{1}{4} v^2 \qquad\qquad (4.5.8)$$

$$\kappa_1^5 + \kappa_2^5 + \kappa_3^5 = \frac{v^2(2v-1)}{12}$$

Solution of this set of equations is equivalent to the solution of a single cubic equation. The three equations in (4.5.8) may be viewed as symmetric functions of the roots of this cubic equation and results in the theory of equations (Cajori, 1904, Chap. 7) may be used to advantage. The locus of roots for (4.5.7) and (4.5.8) are shown in Figure 4.5. The points are exact results obtained from (4.4.17). Results for $v = 2$, 6 and 12 correspond to pure soliton solutions and are seen to be on the curves. The approximate agreement for $v = 3.75$ and 8.75 results from ignoring the nonsoliton part of the solution.

## 4.6   THE INITIAL PULSE PROFILE $\delta'(x)$—SIMILARITY SOLUTION

In Ex. 5 of Chapter 2 a limiting procedure was used to show that the reflection coefficient $R_L(k)$ was equal to unity for a potential proportional to $\delta'(x)$. By using the relation between $R_R(k)$ and $R_L(k)$ in (2.8.16) before this limiting procedure is carried out and then proceeding to the limit, we find that the other reflection coefficient $R_R(k)$ is equal to $-1$. This might be considered to be the simplest reflection coefficient that we could use as an initial pulse profile. Unfortunately, the function $\delta'(x)$ is too singular to be used as an initial pulse profile. As noted in Section 3.1, the requirement that $\int_{-\infty}^{\infty} dx(1+|x|)|u(x,0)| < \infty$ is violated. Nevertheless, it is possible to extract interesting results concerning this initial profile. (We may think of the inverse formalism as being applied to a profile that approaches $\delta'(x)$ as a limit, e.g., $\delta'(x) = -\lim_{a \to 0}[(x/\pi^{1/2}a^3)\exp(-x^2/a^2)]$.)

Since $R_R(k,0) = -1$, we have

$$\Omega_R(x,t) = -\int_{-\infty}^{\infty} \frac{dk}{2\pi} e^{i(kx+8k^3t)} \tag{4.6.1}$$

This integral may be written as the integral that defines the Airy function $\text{Ai}(x)$ which is related to Bessel functions of one-third order (Abramowitz and Stegun, 1964). Introducing the new integration variable $k = \alpha\theta$ and then setting $8\alpha^3 t = \frac{1}{3}$, we find that

$$\Omega_R(x,t) = -\alpha\int_{-\infty}^{\infty} \frac{d\theta}{2\pi} e^{i[\theta\alpha x + (1/3)\theta^3]}$$

$$= -\alpha\,\text{Ai}(\alpha x) \tag{4.6.2}$$

Because of the convergence difficulty mentioned above, we will not attempt to solve the Marchenko equation with this form for $\Omega_r(x,t)$. Instead, we note that since $\alpha x = x(24t)^{-1/3}$, the space and time dependence of $\Omega_R(x,t)$ in (4.6.2) is of the form $\Omega_r(x,t) = t^{-1/3}\omega(\zeta)$, where $\zeta = xt^{-1/3}$. The interesting result is that the space-time dependence of the subsequent evolution of the initial profile $u(x,0) = \delta'(x)$ may also be expressed as a function of $\zeta$. This may be seen by setting $A_1(x,y) = t^{-1/3}\mathcal{C}(\zeta,\eta)$, where $\eta = yt^{-1/3}$. We find that the Marchenko equation becomes

$$\omega(\zeta + \eta) + \mathcal{C}(\zeta,\eta) + \int_{-\infty}^{\zeta} d\zeta'\,\omega(\zeta' + \eta)\mathcal{C}(\zeta,\zeta') = 0 \tag{4.6.3}$$

The pulse profile is now given by

$$u(x,t) = 2\frac{\partial}{ix}A_1(x,x;t) = 2t^{-1/3}\frac{\partial}{\partial x}\mathcal{C}(\zeta,\zeta) \tag{4.6.4}$$

Since $\partial/\partial x = t^{-1/3}d/d\zeta$, we obtain

$$u(x,t) = 2t^{-2/3}\mathcal{C}'(\zeta,\zeta) = t^{-2/3}f(\zeta) \tag{4.6.5}$$

Assuming $u(x,t)$ to be of this form and noting that

$$\frac{\partial}{\partial t} = -\frac{\zeta}{3t}\frac{d}{d\zeta} \quad \text{so that} \quad \frac{\partial u}{\partial t} = -\tfrac{1}{3}t^{-5/3}(2f + \zeta f')$$

we find that the Korteweg–deVries equation reduces to a nonlinear ordinary differential equation for the function $f(\zeta)$. The numerical coefficients in this equation are somewhat simplified if we set $g(\zeta) = 6f(\zeta)$. The equation for $g(\zeta)$ is then

$$3g''' - 3gg' - (2g + \zeta g') = 0 \tag{4.6.6}$$

For values of $g$ that are small enough so that the nonlinear term is negligible, we may set $g = h'$ and integrate (4.6.6) twice to obtain $h'' - \tfrac{1}{3}\zeta h = 0$. Integration constants have been neglected since attention is being confined to solutions that vanish before the pulse arrives. The equation for $h$ is satisfied by Bessel functions of one-third order. The solution that vanishes as $\zeta \to +\infty$ is proportional to $\zeta^{1/2}K_{1/3}[2(-\zeta/3)^{3/2}]$. Using the asymptotic form for the Bessel function we see that as $\zeta \to +\infty$, $g \sim c\zeta^{1/4}e^{-2(\zeta/3)^{3/2}}$. A numerical solution of the full nonlinear equation (4.6.6) that reduces to this small-amplitude form is shown in Figure 4.6 for $c = 0.15$. As noted above, the function $\delta'(x)$ is too singular to be a valid initial condition for a pulse profile. This is borne out in the figure, since the oscillatory solution continues to grow as $\zeta \to -\infty$. Further information on this solution may be obtained by consulting Karpman (1975).

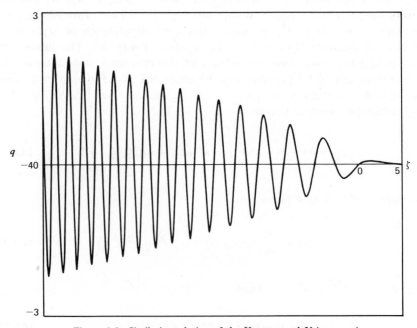

**Figure 4.6**  Similarity solution of the Korteweg–deVries equation.

## 4.7   ALTERNATIVE APPROACH TO THE LINEAR EQUATIONS FOR THE KORTEWEG–deVRIES EQUATION

The solution of the Korteweg–deVries equation may also be phrased in terms of a two-component linear system similar to that considered in Section 2.11. Before beginning a consideration of other evolution equations which are more naturally treated in a two-component framework, we briefly consider a two-component method for solving the Korteweg–deVries equation. This is accomplished by transforming the linear equations for the Korteweg–deVries equation into an equivalent two-component system.

To introduce the two-component method, we begin with the linear equations that were obtained in Section 1.2. According to (1.2.2) and (1.2.10),

$$y_{xx} = (\lambda + u)y \tag{4.7.1a}$$

$$y_t = -4y_{xxx} + 6uy_x + 3u_x y \tag{4.7.1b}$$

We now follow the standard procedure for replacing a second-order differential equation by a pair of first-order equations. Setting

$$z = y_x \tag{4.7.2a}$$

we find that the Schrödinger equation is equivalent to

$$z_x = (\lambda + u)y \tag{4.7.2b}$$

Also, the expression for $y_t$ in (4.7.1b) may be written as a linear combination of $z$ and $y$ when the derivative of (4.7.1a) is used to eliminate $y_{xxx}$ from (4.7.1b). We obtain

$$y_t = -u_x y + (2u - 4\lambda)4x \tag{4.7.3a}$$

Finally, an expression for $z_t$ that is a linear combination of $y$ and $z$ may be constructed by writing $z_t = (y_t)_x$ and then using (4.7.1b). The result, after using (4.7.1a), is

$$z_t = (2u^2 - u_{xx} - 2\lambda u - 4\lambda^2)y + u_x z \tag{4.7.3b}$$

The pair of linear equations (4.7.2) may be brought to a form somewhat similar to the linear equation considered in Section 2.11 by writing $y = v_2, z = -v_1 + i\zeta v_2$, and $\lambda = -\zeta^2$. We then obtain

$$v_{1x} + i\zeta v_1 = -uv_2$$
$$v_{2x} - i\zeta v_2 = -v_1 \tag{4.7.4}$$

When the same transformation is applied to (4.7.3), we obtain equations of the form

$$v_{1t} = Av_1 + Bv_2$$
$$v_{2t} = Cv_1 + Dv_2$$

(4.7.5)

where

$$A = -4i\zeta^3 - 2i\zeta u + u_x$$
$$B = -4\zeta^2 u - 2i\zeta u_x - 2u^2 + u_{xx}$$
$$C = -4\zeta^2 - 2u$$
$$D = -A$$

(4.7.6)

Solution of the Korteweg–deVries equation could now be recovered from these linear equations. The details of this approach will be considered in Chapter 5 when it is applied to other evolution equations.

Of importance as far as analyzing the new equations in Chapter 5 is concerned is the fact that equations of the form of (4.7.5) provide a starting point for introducing the time dependence for the functions $v_1$ and $v_2$.

## CHAPTER 5

## Some Evolution Equations
## Related to a
## Two-Component Linear System

In Chapter 4 we saw how the linear scattering problem associated with the Schrödinger equation could be used to solve the nonlinear partial differential equation known as the Korteweg–deVries equation. Certain other nonlinear evolution equations have been solved in a similar way by relating them to the pair of first-order linear differential equations

$$v_{1x} + i\zeta v_1 = qv_2$$
$$v_{2x} - i\zeta v_2 = -q^* v_1$$

These linear equations are a special case of the more general equations given in (1.5.4). In this chapter we consider how this two-component linear problem may be used to solve the three nonlinear evolution equations known as the modified Korteweg–deVries equation, the sine–Gordon equation, and the cubic or nonlinear Schrödinger equation. Following is a standard form for each of these equations:

$$u_t \pm 6u^2 u_x + u_{xxx} = 0 \qquad \text{modified Korteweg–deVries}$$
$$\text{equation } (q = u)$$

$$\sigma_{xt} = \sin \sigma \qquad \text{sine–Gordon equation}$$
$$(q = -\tfrac{1}{2}\sigma_x)$$

$$iq_t + q_{xx} + 2|q|^2 q = 0 \qquad \text{cubic Schrödinger}$$
$$\text{equation}$$

For the first two of these evolution equations the function $q$ in the above linear equations is real. As we will see below, the pair of first-order linear equations associated with the first two evolution equations is equivalent to a single second-order equation of Schrödinger type in which the potential Las either of the complex forms $q^2 \pm iq_x$.

When the relation of the Schrödinger equation to the Korteweg–deVries equation was established in Chapter 1, we found that we also had to require that the time dependence of the eigenfunctions be given by $y_t = By$, where $B$ is the third-order differential operator given in (1.2.10). As we shall see below, the modified Korteweg–deVries equation arises if that same procedure

**133**

is applied when either of the complex potentials $u^2 \pm iu_x$ is used in the Schrödinger equation and in the operator $B$.

The fact that the modified Korteweg–deVries equation is related to a Schrödinger equation, albeit with a complex potential, provides a very direct way of introducing this evolution equation and the linear equations associated with it. However, the final form of the linear equations for the modified Korteweg–deVries equation will suggest a systematic way of obtaining linear equations for other evolution equations. This more systematic approach, the two-component approach, will then be used for the sine–Gordon and for the cubic Schrödinger equation.

## 5.1  MODIFIED KORTEWEG–deVRIES EQUATION

Of the three equations to be introduced in this chapter, the modified Korteweg–deVries equation $u_t \pm 6u^2 u_x + u_{xxx} = 0$ is the one that occurs least in applications. However, it is the most closely related to the Korteweg–deVries equation and thus provides the most natural transition from the results of Chapter 4.

As noted previously, for the modified Korteweg–deVries equation, we shall be concerned with the two-component linear equations in the form

$$\begin{aligned} v_{1x} + i\zeta v_1 &= uv_2 \\ v_{2x} - i\zeta v_2 &= -uv_1 \end{aligned} \tag{5.1.1}$$

where $u$ is real. We can make contact with the Schrödinger equation and with previous results for the Korteweg–deVries equation by introducing the new variables

$$\begin{aligned} w_1 &= v_1 + iv_2 \\ w_2 &= v_1 - iv_2 \end{aligned} \tag{5.1.2}$$

These new variables are readily found to satisfy the system

$$w_{1x} + iuw_1 = -i\zeta w_2 \tag{5.1.3a}$$

$$w_{2x} - iuw_2 = -i\zeta w_1 \tag{5.1.3b}$$

which, as noted in Section 1.5, is equivalent to a pair of Schrödinger equations with complex potentials. Combining (5.1.1), we find that

$$w_{1xx} + \left[ \zeta^2 + (u^2 + iu_x) \right] w_1 = 0 \tag{5.1.4a}$$

$$w_{2xx} + \left[ \zeta^2 + (u^2 - iu_x) \right] w_2 = 0 \tag{5.1.4b}$$

Let us consider the complex potential $\mathcal{V} = -(u^2 + iu_x)$ that occurs in the equation for $w_1$. From previous considerations we know that if $\mathcal{V}$ contains a parameter $t$, then $\zeta$ will be independent of $t$ provided that $\mathcal{V}$ satisfies any one of a certain sequence of partial differential equations. The first two of these equations are $\mathcal{V}_t + c\mathcal{V}_x = 0$ and the Korteweg–deVries equation $\mathcal{V}_t - 6\mathcal{V}\mathcal{V}_x + \mathcal{V}_{xxx} = 0$. We can now ask what these equations for $\mathcal{V}$ imply as far as the function $u$ is concerned. The first of the two partial differential equations for $\mathcal{V}$ leads to

$$2u(u_t + cu_x) + i(u_t + cu_x)_x = 0 \qquad (5.1.5)$$

A simple way to guarantee that this equation will be satisfied is to choose $u_t + cu_x = 0$, that is, to have $u$ satisfy the same equation as $\mathcal{V}$. Thus no new equation is encountered. However, substitution of $\mathcal{V} = -(u^2 + iu_x)$ into the second equation for $\mathcal{V}$, that is, the Korteweg–deVries equation, yields

$$2u(u_t + 6u^2u_x + u_{xxx}) + i(u_t + 6u^2u_x + u_{xxx})_x = 0 \qquad (5.1.6)$$

This equation will be satisfied if we set

$$u_t + 6u^2u_x + u_{xxx} = 0 \qquad (5.1.7)$$

which is known as the modified Korteweg–deVries equation. It should be noted that if $u(x,t)$ is a solution of this equation, then $-u(x,t)$ is also a solution. A similar consideration of the equation for $w_2$ again leads to the modified Korteweg–deVries equation in the form $u_t + 6u^2u_x + u_{xxx} = 0$.

We have thus found that if $u$ satisfies the modified Korteweg–deVries equation in the form (5.1.7), then $\mathcal{V} = -(u^2 \pm iu_x)$ satisfies the Korteweg–deVries equation in the form (4.1.1). This relation between $\mathcal{V}$ and $u$ is known as the Miura transformation (Miura, 1968).

## The Linear Equations

Since the linear equations for the Korteweg–deVries equation have already been determined, those for the modified Korteweg–deVries equation are readily obtained by exploiting the close relation just noted between these two equations. The linear equations involving spatial derivatives are the equations for $v_1$ and $v_2$ given in (5.1.1). For the time dependence we need only recall that from the results for the Korteweg–deVries equation in (4.7.2a) and (4.7.3a) we may write

$$w_{1t} = (4\zeta^2 + 2\mathcal{V})w_{1x} - \mathcal{V}_x w_1$$
$$w_{2t} = (4\zeta^2 + 2\mathcal{V}^*)w_{2x} - \mathcal{V}_x^* w_2 \qquad (5.1.8)$$

We now form equations for $v_1$ and $v_2$ by addition and subtraction of these equations for $w_{1t}$ and $w_{2t}$. From (5.1.2) we have $v_1 = \frac{1}{2}(w_1 + w_2)$ and $v_2 =$

$-\frac{1}{2}i(w_1 - w_2)$. When $\mathcal{V}$ is replaced by $-(u^2 + iu_x)$, the equations for the sum and difference of $w_{1t}$ and $w_{2t}$ yield

$$v_{1t} = 2i\zeta(u^2 - 2\zeta^2)v_1 + (4\zeta^2 u + 2i\zeta u_x - 2u^3 - u_{xx})v_2$$
$$v_{2t} = (-4\zeta^2 u + 2i\zeta u_x \div 2u^3 + u_{xx})v_1 - 2i\zeta(u^2 - 2\zeta^2)v_2$$

$$(5.1.9)$$

Equations 5.1.1 and 5.1.9 are the linear equations that are customarily used in solving the modified Korteweg–deVries equation by the two-component inverse scattering method.

### Solution by Inverse Scattering

The modified Korteweg–deVries equation is solved by the same procedure as that used in the last chapter for the Korteweg–deVries equation. At large distances from the solution where $u \to 0$, the time dependence of the linear equations is governed by (5.1.9) in the limiting form

$$v_{1t} = -4i\zeta^3 v_1$$
$$v_{2t} = 4i\zeta^3 v_2 \tag{5.1.10}$$

The spatial dependence is now expressed in terms of $\varphi$ and $\psi$, the fundamental solutions of the two-component system (5.1.1) that were introduced in Section 2.11 (with $k$ replaced by $\zeta$). Let us consider the solution $v = \begin{pmatrix} v_1 \\ v_2 \end{pmatrix}$ that is proportional to $\varphi$ as $x$ approaches $-\infty$. Then, as $x$ approaches $-\infty$, $v(x,t) = f(t)\varphi(x) \to f(t)e^{-i\zeta x}\begin{pmatrix} 1 \\ 0 \end{pmatrix}$. Using $v_1 = f(t)e^{-i\zeta x}$ in the first of (5.1.10) yields $f(t) = f(0)\exp(-4i\zeta^3 t)$. The form of $\varphi$ as $x \to +\infty$ is obtained from (2.11.19a). We have

$$v = f(t)\varphi \to f(0)e^{-4i\zeta^3 t}\left[ c_{11}(\zeta,t)e^{i\zeta x}\begin{pmatrix} 0 \\ 1 \end{pmatrix} \right.$$
$$\left. + c_{12}(\zeta,t)e^{-i\zeta x}\begin{pmatrix} 1 \\ 0 \end{pmatrix} \right] \tag{5.1.11}$$

Again substituting into (5.1.10) and equating coefficients, we obtain $\dot{c}_{12} = 0$ and $\dot{c}_{11} = 8i\zeta^3 c_{11}$, so that

$$c_{12}(\zeta,t) = c_{12}(\zeta,0)$$
$$c_{11}(\zeta,t) = c_{11}(\zeta,0)e^{8i\zeta^3 t} \tag{5.1.12}$$

A similar calculation based upon a solution $v$ that is proportional to $\psi$ yields $c_{22}(\zeta,t) = c_{22}(\zeta,0)e^{-8i\zeta^3 t}$. This result also follows from $c_{22}(\zeta) = c_{11}(-\zeta)$. Since $c_{12}$ is again the reciprocal of the transmission coefficient, we see that the transmission coefficient is again independent of time.

For the localized solutions associated with the zeros of $c_{12}(\zeta,0)$ in the upper half-plane at points $\zeta=\kappa_l$, we may use (2.11.37) to write

$$m_{Rl}(\kappa_l,t)=-i\frac{c_{11}(\kappa_l,t)}{\dot{c}_{12}(\kappa_l,0)}=m_{Rl}(\kappa_l,0)e^{8i\kappa_l^3 t} \qquad (5.1.13)$$

and

$$m_{Ll}(\kappa_l,t)=-i\frac{c_{22}(\kappa_l,t)}{\dot{c}_{12}(\kappa_l,0)}=m_{Ll}(\kappa_l,0)e^{-8i\kappa_l^3 t} \qquad (5.1.14)$$

Solution of the modified Korteweg–deVries equation is approached by first constructing the function

$$\Omega_L(z,t)=\int_{-\infty}^{\infty}\frac{d\zeta}{2\pi}R_L(\zeta,0)e^{-8i\zeta^3 t-i\zeta z}$$

$$+\sum_{l=1}^{N}m_{Ll}(\kappa_l,0)e^{-8i\kappa_l^3 t-i\kappa_l z} \qquad (5.1.15)$$

where $R_L(\zeta,0)=-c_{22}(\zeta)/c_{21}(\zeta)$. We then consider the pair of coupled Marchenko equations

$$A_1(x,y)-\int_{-\infty}^{x}dx'\Omega_L(x'+y)A_2(x,x')=0$$
$$A_2(x,y)+\Omega_L(x+y)+\int_{-\infty}^{x}dx'\Omega_L(x'+y)A_1(x,x')=0 \qquad (5.1.16)$$

According to the results of Section 3.6, the solution of the modified Korteweg–deVries equation is given by

$$u(x,t)=-2A_2(x,x,t) \qquad (5.1.17)$$

Solution of the integral equations and determination of the pulse evolution is a simple matter only when the reflection coefficient $R_L(\zeta,0)$ for the initial pulse profile is zero. The result is then one of the pure multisoliton solutions for the modified Korteweg–deVries equation. Using the results of Section 3.7, we immediately find that the solution is

$$u(x,t)=2\frac{\partial}{\partial x}\tan^{-1}\left[\frac{\mathrm{Im}\det(1-iM)}{\mathrm{Re}\det(1-iM)}\right] \qquad (5.1.18)$$

where $1$ is a unit matrix of order $N$, where $N$ is the number of zeros of $c_{12}(\zeta,0)$ in the upper half-plane and $M$ is an $N\times N$ matrix with elements

$$M_{\alpha\beta}=i\frac{m_{L\beta}(\kappa_\beta,t)}{\kappa_\alpha+\kappa_\beta}e^{-i(\kappa_\alpha+\kappa_\beta)x} \qquad (5.1.19)$$

As noted in Section 2.11, the zeros of $c_{12}$ may be either on the imaginary axis or located symmetrically in pairs about the imaginary axis.

For the simplest case, $N = 1$, the matrix $M$ reduces to the scalar $M = i(m_1/2\kappa_1)\exp(-2i\kappa_1 x)$. With $\kappa = ik$ and the time dependence given by (5.1.12), we have

$$M = \frac{m_1(0)}{2k} e^{2kx - 8k^3 t} \tag{5.1.20}$$

As shown in Section 2.11, $m_1(0)$ is real. Thus (5.1.18) reduces to

$$u(x, t) = -2\frac{\partial}{\partial x}\tan^{-1}\left[\frac{m_1(0)}{2k} e^{2(kx - 4k^3 t)}\right] \tag{5.1.21}$$

As noted in (2.11.51), the sign of $m_1(0)$ may be either positive or negative. Finally, the single-soliton solution for the modified Korteweg–deVries equation is

$$u(x, t) = \pm 2k\,\text{sech}(2kx - 8k^3 t + \delta) \tag{5.1.22}$$

where $\delta = \ln[|m_1(0)|/2k]$. The upper sign is to be used for $m_1(0) < 0$ and the lower sign for $m_1(0) > 0$.

## Breather Solution

We now examine a new type of localized solution that may be obtained for the modified Korteweg–deVries equation. It is associated with the fact that zeros of $c_{12}(\zeta)$ can occur in pairs located symmetrically about the imaginary axis of the $\zeta$ plane. The pulse profile is in the form of an oscillatory solution that is modulated by an envelope having the shape of a hyperbolic secant. The oscillations and the envelope move at different velocities. As a result of the undulations in the profile that take place as this pulse propagates, it is frequently referred to as a breather solution. To obtain the analytical expression for this solution, we set $\kappa_1 = \alpha + i\beta$ and $\kappa_2 = -\alpha + i\beta = -\kappa_1^*$. Then using (5.1.19) to determine the elements of the associated $2\times 2$ matrix, we obtain

$$\text{Re}[\det(1 - iM)] = 1 + \frac{\alpha^2 |m_1|^2 e^{4\beta x}}{4\beta^2(\alpha^2 + \beta^2)} \tag{5.1.23}$$

and

$$i\,\text{Im}[\det(1 - iM)] = \frac{e^{2\beta x}}{2(\alpha^2 + \beta^2)}\left[(\alpha - i\beta)m_1 e^{-2i\alpha x} - \text{c.c.}\right] \tag{5.1.24}$$

Also,

$$m_1(\kappa, t) = m_1(\kappa_1, 0)e^{-8i\kappa_1^3 t} = m_1(\kappa_1, 0)e^{(\gamma - i\delta)t} \tag{5.1.25}$$

where $\gamma = 8\beta(3\alpha^2 - \beta^2)$ and $\delta = 8\alpha(\alpha^2 - 3\beta^2)$. Then, after some algebraic reduction, we find that

$$u = -2\frac{\partial}{\partial x}\tan^{-1}\left[\frac{\beta}{\alpha}\frac{\sin(2\alpha x + \delta t - \varphi)}{\cosh(2\beta x + \gamma t + \psi)}\right] \qquad (5.1.26)$$

where we have set $m_1(0) = |m_1(0)|e^{i\varphi}$ and $e^\psi = (\alpha/2\beta)|m_1(0)|(\alpha^2 + \beta^2)^{-1/2}$. The differentiation may be put in the form

$$u(x,t) = -4\beta\,\text{sech}\,\Psi\left[\frac{\cos\Phi - (\beta/\alpha)\sin\Phi\tanh\Psi}{1 + (\beta/\alpha)^2\sin^2\Phi\,\text{sech}^2\Psi}\right] \qquad (5.1.27)$$

where

$$\Phi \equiv 2\alpha x + \delta t - \varphi + \theta$$

$$\Psi \equiv 2\beta x + \gamma t + \psi \qquad (5.1.28)$$

and $\theta = \tan^{-1}(b/\alpha)$.

The expression for $u(x,t)$ can be seen to have the structure described above. The envelope and phase velocities are found from (5.1.28) to be $v_e = \gamma/2\beta$ and $v_{\text{ph}} = \delta/2\alpha$, respectively. A similar type of solution, but moving with different velocities will be obtained for the sine–Gordon equation in the next section. A graph of the breather solution for the modified Korteweg–deVries equation is shown in Figure 5.1.

The breather solution is a localized entity that also has the essential feature of the soliton; that is, it interacts with another breather solution [or a soliton of the type given by (5.1.22)] in an elastic fashion. A figure showing the interaction of two breather solutions is shown in Figure 5.2. The analytical expression for this solution could be obtained by considering the solution (5.1.18) for $N = 4$ with $\kappa_2 = -\kappa_1^*$ and $\kappa_4 = -\kappa_3^*$. However, the solution was actually obtained by using a method to be described in Section 8.2.

For $\beta \ll 1$, the breather solution (5.1.27) is a *small-amplitude* solution. If, in addition, we assume that $\beta/\alpha \ll 1$, this solution may be approximated by writing

$$u \approx -4\beta\,\text{sech}\,\Psi\cos\Phi$$

$$= -2\beta e^{-i\Phi}\,\text{sech}\,\Psi + \text{c.c.} \qquad (5.1.29)$$

The phase term is composed of slowly and rapidly varying terms and may be written as $\Phi = \phi_s + \phi_f$, where

$$\phi_s = -24\alpha\beta^2 t$$

$$\phi_f = 2\alpha x + 8\alpha^3 t \qquad (5.1.30)$$

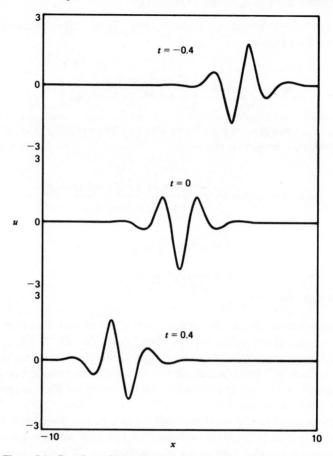

**Figure 5.1** Breather solution for modified Korteweg–deVries equation.

Let us now determine the equation that is satisfied by the small-amplitude, slowly varying phase term $F = -2\beta e^{24i\alpha\beta^2 t} \operatorname{sech} \Psi$. Writing

$$u = Fe^{-i\phi_f} + \text{c. c.} \tag{5.1.31}$$

and substituting into the modified Korteweg—deVries equation (5.1.7), we obtain the approximate equation

$$F_t - 12\alpha^2 F_x - 6i\alpha F_{xx} - 12i\alpha|F|^2 F = 0 \tag{5.1.32}$$

In obtaining this result, we have neglected the terms $|F|^2 F_x$ and $F_{xxx}$, which, according to (5.1.29), are proportional to $\beta^4$. In terms of the independent variables $x' = x + 12\alpha^2 t$ and $t' = 6\alpha t$, the equation for $F$ becomes

$$iF_{t'} + F_{x'x'} + 2|F|^2 F = 0 \tag{5.1.33}$$

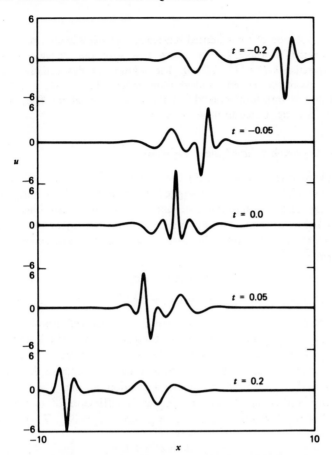

**Figure 5.2** Interaction of two breather solutions of the modified Korteweg–deVries equation.

This equation will be referred to as the cubic Schrödinger equation.* It is the third of the nonlinear evolution equations that are being examined in this chapter and will be considered in Section 5.3.

In terms of the independent variables $x'$ and $t'$, the solution for $F$ given above is

$$F(x',t') = -2\beta e^{4i\beta^2 t'} \operatorname{sech} 2\beta x' \qquad (5.1.34)$$

where a higher-order term proportional to $\beta^3$ has been neglected in the argument of the hyperbolic secant term.

---

*The name follows from the fact that the Schrödinger equation for time-dependent problems in quantum mechanics has the form $i\hbar\psi_t + (\hbar^2/2m)\psi_{xx} - V\psi = 0$, which is of the form of (5.1.33) if $V \sim |\psi|^2$.

*Exercise 1*

The similarity solution of the modified Korteweg–deVries equation may be obtained by setting $u = t^\alpha f(\zeta)$, where $\zeta = t^\beta x$. Show that with the choice $\alpha = \beta = -\frac{1}{3}$, the function $f(\zeta)$ satisfies $f'' + 2f^3 - \frac{1}{3}\zeta f = 0$. The solution of this equation may be expressed in terms of the second Painlevé transcendent (Ince, 1926, p. 345). With the help of the Miura transformation, the similarity solution of the Korteweg–deVries equation may also be related to these functions.

## Alternative Approach to the Linear Equations

The form of the linear equations for the modified Korteweg–deVries equation that appear in (5.1.1) and (5.1.9) suggest an orderly and unifying procedure for obtaining linear equations associated with other nonlinear evolution equations (Ablowitz et al., 1974a). We begin with the linear equations (5.1.1) and also assume that the time derivatives of $v_1$ and $v_2$ are expressible in the general form

$$
\begin{aligned}
v_{1t} &= Av_1 + Bv_2 \\
v_{2s} &= Cv_1 + Dv_2
\end{aligned}
\tag{5.1.35}
$$

as is the case in (5.1.9). We now equate the mixed second derivatives $v_{1xt} = v_{1tx}$, $v_{2xt} = v_{2tx}$ and in the resulting equalities use the linear equations (5.1.1) and (5.1.35) to eliminate time and space derivatives of $v_1$ and $v_2$. As in our treatment of the Korteweg–deVries equation, we impose the requirement that the eigenvalue parameter be time-independent so that $\zeta_t = 0$. Since $v_1$ and $v_2$ are linearly independent, we may equate coefficients of $v_1$ and $v_2$ in the equations that result from setting $v_{1xt} = v_{1tx}$ and $v_{2xt} = v_{2tx}$. We obtain

$$
A_x = q(B + C)
\tag{5.1.36a}
$$

$$
B_x + 2i\zeta B = -2qA + q_t
\tag{5.1.36b}
$$

$$
C_x - 2i\zeta C = -2qA - q_t
\tag{5.1.36c}
$$

and $D_x = -A_x$. The coefficient $D$ thus differs from $-A$ by a function of time which, as in our treatment of the Korteweg–deVries equation in Section 1.2, may be ignored.

The coefficients $A$, $B$, and $C$ may now be determined by observing from (5.1.9) that they are polynomials in $\zeta$. Setting $A = \sum_0^3 a_n \zeta^n$, $B = \sum_0^3 b_n \zeta^n$, and $C = \sum_0^3 c_n \zeta^n$, substituting into (5.1.36) and equating powers of $\zeta$, we obtain a set of simple differential equations from which the coefficients $a_n$, $b_n$, and $c_n$ may be determined. We find that $a_3 = a_3(t)$, $b_3 = c_3 = 0$ and

$$
\begin{aligned}
&a_2 = a_2(t), && b_2 = -c_2 = ia_3 q \\
&a_1 = -\tfrac{1}{2}a_3 q^2, && b_1 = -\tfrac{1}{2}a_3 q_x + ia_2 q,\; c_1 = -\tfrac{1}{2}a_3 q_x - ia_2 q \\
&a_0 = -\tfrac{1}{2}a_2 q^2, && b_0 = c_0 = \frac{i}{4}a_3(q_{xx} + 2q^3) - \tfrac{1}{2}a_2 q_x
\end{aligned}
\tag{5.1.37}
$$

$$
c_0 = \frac{i}{4}a_3(q_{xx} + 2q^3) - \tfrac{1}{2}a_2 q_x
$$

Finally, the terms that are independent of $\zeta$ in the last two of equations (5.1.36) yield $q_{0t} = b_{0x} + 2a_0q$ which is the evolution equation. With the expressions for $a_0$ and $b_0$ obtained above, we find that

$$q_t = -\tfrac{1}{4}ia_3\left(6q^2q_x + q_{xxx}\right) - a_2\left(q^3 + \tfrac{1}{2}q_{xx}\right) \tag{5.1.38}$$

Setting $a_2 = 0$ and $a_3 = 4i$, we obtain the modified Korteweg–deVries equation in the form given in (5.1.7). As expected, the coefficients $A$, $B$, and $C$ reduce to those given in (5.1.9).


## 5.2   SINE–GORDON EQUATION

The method for obtaining linear equations that was used at the end of the previous section provides a very simple way to construct linear equations for other common nonlinear evolution equations. In the present section we consider the use of this method to obtain linear equations for the so-called sine–Gordon equation,* which may be written

$$\sigma_{xt} = \sin\sigma \tag{5.2.1}$$

This is the evolution equation that results when each of the coefficients $A$, $B$, and $C$ in (5.1.36) is chosen so as to contain a single term that is proportional to $\zeta^{-1}$. We thus set

$$A = \frac{a}{\zeta}, \qquad B = \frac{b}{\zeta}, \qquad C = \frac{c}{\zeta} \tag{5.2.2}$$

Substituting into (5.1.36) and separating terms according to powers of $\zeta$, we find that

$$a_x = q(b + c) \tag{5.2.3a}$$

$$2ib = q_t, \qquad b_x = -2qa \tag{5.2.3b}$$

$$2ic = q_t, \qquad c_x = -2qa \tag{5.2.3c}$$

Then, $ia_x = iq(b = c) = 2qq_t = 2iqb$ and we have

$$a_x = 2qb$$

$$b_x = -2qa \tag{5.2.4}$$

Multiplying by $a$ and $b$, respectively, and adding, we may obtain the integrated form $a^2 + b^2 = f^2(t)$, which implies the parametric representation

---

*The nomenclature is somewhat facetious and is related to the form taken by the equation in the coordinate system $\xi = x + t$, $\tau = x - t$, namely $\sigma_{\xi\xi} - \sigma_{\tau\tau} = \sin\sigma$. The linear approximation to this equation, $\sigma_{\xi\xi} - \sigma_{\tau\tau} - \sigma = 0$, is known in quantum theory as the Klein–Gordon equation.

$a = f(t)\cos\sigma$, $b = f(t)\sin\sigma$. We also find that $q = -\frac{1}{2}\sigma_x$. The first of (5.2.3b) then becomes $\sigma_{xt} = -4if(t)\sin\sigma$. The choice $f(t) = \frac{1}{4}i$ leads to the sine–Gordon equation in the form given in (5.2.1). From (5.1.35) we have the linear equations

$$v_{1t} = \frac{i}{4\zeta}\big[(\cos\sigma)v_1 + (\sin\sigma)v_2\big]$$

$$v_{2t} = \frac{i}{4\zeta}\big[(\sin\sigma)v_1 + (\cos\sigma)v_2\big]$$

$$(5.2.5)$$

The spatial derivatives of $v_1$ and $v_2$ are, of course, still given by (5.1.1) with $u = q = -\frac{1}{2}\sigma_x$.

## Some Simple Solutions

Before using the linear equations to solve the sine–Gordon equation by inverse scattering techniques, let us first consider some simple solutions of this equation that have been known for many years. They will also be the simplest of the solutions that arise when the inverse method is used.

The equation $\sigma_{xt} = \sin\sigma$ has a long history that begins in the latter part of the nineteenth century when this equation was found to occur in differential geometry (Eisenhart, 1909). Various techniques for obtaining particular solutions of this equation were developed at that time. One of these, the method of Bäcklund transformations, will be described in Chapter 8. Another approach to the simpler solutions is to express them as functions of the independent variables $u = ax + t/a$ and $v = ax - t/a$ in the form

$$\sigma(u,v) = 4\tan^{-1}\left[\frac{U(u)}{V(v)}\right] \qquad (5.2.6)$$

We shall assume that $a > 0$. Anticipation of a solution in terms of an inverse tangent is now quite natural in view of the results of inverse scattering theory, such as (3.7.16).

In the $u$–$v$ coordinate system the sine–Gordon equation is

$$\sigma_{uu} - \sigma_{vv} = \sin\sigma \qquad (5.2.7)$$

In certain applications of this equation to nonlinear wave propagation, it appears in the form given in (5.2.7), with $u$ and $v$ playing the role of space and time, respectively.

Using the identity $\sin 4\theta = 4\tan\theta(1 - \tan^2\theta)/(1 + \tan^2\theta)^2$, with $\theta = \sigma/4$, we find that the assumed form of the solution given in (5.2.6) converts the sine–Gordon equation into

$$(U^2 + V^2)\left(\frac{U''}{U} + \frac{V''}{V}\right) - 2(U')^2 - 2(V')^2 = V^2 - U^2 \qquad (5.2.8)$$

where the primes indicate differentiation of the functions with respect to their argument. Successive differentiation of this result with respect to both $u$ and $v$ enables us to separate variables. We find that

$$\frac{1}{UU'}\left(\frac{U''}{U}\right)' = -\frac{1}{VV'}\left(\frac{V''}{V}\right)' = -4k^2 \qquad (5.2.9)$$

where the form chosen for the separation constant is determined by future convenience. Each of these ordinary differential equations may be integrated twice to yield $(U')^2 = -k^2 U^4 + \mu_1 U^2 + v_1$ and $(V')^2 = k^2 V^4 + \mu_2 V^2 + v_2$, where the $\mu_i$ and $v_i$ are integration constants. Substitution of these equations into (5.2.8) leads to the requirements $\mu_1 - \mu_2 = 1$ and $v_1 + v_2 = 0$. Setting $\mu_1 = m^2$ and $v_1 = n^2$, we finally obtain

$$
\begin{aligned}
(U')^2 &= -k^2 U^4 + m^2 U^2 + n^2 \\
(V')^2 &= k^2 V^4 + (m^2 - 1) V^2 - n^2
\end{aligned}
\qquad (5.2.10)
$$

In general, the solutions for $U(u)$ and $V(v)$ will be expressed in terms of integrals of the square root of a biquadratic expression and thus will involve elliptic functions. An exhaustive cataloguing of these solutions has been given by Steuerwald (1936).

Certain solutions in terms of elementary functions may be obtained by specializing the constants $k$, $m$, and $n$ in (5.2.10). These solutions will arise again when we consider the multisoliton solutions from the standpoint of inverse scattering. We now examine some of these special cases.

1. $k=0$, $m>1$, $n=0$. Here (5.2.10) yields $U = \gamma_1 e^{\pm mu}$ and $V = \gamma_2 e^{\pm \sqrt{m^2-1}\,v}$, where $\gamma_1$ and $\gamma_2$ are constants of integration. The solution of the sine–Gordon equation in this instance has the shape of a shelf and according to (5.2.6), is given by

$$\sigma = 4 \tan^{-1}\left[\gamma \exp\left(\pm \frac{u \pm \beta v}{\sqrt{1-\beta^2}}\right)\right] \qquad (5.2.11)$$

where $\gamma = \gamma_1/\gamma_2$ and $\beta = \sqrt{m^2-1}\,/m$. All four sign combinations are possible. As an example, if we choose the upper sign in both instances, we see that $\sigma$ is a solution that increases monotonically from 0 to $2\pi$ as $u$ proceeds from $-\infty$ to $+\infty$. This shelf moves with velocity $\beta$ in the negative $u$ direction. This solution, as well as the solution with the sign choice leading to

$$\sigma = 4 \tan^{-1}\left[\gamma \exp\left(\frac{u - \beta v}{\sqrt{1-\beta^2}}\right)\right]$$

are the single-soliton solutions of the sine–Gordon equation. The latter choice represents a shelf that moves in the positive $u$ direction with velocity $\beta$.

Pulselike solutions may be obtained as derivatives of these solutions. We find that

$$\sigma_u = \pm 2m \, \text{sech}\left[ m(u \pm \beta v) + \ln \gamma \right]$$

and

$$\sigma_v = \pm 2\sqrt{m^2 - 1} \, \text{sech}\left[ m(u \pm \beta v) + \ln \gamma \right]$$

Frequently, it is these expressions that are referred to as the soliton solutions of the sine–Gordon equation. The two other choices in sign, which yield solutions that vary from $2\pi$ to $0$ as $v$ proceeds from $-\infty$ to $+\infty$, are sometimes referred to as antisolitons.

Let us briefly consider the solution in terms of the $x$ and $t$ coordinates used in (5.2.1). It is sufficient to consider $m=1$ so that $\beta=0$. The solution is then $\sigma = 4\tan^{-1}\{\gamma \exp[\pm(ax + t/a)]\}$, which is a shelf that moves with velocity $a^{-2}$ in the negative $x$ direction. Pulse solutions are again obtainable from $\sigma_x$ or $\sigma_t$. The pulse $\sigma_x$ has an area $\int_{-\infty}^{\infty} dx \, \sigma_x(x,t) = \sigma(\infty,t) - \sigma(-\infty,t) = 2\pi$. In certain applications of the sine–Gordon equation that will be discussed in Chapter 7, this solution (with space and time coordinates interchanged) will be referred to as a $2\pi$ pulse.

Although the expressions for $\sigma$ that have just been obtained are the single-soliton solutions of the sine–Gordon equation, the soliton nature of these solutions is not, of course, evident until the interaction of two such pulses has been investigated. This may be done by examining the following choice of constants.

**2.** $k=0$, $m>1$, $n\neq0$. The integrations involved in solving (5.2.10) are again given in terms of elementary functions. We find $U(u) = \pm(n/m)\sinh(mu + c_1)$ and $V(v) = (n/\sqrt{m^2 - 1})\cosh(\sqrt{m^2 - 1}\ v + c_2)$, where $c_1$ and $c_2$ are constants of integration that may be used to set the origin of the $u$–$v$ coordinate system. Then, from (5.2.6), we have

$$\sigma = \pm 4\tan^{-1}\left[ \frac{\sqrt{m^2 - 1}}{m} \frac{\sinh(mu + c_1)}{\cosh(\sqrt{m^2 - 1}\ v + c_2)} \right] \qquad (5.2.12)$$

The result is independent of $n$ since only the ratio $U/V$ occurs in the expression for $\sigma$. Let us consider the case with the upper sign in (5.2.12) and examine the meaning of this solution when $u$ and $v$ refer to space and time, respectively. We also set $c_1 = c_2 = 0$. The solution is found to represent the collision of two solitons. In the remote past as $v \to -\infty$ and as $u \to -\infty$, we find that $\sigma \to -\tan^{-1}[\beta e^{-m(u - \beta v)}]$, where again $\beta = \sqrt{m^2 - 1}\ /m$. The solution increases from $-2\pi$ to $0$ as $u$ passes through the value of $\beta v$. It thus represents a shelf that moves in the positive $u$ direction. As $u \to +\infty$ and as

$v \rightarrow -\infty$, we find $\sigma \rightarrow 4\tan^{-1}[\beta e^{m(u+\beta v)}]$, which represents a shelf in which $\sigma$ rises from 0 to $2\pi$ as $u$ increases through the value $-\beta v$. This shelf moves in the negative $u$ direction. At $v=0$ the two shelves collide. As $v \rightarrow +\infty$ we obtain two receding disturbances, $\sigma \rightarrow -4\tan^{-1}[\beta e^{-m(u+\beta v)}]$ as $u \rightarrow -\infty$ and $\sigma \rightarrow 4\tan^{-1}[\beta e^{m(u-\beta v)}]$ as $u \rightarrow +\infty$. Since $\sigma$ varies from $-2\pi$ to $+2\pi$ as $u$ varies from $-\infty$ to $+\infty$, a solution of this type is sometimes referred to as a $4\pi$ pulse.

Let us now consider the solution in $x-t$ space. We again examine (5.2.12) with the positive sign and set $c_1 = c_2 = 0$, as before. As $x \rightarrow +\infty$ both $u$ and $v \rightarrow +\infty$ and $\sigma$ approaches $2\pi$. Similarly, as $x \rightarrow -\infty$, $\sigma$ approaches $-2\pi$. It is sometimes convenient to have a solution that varies from 0 to $+4\pi$ as $u$ varies from $-\infty$ to $+\infty$. Such a solution is readily constructed by merely setting $\tilde{\sigma} = \sigma + 2\pi$ and noting that $\tilde{\sigma}$ still satisfies the sine–Gordon equation and also varies from 0 to $4\pi$. If we use the identity $\tan^{-1}\theta + \pi/2 = -\tan^{-1}(1/\theta)$, (5.2.12) yields

$$\tilde{\sigma} = \pm 4\tan^{-1}\left[\frac{m}{\sqrt{m^2-1}}\frac{\cosh(\sqrt{m^2-1}\,v+c_2)}{\sinh(mu+c_1)}\right] \qquad (5.2.13)$$

The two-soliton nature of the result is readily made evident. This aspect of the solution may be understood analytically by first setting $m=(a_1+a_2)/2\sqrt{a_1 a_2}$ and $a=\sqrt{a_1 a_2}$. We then obtain

$$\tilde{\sigma} = -4\tan^{-1}\left\{\frac{a_1+a_2}{a_1-a_2}\frac{\cosh\left[\frac{1}{2}(u_1-u_2)+c_2\right]}{\sinh\left[\frac{1}{2}(u_1+u_2)+c_1\right]}\right\} \qquad (5.2.14)$$

where $u_1 = a_1 x + t/a_1$ and $u_2 = a_2 x + t/a_2$. Again considering the spatial derivative, we find that

$$\tilde{\sigma}_x = 2\frac{a_1^2 - a_2^2}{a_1^2 + a_2^2}\frac{a_1\operatorname{sech}u_1 + a_2\operatorname{sech}u_2}{1+2\dfrac{a_1 a_2}{a_1^2+a_2^2}(\operatorname{sech}u_1\operatorname{sech}u_2 - \tanh u_1\tanh u_2)} \qquad (5.2.15)$$

As $x \rightarrow \pm\infty$, the pulses become well separated and considerations similar to those employed in obtaining (1.3.21) can be introduced. The denominator now simplifies and for $a_1 > a_2$ we can put (5.2.15) in the form

$$\tilde{\sigma}_x = 2a_1\operatorname{sech}(u_1 \mp \delta) + 2a_2\operatorname{sech}(u_2 \pm \delta) \qquad (5.2.16)$$

where the choice of signs refers to $t \rightarrow \mp\infty$ and $\delta = \tanh^{-1}[2a_1 a_2/(a_1^2 + a_2^2)]$. The solution has the form of two pulses with a phase shift that is due to their interaction at $t=0$, and is qualitatively similar to the two-soliton interaction for the Korteweg–deVries equation that was shown in Figure 1.2. This justifies the interpretation of the pulses as solitons.

**3.** $k \neq 0$, $n = 0$. Here three subcases are of interest and they can all be expressed in terms of elementary functions.

**3a.** $m^2 > 1$. The result is similar to that given in (5.2.12). We find that

$$\sigma = -4 \tan^{-1}\left[ \frac{m}{\sqrt{m^2-1}} \frac{\sinh\left(\sqrt{m^2-1}\ v + c_2\right)}{\cosh(mu + c_1)} \right] \tag{5.2.17}$$

Considerations similar to those of Case 2 show that this solution represents the collision of a soliton with an antisoliton. A graph of $\sigma_u$ is shown in Figure 5.3. The area under the whole pulse is zero for all values of $v$.

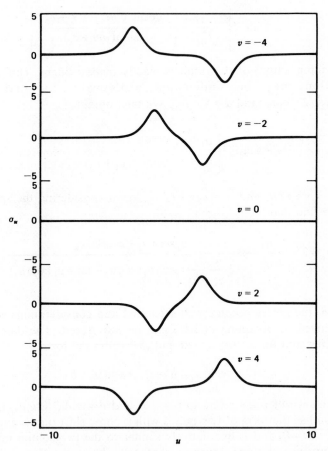

**Figure 5.3**  Collision of soliton and antisoliton solutions of the sine–Gordon equation.

*Exercise 2*

Show that

$$\sigma_u = -\frac{A(\operatorname{sech}\varphi - \operatorname{sech}\psi)}{B - \operatorname{sech}\varphi \operatorname{sech}\psi - \tanh\varphi \tanh\psi}$$

where $\varphi = m(u - \beta v)$, $\psi = m(u + \beta v)$, $A = 4\beta m/(1 - \beta^2)$, and $B = (1 + \beta^2)/(1 - \beta^2)$.

**3b.** $m = 1$. This is the limiting case in which the relative velocity of the two pulses in Figure 5.3 becomes zero. The analytical expression is found immediately from (5.2.17) to be

$$\sigma = -4\tan^{-1}\left[(v + c_2)\operatorname{sech}(u + c_1)\right] \qquad (5.2.18)$$

**3c.** $m^2 < 1$. This is another zero area pulse and perhaps the most interesting. The analytical expression may be obtained immediately from (5.2.17) by setting $\sqrt{m^2 - 1} = i\sqrt{1 - m^2}$. We find that

$$\sigma = -4\tan^{-1}\left[\frac{m}{\sqrt{1 - m^2}} \frac{\sin\left(\sqrt{1 - m^2}\, v + c_2\right)}{\cosh(mu + c_1)}\right] \qquad (5.2.19)$$

This is the so-called breather solution for the sine–Gordon equation and should be compared with the corresponding solution for the modified Korteweg–deVries equation that was given in (5.1.26). A graph of the sine–Gordon breather solution as a function of $u$ for $m = 0.8$ is shown in Figure 5.4. As a function of $v$ this solution is periodic.

For $m \ll 1$ the breather is a *small-amplitude* solution. It may then be approximated by writing (5.2.19) as

$$\sigma = 2i\epsilon(\operatorname{sech}\epsilon u)e^{i(1 - \epsilon^2/2)v} + \text{c.c.} \qquad (5.2.20)$$

where we have set $m = \epsilon$ and retained only the first term in the power series expansion for the inverse tangent. The solution is now of the form

$$\sigma(u, v) = F(u, v)e^{iv} + \text{c.c.} \qquad (5.2.21)$$

where $F(u, v) = 2i\epsilon(\operatorname{sech}\epsilon u)e^{-\epsilon^2 v/2}$. Substituting this form of $\sigma$ into the sine–Gordon equation in the form $\sigma_{uu} - \sigma_{vv} = \sin\sigma \approx \sigma - \sigma^3/6$, we find that $F$ satisfies the cubic Schrödinger equation

$$2iF_v - F_{uu} - \tfrac{1}{3}|F|^2 F = 0 \qquad (5.2.22)$$

The term $F_{vv}$ has been neglected since it is $O(\epsilon^4)$. This result is similar to that obtained for the modified Korteweg–deVries equation in (5.1.33).

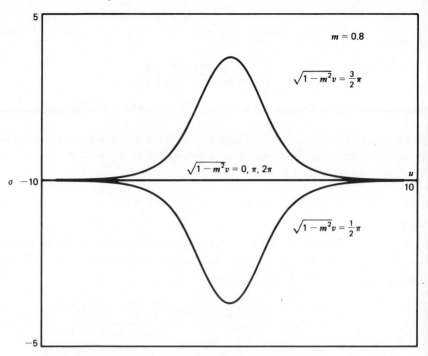

**Figure 5.4**  Breather solution of sine–Gordon equation $\sigma_{uu} - \sigma_{vv} = \sin \sigma$.

### Energy Considerations

It is instructive to calculate the energy that may be associated with the solutions obtained here. This is readily accomplished by using the Hamiltonian density for the sine–Gordon equation. An expression for the energy may then be obtained in the same way that the energy associated with waves on a string was determined in Section 2.1. For the sine–Gordon equation in the form given in (5.2.7), a Lagrangian density may be written as $\mathcal{L} = \frac{1}{2}[(\sigma_v)^2 - (\sigma_u)^2] - (1 - \cos \sigma)$. Substitution into the Euler–Lagrange equation (2.1.17), with $x$ and $t$ playing the role of $u$ and $v$, respectively, yields the sine–Gordon equation (5.2.7). The Hamiltonian density, which is $\mathcal{H} = \sigma_v(\partial \mathcal{L}/\partial \sigma_v) - \mathcal{L}$, takes the form

$$\mathcal{H} = \frac{1}{2}\left[(\sigma_u)^2 + (\sigma_v)^2\right] + (1 - \cos \sigma) \tag{5.2.23}$$

For the single-soliton solution $\sigma = 4 \tan^{-1}\{\exp[m(u - \beta v)]\}$, where $m = (1 - \beta^2)^{-1/2}$, we find that $\mathcal{H} = 4m^2 \operatorname{sech}^2[m(u - \beta v)]$ and $E = \int_{-\infty}^{\infty} du\, \mathcal{H} = 8m$ with $m > 1$. The same energy is obtained for the antisoliton $\sigma = 4 \tan^{-1}\{\exp[-m(u - \beta v)]\}$.

For the zero area or soliton–antisoliton solution given in (5.2.17), we find that

$$\mathcal{H} = \frac{8m^2}{k^2} \frac{\operatorname{sech}^2 mu (p + q \tanh^2 mu)}{(1 + r \operatorname{sech}^2 mu)^2} \tag{5.2.24}$$

where $k^2 = m^2 - 1$, $p = k^2 + (1 + k^2)\sinh^2 kv$, $q = m^2 \sinh^2 kv$, and $r = (m/k)^2 \sinh^2 kv$. When the integral $\int_{-\infty}^{\infty} du\, \mathcal{H}$ is evaluated, we obtain $E = 16m$ with $m > 1$. This is an expected result since the solution represents the elastic interaction of a soliton and an antisoliton each of which, in isolation, has energy $8m$.

For the breather solution (5.2.18) we again obtain $E = 16m$, where now $m < 1$. The energy is thus less than that of two free solitons (or antisolitons) and leads to the interpretation of the breather as a bound state composed of a soliton and an antisoliton pair.

## Solution by Inverse Scattering

Returning to the linear equations given in (5.1.1) and (5.2.5), we may proceed to the solution of the sine–Gordon equation by the inverse scattering procedure. Since the dependent variable $\sigma$ is related to $u$ in the potential through $\sigma(x, t) = -2\int_{-\infty}^{x} dx'\, u(x', t)$, we see that $\sigma$ itself is no longer localized. The integral over a localized solution $u(x, t)$ will lead to $\sigma(x, t)$ in the form of a step or shelf, as noted in our previous consideration of the single-soliton solution. Thus, although $\sigma$ reduces to zero as $x \to -\infty$, and (5.2.5) becomes

$$v_{1t} \to \frac{i}{4\zeta} v_1$$

$$x \to -\infty \qquad\qquad (5.2.25)$$

$$v_{2t} \to -\frac{i}{4\zeta} v_2$$

the form of linear equations (5.2.5) in the limit $x \to +\infty$ will depend upon the value of $\sigma$ as $x \to +\infty$. The only cases that can be treated with ease are those in which $\sigma \to 2n\pi$ as $x$ approaches $+\infty$.

From (5.2.25) we see that as $x \to -\infty$, $v_1$ may either vanish or approach $e^{it/4\zeta}$. Similarly, $v_2$ may either vanish or approach $e^{-it/4\zeta}$. Let us now consider in some detail the solution that is proportional to the fundamental solution $\varphi$ given in (2.11.6). Then

$$v = \begin{pmatrix} v_1 \\ v_2 \end{pmatrix} \to e^{it/4\zeta} \begin{pmatrix} 1 \\ 0 \end{pmatrix} e^{-i\zeta x}, \qquad x \to -\infty \qquad (5.2.26)$$

According to (2.11.19a), the form of the solution as $x \to \infty$ will then be

$$v \to e^{it/4\zeta} \left[ c_{11} \begin{pmatrix} 0 \\ 1 \end{pmatrix} e^{i\zeta x} + c_{12} \begin{pmatrix} 1 \\ 0 \end{pmatrix} e^{-i\zeta x} \right], \qquad x \to \infty \qquad (5.2.27)$$

Consequently, in this limit $v_1 \sim c_{12}(\zeta, t) e^{i(t/4\zeta) - i\zeta x}$ and $v_2 \sim c_{12}(\zeta, t) e^{i(t/4\zeta) + i\zeta x}$. The time dependence of $c_{11}$ and $c_{12}$ is obtained from the form of the linear equations (5.2.5) in the limit of $x \to +\infty$. As noted above, we shall consider the case in which $\sigma \to 2n\pi$ as $x \to +\infty$. Then

$$v_{1t} \to \frac{i}{4\zeta} v_1$$
$$\qquad\qquad\qquad\qquad x \to +\infty \qquad\qquad\qquad (5.2.28)$$
$$v_{2t} \to -\frac{i}{4\zeta} v_2$$

Following the procedure used for the modified Korteweg–deVries equation in (5.1.10), we arrive at

$$c_{11}(\zeta, t) = c_{11}(\zeta, 0) e^{-it/2\zeta}$$
$$c_{12}(\zeta, t) = c_{1s}(\zeta, 0) \qquad\qquad\qquad\qquad (5.2.29)$$

Recalling that $R_R(\zeta, t) = c_{11}(\zeta, t)/c_{12}(\zeta, t)$ and $m_{RI}(k_l, t) = -ic_{11}(k_l, t)/\dot{c}_{12}(k_l, t)$, we find that

$$\Omega_R(z, t) = \int_{-\infty}^{\infty} \frac{d\zeta}{2\pi} R_R(\zeta, 0) e^{-i(t/2\zeta) + i\zeta z}$$
$$\qquad - \sum m_{RI}(k_l, 0) e^{-i(t/2k_l) + ik_l z} \qquad\qquad (5.2.30)$$

Also, using the relations between the $c_{ij}(\zeta, t)$ given in (2.11.23), we obtain

$$\Omega_L(z, t) = \int_{-\infty}^{\infty} \frac{d\zeta}{2\pi} R_L(\zeta, 0) e^{i(t/2\zeta) - i\zeta z}$$
$$\qquad - \sum m_{LI}(k_l, 0) e^{i(t/2k_l) - ik_l z} \qquad\qquad (5.2.31)$$

   Solutions of the sine–Gordon equation are now obtained by the procedure used for the modified Korteweg–deVries equation. Again, it is only the pure soliton solutions that can be obtained with ease. Since $u = -\frac{1}{2}\sigma_x$, we see from (5.1.18) that the multisoliton solutions will be

$$\sigma = -4 \tan^{-1} \left[ \frac{\operatorname{Im} \det(1 - iM)}{\operatorname{Re} \det(1 - iM)} \right] \qquad\qquad (5.2.32)$$

The elements of $M$ are again given by (5.1.19) but with the time dependence

$$m_{Lj}(k_j, t) = m_{Lj}(k_j, 0) e^{it/2k_j} \qquad\qquad (5.2.33)$$

   The simplest case of a single pole on the imaginary axis at a point $k_1 = ia_1/2$ yields $m_{L1}(k_1, t) = m_{L1}(k_1, 0) e^{t/a_1}$ and

$$m_L = \frac{m_{L1}(k_1, 0)}{a} e^{a_1 x + t/a_1} \qquad\qquad (5.2.34)$$

The expression for $\sigma$ is found from (5.2.32) to be

$$\sigma = 4\tan^{-1}(e^{a_1 x + t/a_1 + \delta}) \tag{5.2.35}$$

where $\delta = \ln[m_{L1}(k_1,0)/a_1]$. This solution is the same as the single-soliton result given in (5.2.11).

## Two-Soliton Solution

The solution of the sine–Gordon equation that describes the interaction of two solitons is obtained by considering the case in which there are two poles on the imaginary axis. The situation is very much the same as that described in Section 4.4 for the Korteweg–deVries equation. Setting $k_1 = ia_1/2$ and $k_2 = ia_2/2$, we find that $\det(1 - iM_L)$ takes the form

$$\text{Re}[\det(1 - iM)] = 1 - \frac{m_{L1}m_{L2}}{a_1 a_2}\left(\frac{a_1 - a_2}{a_1 + a_2}\right)^2 e^{(a_1 + a_2)x}$$

$$\text{Im}[\det(1 - iM)] = -\left(\frac{m_{L1}}{a_1}e^{a_1 x} + \frac{m_{L2}}{a_2}e^{a_2 x}\right) \tag{5.2.36}$$

The expression for $\sigma$ given in (5.2.32) can be put into a form that agrees with (5.2.14). We obtain

$$\sigma = -4\tan^{-1}\left[\frac{a_1 + a_2}{a_1 - a_2}\frac{\cosh\frac{1}{2}(a_1 - a_2)}{\sinh\frac{1}{2}(a_1 + a_2)}\right], \qquad a_1 > a_2 \tag{5.2.37}$$

where

$$u_i = a_i x + \frac{t}{a_i} + \delta_i \quad \text{and} \quad \exp\delta_i = \frac{a_1 - a_2}{a_1 + a_2}\left[\frac{m_{Li}(k,0)}{a_i}\right]$$

The solutions treated here are those special solutions that evolve into pure multisoliton solutions. The simplest way of estimating the amplitude of other initial profiles is to use the conservation laws obtained from (2.11.57). The procedure is similar to that developed for the Korteweg–deVries equation in Section 4.5. As long as the initial pulse profile is smoothly varying and no zero area pulses (breathers) are involved, we can expect reasonably accurate results by this method.

## $\pi$ Pulse and the Similarity Solution

The obvious invariance of the sine–Gordon equation under the transformation $x' = ax$, $t' = t/a$ implies the existence of a similarity variable $\eta = xt$ (Ames, 1965, p. 133). This may be seen in a direct way by using the corresponding differential expressions $\partial/\partial x = t\, d/d\eta$ and $\partial/\partial t = x\, d/d\eta$ in the sine–Gordon equation (5.2.1). We then obtain the ordinary differential equation (Amsler, 1955)

$$\eta\sigma'' + \sigma' - \sin\sigma = 0 \qquad (5.2.38)$$

where the prime indicates differentiation with respect to $\eta$. A new dependent variable $w$, related to $\sigma$ by $w = \exp(i\sigma)$, is readily shown to satisfy the equation

$$w'' - \frac{(w')^2}{w} + \frac{2w' - w + 1}{2\eta} = 0 \qquad (5.2.39)$$

which is a special case of the equation that defines the third Painlevé transcendent (Ince, 1926, p. 345; Davis, 1960, p. 185). Since these functions are not available in any convenient form, it is preferable to resort to numerical integration of (5.2.38). An example of such a solution is shown in Figure 5.5, which also includes the result for $\sigma' = -2u/t$ and a phase plane diagram of the solution. The example shown in the figure satisfies the conditions $\sigma(0) = 0.1$ as well as $\sigma'(0) = \sin\sigma(0)$, which yields a solution that is finite for $\eta = 0$.

Since $\sigma$ increases from 0 to $\pi$ as $\eta$ varies from $-\infty$ to $+\infty$, we see that the solution is outside the class of solutions treated above by the inverse method. When the time dependence of the solution was calculated previously, it was assumed that $\sigma$ approached some integral multiple of $2\pi$. The corresponding

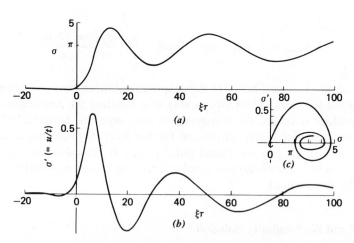

**Figure 5.5** Similarity solution of sine–Gordon equation (*a*) Numerical solution of (5.2.38). (*b*) Derivative of solution shown in (*a*). (*c*) Phase-plane diagram of solution. (With permission of American Institute of Physics.)

calculation for $\sigma$ approaching $\pi$ yields $c_{11}(k,t)=c_{11}(k,0)e^{it/2k}$ and $c_{12}(k,t)=c_{12}(k,0)$. The reflection coefficients in this case are then

$$R_R(k,t)=c_{11}(k,t)/c_{12}(k,0)=R_R(k,0)e^{it/2k}$$

and

$$R_L(k,t)=c_{11}(-k,t)/c_{12}(k,0)=R_L(k,0)e^{-it/2k}.$$

As an example, let us consider the reflection coefficient associated with the truncated pulse profile $u(x,0)=\theta(x-x_1)\mu\operatorname{sech}\mu x$ that was given in (2.11.61). For $\mu>0$ and $x_1>0$ there is no pole in the upper half-plane, and we obtain

$$\Omega_L(2z,t)=\tfrac{1}{2}\mu\operatorname{sech}\mu x_1\int_{-\infty}^{\infty}\frac{dk}{2\pi}\frac{e^{-2ik(z-x_1)-it/2k}}{k+i(\mu/2)\tanh\mu x_1} \tag{5.2.40}$$

Introducing the new integration variable $\kappa$ through $k=\theta\kappa$ and choosing $4\theta^2=t/(z-x_1)$, we obtain

$$\Omega_L(2z,t)=\frac{\mu}{2}\operatorname{sech}\mu x_1\int_{-\infty}^{\infty}\frac{d\kappa}{2\pi i}\frac{e^{-i\sqrt{t(x-x_1)}\,(\kappa+1/\kappa)}}{\kappa+i\mu\sqrt{(x-x_1)/t}\,\tanh\mu x_1} \tag{5.2.41}$$

In general, then, $\Omega_L(2z,t)$ is a function of both $x$ and $t$ as well as $x_1$. If we now consider the case in which $x_1=0$, so that the area under the curve $u(x,0)$ equals $\pi/2$, $\Omega_L(2z,t)$ will be a function of only the similarity variable $\eta=xt$. Hence we can expect that the similarity solution will be directly related to this case. This example will be considered again in Section 7.11.

## 5.3 CUBIC SCHRÖDINGER EQUATION

We have seen in Section 5.1 that the breather solution of the modified Korteweg–deVries equation exhibits solitonlike properties when interacting with another breather solution. The same is found to be the case for the breather solution of the sine–Gordon equation. For small-amplitude breather solutions it was also seen that the small-amplitude, slowly varying phase part of the breather solution satisfies an equation of the type

$$iu_t+u_{xx}+2|u|^2u=0 \tag{5.3.1}$$

which is referred to as a cubic Schrödinger equation. We now investigate this equation and show that, even for large-amplitude solutions, it has the properties expected of soliton equations except that the most elementary soliton solution is now a localized oscillatory solution similar to the breather solution considered previously.

**Linear Equations**

The dependent variable in the cubic Schrödinger equation is complex. To obtain the appropriate linear equations for solution by inverse scattering methods, we begin by using the system

$$v_{1x} + i\zeta v_1 = uv_2$$
$$v_{2x} - i\zeta v_2 = -u^* v_1 \tag{5.3.2}$$

The time derivatives of $v_1$ and $v_2$ are given by (5.1.35) with appropriate coefficients $A$, $B$, $C$, and $D$ $(= -A)$. To obtain these coefficients, we first equate mixed second derivatives of $v_1$ and $v_2$ as in obtaining (5.1.36). The result is now

$$A_x = u^* B + uC$$
$$B_x + 2i\zeta B = -2Au + u_t \tag{5.3.3}$$
$$C_x - 2i\zeta C = -2Au^* - u_t^*$$

Since the cubic Schrödinger equation is of second order in $x$, we may expect that the series expansions in $\zeta$ for $A$, $B$, and $C$ will include only terms through $\zeta^2$. We therefore write $A = \sum_0^2 A_n \zeta^n$, $B = \sum_0^2 B_n \zeta^n$, and $C = \sum_0^2 C_n \zeta^n$. Substituting these expressions into (5.4.3) and equating terms in each power of $\zeta$, we find that $B_2 = C_2 = 0$, $A_2 = a_2(t)$ and $B_1 = ia_2 u$, $C_1 = -ia_2 u^*$, $A_1 = a_1(t)$ as well as $2iB_0 = -2a_1 u - ia_2 u_x$, $2ic_0 = 2a_1 u^* - ia_2 u_x^*$ and $A_0 = -\frac{1}{2} a_2 |u|^2$. The evolution equation for $u$ is obtained from the terms that do not contain $\zeta$ in the second of equations (5.3.3). We find that

$$u_t - ia_1 u_x + \frac{1}{2} a_2 (u_{xx} + 2|u|^2 u) = 0 \tag{5.3.4}$$

Setting $a_1 = 0$ and $a_2 = -2i$, we obtain the cubic Schrödinger equation in the form given in (5.3.1). The expressions for $A$, $B$, and $C$ obtained above yield the linear equations in the form

$$v_{1t} = i(|u|^2 - 2\zeta^2)v_1 + (iu_x + 2\zeta u)v_2$$
$$v_{2t} = (iu_x^* - 2\zeta u^*)v_1 - i(|u|^2 - 2\zeta^2)v_2 \tag{5.3.5}$$

**Solution by Inverse Scattering**

Following our previous treatments of the other soliton equations, we now consider a solution $(v_1, v_2)$ of the system (5.3.2) that is proportional to one of the fundamental solutions $(\varphi_1, \varphi_2)$ or $(\psi_1, \psi_2)$ that were introduced in Section 2.11. We shall concentrate on the latter solution and write

$$v = \begin{pmatrix} v_1 \\ v_2 \end{pmatrix} = f(t)\psi = f(t)\begin{pmatrix} \psi_1 \\ \psi_2 \end{pmatrix} \tag{5.3.6}$$

According to (2.11.7), we thus have $v_2(x,t) \to f(t)e^{i\zeta x}$ as $x$ approached $+\infty$. From the first of (5.3.5) we obtain, assuming that $u$ is localized, $f_t = 2i\zeta^2 f$ or $f(t) = f(0)e^{2i\zeta^2 t}$.

As $x$ approaches $-\infty$, we use (3.9.8b) to write

$$v = f(t)\psi \to f(0)e^{2i\zeta^2 t}\left[c_{21}\begin{pmatrix}0\\-1\end{pmatrix}e^{i\zeta x} + c_{22}\begin{pmatrix}1\\0\end{pmatrix}e^{-i\zeta x}\right] \tag{5.3.7}$$

Thus $v_1 \to f(0)c_{22}(\zeta,t)e^{-i\zeta x + 2i\zeta^2 t}$ and $v_2 \to f(0)c_{21}(\zeta,t)e^{i\zeta x + 2i\zeta^2 t}$ in this limit. Substitution into (5.3.5) with $u$ negligible yields

$$
\begin{aligned}
c_{21}(\zeta,t) &= c_{21}(\zeta,0) \\
c_{22}(\zeta,t) &= c_{22}(\zeta,0)e^{-4i\zeta^2 t}
\end{aligned}
\tag{5.3.8}
$$

From (2.11.37b), the time dependence of the normalization constant is

$$m_{Li}(k_i,t) = -i\frac{c_{22}(k_i,t)}{\dot{c}_{21}(k_i,t)} = m_{Li}(k_i,0)e^{-4ik_i^2 t} \tag{5.3.9}$$

From either a repetition of the foregoing calculation for the fundamental solution $\varphi$ or from the relations among the various coefficients $c_{ij}(\zeta,t)$ given in (3.9.10), we also have $c_{11}(\zeta,t) = c_{11}(\zeta,0)e^{4i\zeta^2 t}$, $c_{12}(\zeta,t) = c_{12}(\zeta,0)$, and

$$m_{Ri}(k_i,t) = i\frac{c_{11}(k_i,t)}{\dot{c}_{12}(k_i,t)} = m_{Ri}(k_i,0)e^{4ik_i^2 t} \tag{5.3.10}$$

To obtain solutions of the cubic Schrödinger equation, we now employ the results on inverse scattering for complex potentials that were obtained in Section 3.9. The only solutions that can be expressed with ease are the multisoliton solutions associated with reflectionless potentials. Even here the explicit evaluations entail considerable algebraic complexity. The simplest case is, of course, the single-soliton solution, since the matrix $M$ given in (3.9.19) then has only one element. For a single pole located at $k_1 = \alpha + i\beta$ ($\beta > 0$), we find that

$$M = \frac{m_{R1}(k_1,0)}{2\beta}e^{-2\beta x + 4i(\alpha + i\beta)^2 t} \tag{5.3.11}$$

From (3.9.20) we have

$$u(x,t) = \frac{2m_{R1}^*(k_1^*,0)}{1+|M|^2}e^{-2ik_1^* x - 4ik_1^{*2} t} \tag{5.3.12}$$

Setting $m_{R1}(k_1,0)/2\beta = e^{\delta + i\theta}$, this single-soliton solution may be written

$$u(x,t) = 2\beta\,\mathrm{sech}(2\beta x + 8\alpha\beta t - \delta)e^{-2i\alpha x - 4i(\alpha^2 - \beta^2)t - i\theta} \tag{5.3.13}$$

For $\alpha = 0$ this result reduces to the special case obtained for the small-amplitude solution in (5.1.34). It should be noted that if $u(x,t)$ is a solution of the cubic Schrödinger equation, then $-u(x,t)$ is also a solution.

### Exercise 3

The linear equations for the modified Korteweg–deVries equation and the cubic Schrödinger equation may also be obtained by the method introduced in Chapter 1 for the Korteweg–deVries equation if we use an operator $L$ in the form of a $2 \times 2$ matrix.

Consider $Ly = \lambda y$, where $\lambda$ is a constant, $y = \begin{pmatrix} y_1 \\ y_2 \end{pmatrix}$, and $L = KD + U$, where $K$ is the constant diagonal matrix $\begin{pmatrix} k_1 & 0 \\ 0 & k_2 \end{pmatrix}$ and $D = \partial/\partial x$. We further set $U = \begin{pmatrix} 0 & u^* \\ u & 0 \end{pmatrix}$ for the cubic Schrödinger equation and $U = \begin{pmatrix} 0 & u \\ u & 0 \end{pmatrix}$, with $u$ real for the modified Korteweg–deVries equation. Assume $y_t = Ay$, where $A$ is a $2 \times 2$ matrix, and show that $\lambda_t = 0$ if $U_t + [L,A] = 0$. Show that the choice $A = aID$, where $a$ is a constant and $I$ is a $2 \times 2$ unit matrix leads to the linear equation

$$U_t - aU_x = 0$$

**A.** *Cubic Schrödinger Equation:* $iu_t + u_{xx} = \kappa |u|^2 u = 0$

Show that this equation follows from the choice $A = aID^2 + B$. In particular:

**(a)** Obtain the matrix partial differential equation $U_t + KB_x - aU_{xx} + [U,B] = 0$ when the requirement $[K,B] - 2aU_x = 0$ is imposed.

**(b)** Determine the elements of $B$ and show that the cubic Schrödinger equation is obtained in the form

$$-u_t + a\left(\frac{k_1 + k_2}{k_1 - k_2}\right)u_{xx} - 2a\frac{(k_1 + k_2)}{k_1 k_2(k_1 - k_2)}|u|^2 u = 0$$

**(c)** Simplify this result with the choices $k_1 = i(1+p), k_2 = i(1-p), a = ip$, and obtain the cubic Schrödinger equation in the form given above with $\kappa = 2/(1-p^2)$ and

$$L = i\begin{pmatrix} 1+p & 0 \\ 0 & 1-p \end{pmatrix}D + \begin{pmatrix} 0 & u^* \\ u & 0 \end{pmatrix}$$

$$A = -i\left\{-p\begin{pmatrix} 1 & 0 \\ 0 & 1 \end{pmatrix}D^2 + \begin{bmatrix} \dfrac{|u|^2}{1+p} & iu_x^* \\ -iu_x & -\dfrac{|u|^2}{1-p} \end{bmatrix}\right\}$$

The operators $L$ and $\tilde{A} = iA$ were first given by Zakharov and Shabat (1972).

**B.** *Modified Korteweg–deVries Equation*: $u_t + 6u^2 u_x + u_{xxx} = 0$.

(a) Choose $A = aID^3 + BD + C$ and show that the operator equation reduces to the matrix partial differential equation $U_t - aU_{xxx} + KC_x - BU_x + [U, C] = 0$ if we choose

$$[K, B] - 3aU_x = 0$$
$$KB_x + [K, C] + [U, B] - 3aU_{xx} = 0$$

(b) Express the elements of $B$ and $C$ in terms of $u$ and its spatial derivatives as well as $k_1$, $k_2$, and $a$. Note that the result is simplified by the choice $k_1 + k_2 = 0$. In particular, for $k_1 = -k_2 = 1$, obtain $a = -4$ and

$$A = -4 \begin{pmatrix} 1 & 0 \\ 0 & 1 \end{pmatrix} D^3 - 6 \begin{pmatrix} u^2 & u_x \\ -u_x & u^2 \end{pmatrix} D - 3 \begin{pmatrix} 2uu_x & u_{xx} \\ -u_{xx} & 2uu_x \end{pmatrix}$$

Similar results have been given by Wadati (1973) and Tanaka (1972).

*Exercise 4*

Show that when $\zeta$ is eliminated from (5.3.5) by means of (5.3.2), the result may be written $v_t = Av$ where $v = \begin{pmatrix} v_1 \\ v_2 \end{pmatrix}$ and $A = MD^2 + ND + C$, in which $D = \partial/\partial x$ and

$$M = 2i \begin{pmatrix} 1 & 0 \\ 0 & -1 \end{pmatrix}, \qquad N = -2i \begin{pmatrix} 0 & u \\ u^* & 0 \end{pmatrix}, \qquad C = -i \begin{pmatrix} -|u|^2 & u_x \\ u_x^* & |u|^2 \end{pmatrix}$$

*Exercise 5*

Determine the proper time dependence for the functions $\zeta$ and $\theta$ that appear in Ex. 5 of Chapter 3 so that the potential $u(x, t)$ will satisfy the cubic Schrödinger equation (5.3.1) and $\varphi$ and $\psi$ will be the appropriate fundamental solutions. Show that the result may be put in the form

$$u = 2\beta e^{-i(\alpha z/\beta + \delta)} \operatorname{sech} z$$

$$\varphi = \frac{e^{-ik\sigma}}{k - k_1^*} \begin{pmatrix} k - \alpha - i\beta \tanh z \\ -i\beta e^{i\theta} \operatorname{sech} z \end{pmatrix}$$

$$\psi = \frac{e^{ik\sigma}}{k_1 - k_1^*} \begin{pmatrix} -i\beta e^{-i\theta} \operatorname{sech} z \\ k - \alpha + i\beta \tanh z \end{pmatrix}$$

where $k = \alpha + i\beta, z = 2\beta(x - \xi)$, and $\xi = -4\alpha\beta t + \xi_0, \sigma = z/2\beta + \xi, \theta = \alpha z/\beta + \delta$, and $\delta = 2\alpha\xi + 4(\alpha^2 - \beta^2)t + \delta_0 = -4(\alpha^2 + \beta^2) + 2\alpha\xi_0 + \delta_0$. This form of the solution will be useful in Chapter 9.

*Exercise 6*

The sine–Gordon equation and the associated linear equations may be obtained by using the method of Chapter 1 if we use the complex potential $\mathcal{V} = -\frac{1}{4}[(\sigma_x)^2 \pm 2i\sigma_{xx}]$

and write the operator $B$ in the integral form $B = f(x, t) \int dx' g(x', t)$, where $f$ and $g$ are to be determined.

(a) Show that the requirement $\mathcal{V}_t + [L, B] = 0$, which must be satisfied for the eigenvalues to be time-independent, leads to $[-\mathcal{V}_t + 2(fg)_x]w + (f_{xx} - \mathcal{V}f) \int dx'(gw) - f \int dx'(g_{xx} - \mathcal{V}g)w = 0$, where $w$ satisfies $(D^2 - \mathcal{V})w = \lambda w$. This result will be devoid of integral operators if we set $f_{xx}/f = g_{xx}/g = \mathcal{V} = \exp(\pm \frac{1}{2} i\sigma)[\exp(\mp \frac{1}{2} i\sigma)]_{xx}$.

(b) Show that the evolution equation is

$$-\tfrac{1}{2}\sigma_x \sigma_{xt} \pm \frac{i}{2}\sigma_{xxt} - 2ik(t)\sigma_x e^{\mp i\sigma} = 0$$

where $k(t)$ arises from integration. Choose $k = \frac{1}{4}$ and show that the real part of this evolution equation is the sine–Gordon equation while the imaginary part is the derivative of the real part.

(c) Show that $w_t = e^{\mp i\sigma/2} \int dx' e^{\pm i\sigma/2} w$. Indicate the upper and lower choices of sign by $w_1$ and $w_2$, respectively, and show that $v_1 = \frac{1}{2}(w_1 + w_2)$ and $v_2 = \frac{1}{2}i(w_1 - w_2)$. Satisfy the linear equations (5.2.5).


## 5.4   A GENERAL CLASS OF SOLUBLE NONLINEAR EVOLUTION EQUATIONS

Before concluding this chapter, we indicate how the class of soluble nonlinear evolution equations can be greatly extended. For a full discussion of this topic, the reader is referred to the research literature (Ablowitz et al., 1974a, Calogero and Degasparis, 1977).

As we have noted, the linear equations that have been associated with the modified Korteweg–deVries equation in (5.1.9), the sine–Gordon equation in (5.2.5), and the cubic Schrödinger equation in (5.3.5) are all of the form

$$\begin{aligned} v_{1t} &= Av_1 + Bv_2 \\ v_{2t} &= Cv_1 - Av_2 \end{aligned} \tag{5.4.1}$$

This result may be used as a starting point for determining linear equations for a fairly broad class of evolution equations to which the inverse method may be applied. We begin by writing equations for the spatial derivatives in the more general form

$$\begin{aligned} v_{1x} + i\zeta v_1 &= qv_2 \\ v_{2x} - i\zeta v_2 &= rv_1 \end{aligned} \tag{5.4.2}$$

The relations $r = -q$ or $r = -q^*$ yield the linear equations considered previously. For the present we make no such restrictions and ultimately derive pairs of *coupled* nonlinear evolution equations satisfied by the two dependent variables, $q$ and $r$. Previous results can then be obtained when the above-mentioned relations between $q$ and $r$ are imposed.

By requiring $\zeta_t = 0$ and the equality of mixed second derivatives, that is, $v_{1xt} = v_{1tx}, v_{2xt} = v_{2tx}$, we find that the following equations must be satisfied since $v_1$ and $v_2$ are linearly independent:

$$A_x = -rB + qC$$
$$B_x + 2i\zeta B = -2qA + q_t \qquad (5.4.3)$$
$$C_x - 2i\zeta C = 2rA + r_t$$

In the evolution equations considered thus far the coefficients $A, B$, and $C$ took the form of series expansions in positive (or negative) powers of $\zeta$. We will no longer consider such a restricted form for these coefficients.

The solubility of all evolution equations treated previously may be traced to the fact that the reflection coefficient in the associated scattering problem has a simple exponential time dependence. The term in the exponential actually turns out to be the dispersion relation for the linear equation that results when the nonlinear term of each of the evolution equations is omitted. More specifically, the linearized version of the modified Korteweg–deVries, cubic Schrödinger, and sine–Gordon equations are $v_t + v_{xxx} = 0, iv_t + v_{xx} = 0$ and $v_{xt} = v$, respectively. The assumption $v = e^{i[kx - \omega(k)t]}$ yields the dispersion relations $\omega = -k^3$, $\omega = k^2$, and $\omega = 1/k$, respectively. These results should be compared with the reflection coefficients for each of these equations. The class of evolution equations for which the reflection coefficient has a simple exponential time dependence is actually much larger than that encountered thus far. We now consider a method for determining such a wider class of evolution equations (Ablowitz et al., 1974a).

In Chapter 2 it was shown that the equations expressing the linear dependence of the various fundamental solutions could be interpreted as solutions to various scattering problems. The reflection coefficients for these scattering problems were then readily identified. We now reexamine this result in a slightly more general context and see what restriction must be introduced in order that the time dependence of the reflection coefficient will be of simple exponential form.

We consider solutions of (5.4.2) that are proportional to fundamental solutions $\varphi_1, \varphi_2$ and $\bar{\varphi}_1, \bar{\varphi}_2$ of the type introduced in (2.11.15). That is

$$\varphi = \begin{pmatrix} \varphi_1 \\ \varphi_2 \end{pmatrix} \to \begin{pmatrix} 1 \\ 0 \end{pmatrix} e^{-i\zeta x} \qquad (5.4.4)$$
$$x \to -\infty$$
$$\bar{\varphi} = \begin{pmatrix} \bar{\varphi}_1 \\ \bar{\varphi}_2 \end{pmatrix} \to \begin{pmatrix} 0 \\ -1 \end{pmatrix} e^{i\zeta x} \qquad (5.4.5)$$

A relation between $\varphi$ and $\bar{\varphi}$ of the type given in (3.9.7) no longer obtains since no relation between $q$ and $r$ is being specified. We will also need solutions $\psi$ and $\bar{\psi}$ similar to those given in (2.11.17). They have the limiting forms

$$\psi = \begin{pmatrix} \psi_1 \\ \psi_2 \end{pmatrix} \to \begin{pmatrix} 0 \\ 1 \end{pmatrix} e^{i\xi x} \qquad (5.4.6)$$

$$x \to +\infty$$

$$\bar{\psi} = \begin{pmatrix} \bar{\psi}_1 \\ \bar{\psi}_2 \end{pmatrix} \to \begin{pmatrix} 1 \\ 0 \end{pmatrix} e^{-i\xi x} \qquad (5.4.7)$$

The relations between these solutions will be written

$$\bar{\varphi} = a\psi + b\bar{\psi}$$

$$\varphi = \bar{b}\bar{\psi} - \bar{a}\psi \qquad (5.4.8)$$

and may be compared with (2.11.19) and (2.11.20) as well as (3.9.8) and (3.9.9). We can also write

$$\psi = -a\bar{\varphi} + \bar{b}\varphi$$

$$\bar{\psi} = b\bar{\varphi} + \bar{a}\varphi \qquad (5.4.9)$$

In terms of a Wronskian defined as in (2.11.18) we find that

$$a = W(\varphi, \psi), \qquad b = -W(\varphi, \bar{\psi}), \qquad \bar{a} = W(\bar{\varphi}, \bar{\psi}), \qquad \bar{b} = W(\bar{\varphi}, \psi)$$

$$(5.4.10)$$

We also find that $W(\varphi, \bar{\varphi}) = -1$ and then from (5.4.8) that $a\bar{a} + b\bar{b} = 1$.

The first of (5.4.9) is a solution that approaches $\begin{pmatrix} 0 \\ 1 \end{pmatrix} e^{i\xi x}$ as $x$ goes to $+\infty$ and approaches $a\begin{pmatrix} 0 \\ 1 \end{pmatrix} e^{i\xi x} + \bar{b}\begin{pmatrix} 1 \\ 0 \end{pmatrix} e^{-i\xi x}$ as $x$ goes to $-\infty$. It thus represents the scattering of a solution $a\begin{pmatrix} 0 \\ 1 \end{pmatrix} e^{i\xi x}$ that is incident from $-\infty$. The reflection coefficient for a *unit* incident amplitude is then $\bar{b}/a$. It is the time dependence of this reflection coefficient that we wish to examine. In particular, we will determine how it may be made to have simple exponential time dependence. As would be expected in the analysis of a scattering problem, we must enter into a detailed investigation of the solutions $v_1$ and $v_2$ as well as the functions $A, B$, and $C$ in the two limits $x \to \pm\infty$.

When (5.4.1) is examined in the limit $x \to -\infty$, we see that $B$ and $C$ must vanish in this limit if $v = (v_1, v_2)^T$ is to be proportional to the fundamental solutions $\varphi$ or $\bar{\varphi}$. We then find from the first of (5.4.3) that $A$ is independent of $x$ in this limit. We shall also assume that $A$ is independent of $t$ as $x \to -\infty$. In this regard, it should be noted that the functions $A$ that occur in the three previous examples are independent of $t$ in this limit. We may therefore write the first of (5.4.3) as $v_{1t} = A_-(\zeta)v_1$, where

$$A_-(\zeta) = \lim_{x \to -\infty} A(x,t,\zeta) \tag{5.4.11}$$

In terms of the fundamental solution $\varphi$, we then have $v = \exp[A_-(\zeta)t]\varphi$. Similarly, writing

$$\bar{v}_{1t} = \bar{A}\bar{v}_1 + \bar{B}\bar{v}_2$$
$$\bar{v}_{2t} = \bar{C}\bar{v}_1 - \bar{A}\bar{v}_2 \tag{5.4.12}$$

and noting the asymptotic form (5.4.5), we find that $\bar{v} = \exp[-A_-(\zeta)t]\varphi$. The linear equations (5.4.1) and (5.4.12) may be written in terms of $\varphi$ and $\bar{\varphi}$ in the matrix form

$$\begin{pmatrix} \varphi_1 \\ \varphi_2 \end{pmatrix}_t = \begin{pmatrix} A - A_- & B \\ C & -A - A_- \end{pmatrix} \begin{pmatrix} \varphi_1 \\ \varphi_2 \end{pmatrix} \tag{5.4.13a}$$

$$\begin{pmatrix} \bar{\varphi}_1 \\ \bar{\varphi}_2 \end{pmatrix}_t = \begin{pmatrix} A + A_- & B \\ C & -A + A_- \end{pmatrix} \begin{pmatrix} \bar{\varphi}_1 \\ \bar{\varphi}_2 \end{pmatrix} \tag{5.4.13b}$$

As $x$ approaches $+\infty$, (5.4.8) shows that

$$\varphi \to \begin{pmatrix} ae^{-i\zeta x} \\ be^{i\zeta x} \end{pmatrix}, \qquad \bar{\varphi} \to \begin{pmatrix} \bar{b}e^{-i\zeta x} \\ -\bar{a}e^{i\zeta x} \end{pmatrix}, \qquad x \to +\infty \tag{5.4.14}$$

When (5.4.13) is examined in the limit $x \to +\infty$, we find that

$$a_t = (A_+ - A_-)a + B_+b$$
$$b_t = C_+a - (A_+ + A_-)b$$
$$\bar{a}_t = -(A_+ - A_-)\bar{a} - C_+\bar{b}$$
$$\bar{b}_t = -B_+\bar{a} + (A_+ + A_-)\bar{b} \tag{5.4.15}$$

where $A_+(\zeta,t) \equiv A(\infty,t,\zeta)$, $B_+ \equiv B(\infty,t,\zeta)e^{2i\zeta x}$, and $C_+ \equiv C(\infty,t,\zeta)e^{-2i\zeta x}$.

Writing $\rho = b/a$ for the reflection coefficient, we find that the first and fourth of (5.4.15) may be combined to yield

$$\rho_t = 2A_-(\zeta)\rho - \frac{B_+}{a^2} \tag{5.4.16}$$

A differential equation for $\rho$ that leads to a purely exponential time dependence for $\rho$ will thus be obtained if either $B_+ = 0$ or, more generally, $B_+ = F(\zeta)\bar{b}a$. The form of $B_+$ must therefore be analyzed.

An expression for $B_+$ may be obtained quite directly by solving (5.4.3) and then taking the limit of this solution as $x \to +\infty$. However, it is clear from an examination of this solution (Ablowitz et al., 1974a) that the final expression for $B_+$ may be constructed quite simply. First, it is found that we require only the equations that specify the temporal variation of $\psi_1$ and $\psi_2$. These are obtained in a manner analogous to that employed in obtaining (5.4.13a) and are

$$\psi_{1t} = (A + A_+)\psi_1 + B\psi_2$$
$$\psi_{2t} = C\psi_1 - (A - A_+)\psi_2 \tag{5.4.17}$$

We differentiate the first of (5.4.2) with respect to time, use the first of (5.4.17), and then multiply the resulting equation by $\psi_2$. We then multiply the second of (5.4.2) by $B\psi_2$ and add to the equation just obtained. After eliminating $A_x$ by the first of (5.4.3), we have

$$-2rB\psi_1\psi_2 + 2Aq\psi_2^2 + (B\psi_2^2)_x = q_t\psi_2^2 \tag{5.4.18}$$

When a similar procedure is applied to the time derivative of the second of (5.4.2) and the result is added to (5.4.18), we obtain

$$(-2\psi_1\psi_2 A - \psi_2^2 B + \psi_1^2 C)_x = -q_t\psi_2^2 + r_t\psi_1^2 \tag{5.4.19}$$

We now integrate this expression from $-\infty$ to $+\infty$. Using (5.4.4) through (5.4.7) and (5.4.9), we readily obtain

$$B_+ = -\int_{-\infty}^{\infty} dx(-q_t\psi_2^2 + r_t\psi_1^2) + 2a\bar{b}A_-(\zeta) \tag{5.4.20}$$

The first-mentioned condition for obtaining exponential time dependence for $\rho$, namely $B_+ = 0$, is seen from (5.4.20) to imply that

$$\int_{-\infty}^{\infty} dx(-q_t\psi_2^2 + r_t\psi_1^2) = 2a\bar{b}A_-(\zeta)$$

$$= -2A_- \int_{-\infty}^{\infty} dx \frac{\partial}{\partial x}(\psi_1\psi_2) \tag{5.4.21}$$

$$= -2A_- \int_{-\infty}^{\infty} dx(q\psi_2^2 + r\psi_1^2)$$

The integral representation for $a\bar{b}$ used here arises in obtaining (5.4.20). We may now write (5.4.21) as

$$\int_{-\infty}^{\infty} dx[(-q_t + 2A_- q)\psi_2^2 + (r_t + 2A_- q)\psi_1^2] = 0 \tag{5.4.22}$$

The ultimate goal in these considerations is the determination of nonlinear evolution equations that are associated with reflection coefficients with exponential time dependence. Equation 5.4.22 incorporates one of the possibilities that leads to such equations, namely $B_+ = 0$. We now consider how the associated evolution equations can be obtained from (5.4.22).

We must, of course, extract from (5.4.22) an expression that is independent of $\zeta$ since the evolution equations themselves are independent of this parameter. This may be accomplished by introducing an eigenvalue problem for the squared eigenfunctions $\psi_1^2$ and $\psi_2^2$ that appear in (5.4.22). We introduce the eigenvector $\Psi \equiv \begin{pmatrix} \Psi_1 \\ \Psi_2 \end{pmatrix} = \begin{pmatrix} \psi_1 2 \\ \psi_2 2 \end{pmatrix}$ and construct an operator $L$ such that $L\Psi = \zeta\Psi$. To exploit this operator we first write (5.4.22) in the scalar product form

$$\int_{-\infty}^{\infty} dx \left[ r_t + 2A_-(\zeta)r, -q_t + 2A_-(\zeta)q \right] \cdot \begin{pmatrix} \Psi_1 \\ \Psi_2 \end{pmatrix} = 0 \qquad (5.4.23)$$

Introducing the vector $u = \begin{pmatrix} r \\ q \end{pmatrix}$ and the matrix $\sigma_3 = \begin{pmatrix} 1 & 0 \\ 0 & -1 \end{pmatrix}$, which is one of the Pauli spin matrices (Schiff, 1949, Sec. 33), we may write (5.4.23) as

$$\int_{-\infty}^{\infty} dx \left\{ \left[ u_t^T \sigma_3 + 2u^T A_-(\zeta) \right] \cdot \Psi \right\} = 0 \qquad (5.4.24)$$

where $T$ indicates the transpose operation. If $A_-(\zeta)$ were merely $\zeta$, we could immediately replace $\zeta\Psi$ by $L\Psi$. Also, since $\zeta^2\Psi = \zeta L\Psi = L^2\Psi$, and in general $\zeta^n\Psi = L^n\Psi$, we can replace $A_-(\zeta)$ by $A_-(L)$ as long as $A_-(\zeta)$ can be expanded in powers of $\zeta$.

Before proceeding, let us determine the operator $L$. We first use (5.4.2) to write

$$\zeta \begin{pmatrix} v_1^2 \\ v_2^2 \end{pmatrix} = \frac{1}{2i} \frac{\partial}{\partial x} \begin{pmatrix} -v_1^2 \\ v_2^2 \end{pmatrix} + i \begin{pmatrix} qv_1v_2 \\ -rv_1v_2 \end{pmatrix} \qquad (5.4.25)$$

Again using the relation $-v_1v_2 = \int_x^{\infty} dy \, (v_1v_2)_y = \int_x^{\infty} dy \, (rv_1^2 + qv_2^2)$ we can write

$$\begin{pmatrix} qv_1v_2 \\ rv_1v_2 \end{pmatrix} = \begin{bmatrix} q \int_x^{\infty} dy (rv_1^2 + qv_2^2) \\ r \int_x^{\infty} dy (rv_1^2 + qv_2^2) \end{bmatrix}$$

$$= - \begin{bmatrix} q \int_x^{\infty} dy \, r & q \int_x^{\infty} dy \, q \\ -r \int_x^{\infty} dy \, r & -r \int_x^{\infty} dy \, q \end{bmatrix} \begin{pmatrix} v_1^2 \\ v_2^2 \end{pmatrix} \qquad (5.4.26)$$

The operator $L$ is now seen from (5.4.25) to be

$$
L = \frac{1}{2i}
\begin{bmatrix}
-\dfrac{\partial}{\partial x} - 2q \displaystyle\int_x^\infty dy\, r(y) & -2q \displaystyle\int_x^\infty dy\, q(y) \\[2ex]
2r \displaystyle\int_x^\infty dy\, r(y) & \dfrac{\partial}{\partial x} + 2r \displaystyle\int_x^\infty dy\, q(y)
\end{bmatrix}
\tag{5.4.27}
$$

We return now to the use of this operator in determining the evolution equations. For simplicity, let us again consider the case $A_-(\zeta) = \zeta$. Then in (5.4.24) we may write

$$
\int_{-\infty}^\infty dx\, u^T A_-(\zeta) \cdot \Psi = \int_{-\infty}^\infty dx\, u^T \cdot L\Psi
$$

$$
= \int_{-\infty}^\infty dx\, \{ r(L_{11}\Psi_1 + L_{12}\Psi_2) + q(L_{21}\Psi_1 + L_{22}\Psi_2) \}
\tag{5.4.28}
$$

where the $L_{ij}$ are the elements of the operator $L$ in (5.4.27) and imply integrals over the $\Psi_i$ as indicated in (5.4.26). In order for the entire integrand in (5.4.24) to be proportional to $\Psi$, we now use partial integration to transfer the operators in the $L_{ij}$ from the $\Psi_i$ to $r$ and $q$. (In effect, we are merely interchanging orders of integration.) As an example, consider

$$
2i \int_{-\infty}^\infty dx\, q L_{21}\Psi_1 = \int_{-\infty}^\infty dx\, q(x) 2r(x) \int_x^\infty dy\, r(y)\Psi(y)
$$

$$
= 2 \int_{-\infty}^\infty dy\, \Psi_1(y) r(y) \int_{-\infty}^x dx\, q(x) r(x)
\tag{5.4.29}
$$

$$
= \int_{-\infty}^\infty dy\, \Psi_1(y) \tilde{L}_{12} r(y)
$$

The operators in $\tilde{L}_{12}$ no longer act upon $\Psi_1$, and the integrand is proportional to $\Psi$ itself. The final form of (5.4.24) with $A_-(\zeta) = \zeta$ can then be written

$$
\int_{-\infty}^\infty dx\, \Psi^T \cdot (\sigma_3 u_t + 2\tilde{L}u) = 0
\tag{5.4.30}
$$

This equation will be satisfied if we set $\sigma_3 u_T + 2\tilde{L}u = 0$. For a more general $A_-(\zeta)$ that possesses a series expansion in $\zeta$, we obtain the final result,

$$
\sigma_3 u_t + 2A_-(\tilde{L})u = 0
\tag{5.4.31}
$$

The other components of $L$ may be treated similarly and the operator $\tilde{L}$ is found to be

$$
\tilde{L} = \frac{1}{2i}
\begin{bmatrix}
\dfrac{\partial}{\partial x} - 2r\displaystyle\int_{-\infty}^{x} dy\, q & 2r\displaystyle\int_{-\infty}^{x} dy\, r \\[2ex]
-2q\displaystyle\int_{-\infty}^{x} dy\, q & -\dfrac{\partial}{\partial x} + 2q\displaystyle\int_{-\infty}^{x} dy\, r
\end{bmatrix}
\tag{5.4.32}
$$

As a simple example, consider the linear form $A_-(\zeta) = i\zeta$. Then (5.4.31) yields

$$
\begin{aligned}
r_t + 2i(\tilde{L}_{11}r + \tilde{L}_{12}q) &= 0 \\
-q_t + 2i(\tilde{L}_{21}r + \tilde{L}_{22}q) &= 0
\end{aligned}
\tag{5.4.33}
$$

From the definition of the $\tilde{L}_{ij}$ in (5.4.32), we find that $2i(\tilde{L}_{11}r + \tilde{L}_{12}q) = r_x$ and $2i(\tilde{L}_{21}r + \tilde{L}_{22}q) = -q_x$. Hence (5.4.33) simplifies to the linear equations $r_t + r_x = 0$ and $q_t + q_x = 0$.

For the quadratic case $A(\zeta) = -2i\zeta^2$, the simplifications that appeared in the previous example again arise and we find that

$$
\begin{aligned}
r_t - 2(\tilde{L}_{11}r_x - \tilde{L}_{12}q_x) &= 0 \\
q_t + 2(\tilde{L}_{21}r_x - \tilde{L}_{22}q_x) &= 0
\end{aligned}
\tag{5.4.34}
$$

On setting $r = -q^*$ we obtain the cubic Schrödinger equation in the form given in (5.3.1).

The alternative choice for obtaining exponential time dependence for $\rho_t$, namely $B_+ = F(\zeta)\bar{b}a$, leads to results similar to those obtained above except that $2A_-(\zeta)$ is replaced by $2A_-(\zeta) - F(\zeta)$. If we write $2A_- - F = 2\Omega_1(\zeta)/\Omega_2(\zeta)$, the evolution equations are of the form

$$
\Omega_2(\tilde{L})\sigma_3 u_t + 2\Omega_1(\tilde{L})u = 0
\tag{5.4.35}
$$

As an example, consider $\Omega_1 = i$ and $\Omega_2 = 4\zeta$; then (5.4.35) becomes $2\tilde{L}\sigma_3 u_t + iu^T = 0$. For $r = -q$ the two equations become

$$
q_{tx} + 2q\int_{-\infty}^{x} dy\,(q^2)_t - \tfrac{1}{2}q = 0
\tag{5.4.36}
$$

This equation may be transformed into the sine–Gordon equation. To see this we first divide by $q$ and differentiate to obtain

$$
\frac{\partial}{\partial x}\left(\frac{q_{tx}}{q}\right) + 4qq_t = 0
\tag{5.4.37}
$$

If we set $q_t = f$, this equation may be written

$$f_{xx} - \left(\frac{q_x}{q}\right)f_x + 4q^2 f = 0 \tag{5.4.38}$$

which has the solution (Kamke, 1971 p. 420) $f = c_1 \cos \sigma + c_2 \sin \sigma$, where $\sigma = 2\int_{-\infty}^{x} dy\, q$. Since $f$ must vanish as $x \to -\infty$, we require that $c_1 = 0$. Also, substitution into (5.4.36) shows that $c_2 = \frac{1}{2}$. Since $f = q_t$ and $q = \frac{1}{2}\sigma_x$, we finally arrive at the sine–Gordon equation $\sigma_{xt} = \sin \sigma$.

### Exercise 7

Show that the assumption $A = \sum_{n=0}^{3} a_n \zeta^n$, with similar expansions for $B$ and $C$, leads to the evolution equations

$$q_t + a_2(t)\left(\tfrac{1}{2}q_{xx} - rq^2\right) + \tfrac{1}{4}ia_3(t)(q_{xxx} - 6rqq_x) = 0$$

$$r_t - a_2(t)\left(\tfrac{1}{2}r_{xx} - r^2 q\right) + \tfrac{1}{4}ia_3(t)(r_{xxx} - 6rqr_x) = 0$$

For $r = -q$, $a_2 = 0$, and $a_3 = -4i$, obtain the modified Korteweg–deVries equation and for $r = -1$ with the same values of $a_2$ and $a_3$, obtain the Korteweg–deVries equation. Show also that when $r = -1$, $v_2$ satisfies $v_{2xx} + (\zeta^2 + q)v_2 = 0$.

### Exercise 8

Use the method described in the text to derive (5.4.19) to obtain the three equations

$$\left(-2\psi_1 \psi_2 A - \psi_2^2 B + \psi_1^2 C\right)_x = -q_t \psi_2^2 + r_t \psi_1^2$$

$$\left(-2\bar{\psi}_1 \bar{\psi}_2 A - \bar{\psi}_2^2 B + \bar{\psi}_1^2 C\right)_x = -q_t \bar{\psi}_2^2 + r_t \bar{\psi}_1^2$$

$$\left[-\left(\bar{\psi}_1 \psi_2 + \psi_1 \bar{\psi}_2\right)A - \psi_2 \bar{\psi}_2 B + \psi_1 \bar{\psi}_1 C\right]_x = -q_t \psi_2 \bar{\psi}_1 + r_t \psi_1 \bar{\psi}_1$$

These equations may now be integrated from $-\infty$ to $x$ and solved for $A$, $B$, and $C$. Show that the determinant of the linear system has magnitude unity.

# CHAPTER 6

# Applications I

The soliton equations arise quite frequently as approximations to the familiar and more exact equations in many fields of physics. Indeed, this is to be expected since the dispersion and nonlinearities incorporated in the various soliton equations can be thought of as the first terms in expansions of more complete expressions for dispersion and nonlinearity. As already seen in Section 1.4, the perturbation procedures used to obtain a soliton equation can be somewhat long and tedious. This chapter contains additional examples taken from hydrodynamics and plasma physics as well as a classical model in solid-state dislocation theory. The main emphasis is placed on obtaining the soliton equations by starting from the standard equations employed in these fields of physics. Readers with a minimum of familiarity with these subjects should find the presentations to be self-contained. Examples of the occurrence of solitons in low-temperature physics, quantum field theory, and other areas that require a more specialized background have not been included.

## 6.1 SHALLOW WATER WAVES AND THE KORTEWEG–deVRIES EQUATION

We consider the motion of plane waves on the surface of a liquid. The maximum amplitude of the disturbance, $a$, is assumed to be small compared to the depth of the liquid, $h$, while the length of the disturbance, $l$, is assumed to be large compared to $h$. The two smallness parameters $\epsilon = a/h$ and $\delta = h/l$ play fundamental roles in the subsequent perturbation development.

The fluid motion is described by a velocity vector $\mathbf{V}(x,y,t) = \mathbf{i}u(x,y,t) + \mathbf{j}v(x,y,t)$, where $\mathbf{i}$ and $\mathbf{j}$ are unit vectors in the horizontal and vertical directions, respectively. It is assumed that the liquid motion is irrotational, so that $\nabla \times \mathbf{V} = 0$. It is also assumed that the density of the fluid, $\rho$, is constant, so that the equation of continuity $\rho_t + \nabla \cdot (\rho V) = 0$ reduces to $\nabla \cdot \mathbf{V} = 0$. We may then set $\mathbf{V} = \nabla \phi$ with $\nabla^2 \phi = 0$.

The fluid is assumed to be bounded below by a hard horizontal bed so that $\phi_y(x,0) = 0$. The upper boundary is the free surface $y = h + \eta(x,t)$, as shown in Figure 6.1. Henceforth all terms evaluated on this surface will be designated by the subscript 1. The boundary conditions at the free surface are nonlinear,

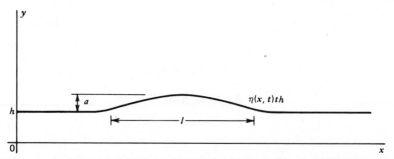

**Figure 6.1** Coordinate system for shallow water wave analysis leading to Korteweg–deVries equation.

and it is here that the complexity of the problem arises. On the free surface $y = y_1 = h + \eta(x, t)$ we may write

$$\frac{dy_1}{dt} = \frac{\partial \eta}{\partial t} + \frac{\partial \eta}{\partial x} \frac{dx_1}{dt} \tag{6.1.1}$$

Since $dy_1/dt = v_1 = \partial \varphi_1/\partial y$ and $dx_1/dt = u_1 = \partial \varphi_1/\partial x$ we have the first boundary condition in either of the forms

$$v_1 = \eta_t + \eta_x u_1 \tag{6.1.2a}$$

or

$$\phi_{1y} = \eta_t + \eta_x \phi_{1x} \tag{6.1.2b}$$

The second boundary condition comes from the momentum conservation relation

$$\frac{d\mathbf{V}}{dt} = \frac{\partial \mathbf{V}}{\partial t} + (\mathbf{V} \cdot \nabla)\mathbf{V} = -\frac{1}{\rho_0} \nabla p - g\mathbf{j} \tag{6.1.3}$$

where $\rho_0$ is the constant density of the liquid, $p$ the pressure at any point in the fluid, and $g$ the gravitational force constant. Since $\frac{1}{2}\nabla(V^2) = \mathbf{V} \times (\nabla \times \mathbf{V}) + (\mathbf{V} \cdot \nabla)\mathbf{V}$ and $\nabla \times \mathbf{V} = 0$ in our considerations, (6.1.3) may be written in terms of the divergence of the expression $\phi_t + \frac{1}{2}(\nabla \phi)^2 + p/\rho_0 - gy$. Integrating and absorbing the arbitrary function of time into $\phi$, we obtain

$$\phi_t + \frac{1}{2}(\nabla \phi)^2 + \frac{p}{\rho_0} + gy = 0 \tag{6.1.4}$$

Evaluating this expression on the upper surface, assuming that $p_1 = 0$, and then differentiating with respect to $x$, we obtain the second boundary condition in the form

$$u_{1t} + u_1 u_{1x} + v_1 v_{1x} + g\eta_x = 0 \tag{6.1.5}$$

The motion of the surface is thus described by obtaining a solution of Laplace's equation that satisfies the nonlinear boundary conditions (6.1.2) and (6.1.5).

Since we are only interested in small values of $y$ (i.e., $h \ll l$), we shall attempt to obtain an adequate solution of Laplace's equation by using the first few terms in an expansion of the form

$$\phi(x,y,t) = \sum_{n=0}^{\infty} y^n \phi_n(x,t) \qquad (6.1.6)$$

Substituting this expansion into Laplace's equation and separating terms according to powers of $y$, we obtain the recurrence relation $\phi_{nxx} + (n+2)(n+1)\phi_{n+2} = 0$. The boundary condition $\phi_y(x,0) = 0$ yields $\phi_1 = 0$. All odd $\phi_n$ thus vanish and we obtain

$$\phi = \phi_0 - \tfrac{1}{2}y^2\phi_{0xx} + \tfrac{1}{24}y^4\phi_{0xxxx} + \cdots \qquad (6.1.7)$$

The velocity components at the free surface are then

$$u_1 = \phi_{x1} = f - \tfrac{1}{2}y_1^2 f_{xx} + \cdots$$

$$v_1 = \phi_{y1} = -y_1 f_x + \tfrac{1}{6}y_1^3 f_{xxx} + \cdots \qquad (6.1.8)$$

where $f(x,t) = \partial\phi_0(x,t)/\partial x$.

Before we proceed with the derivation of the Korteweg–deVries equation, it will be instructive to consider the linear wave equation that is obtained when the product terms in the boundary conditions (6.1.2) and (6.1.5) are neglected. Besides providing insight into the role played by the two small parameters $\epsilon = a/h$ and $\delta = h/l$, it will enable us to determine a dispersion relation for the problem.

If we linearize the boundary conditions (6.1.2) and (6.1.5) by setting $y_1 = h$ and dropping the quadratic terms, we obtain

$$v_1 = \eta_t \qquad (6.1.9a)$$

$$u_1 t = -g\eta_x \qquad (6.1.9b)$$

Initially, we shall use only the first terms in (6.1.8) to express $u_1$ and $v_1$ in terms of $f$. We then have

$$hf_x = -\eta_t \qquad (6.1.10a)$$

$$f_t = -g\eta_x \qquad (6.1.10b)$$

Combining so as to eliminate $f$, we find that $\eta$ satisfies the linear wave equation $\eta_{tt} - c_0^2\eta_{xx} = 0$, where $c_0^2 = gh$. Hence, in this simplest linear approximation we see that the surface can support nondispersive sinusoidal displacements $\eta = a \exp i(kx - \omega t)$, where $\omega = c_0 k$. Returning to (6.1.9b), we find that

$$u_1 = -g \int dt \, \eta_x = \left( \frac{g}{c_0} \right) \eta = \epsilon c_0 \exp[i(kx - \omega t)] \qquad (6.1.11)$$

We thus expect that waves of characteristic amplitude $a$ will have surface velocities $u_1$ of order $\epsilon c_0$. Similarly, from (6.1.9a) we find $v_1 = -i\omega\eta = -ic_0\eta/\lambda$. For disturbances in which the dominant wavelengths $\lambda$ are of magnitude $l$, we see that the magnitude of the vertical velocity $v_1$ is $v_1 \approx c_0(a/h)(h/l) \approx c_0 \epsilon \delta$. This information on the relative sizes of $u_1$ and $v_1$ is useful in establishing a perturbation procedure for dealing with the nonlinear terms.

Before returning to the nonlinear problem, we note that retention of the second terms in (6.1.8) yields the linear equations

$$-hf_x + \tfrac{1}{6}h^3 f_{xxx} = \eta_t$$
$$f_t - \tfrac{1}{2}h^2 f_{xxt} + g\eta_x = 0 \qquad (6.1.12)$$

which can be combined to form the linear equation

$$f_{tt} - c_0^2 f_{xx} + \tfrac{1}{6}c_0^2 h^2 f_{xxxx} - \tfrac{1}{2}h^2 f_{xxtt} = 0 \qquad (6.1.13)$$

The assumption $f \approx \exp i(kx - \omega t)$ now leads to the dispersion relation $\omega \approx c_0 k[1 - (kh)^2/6]$, which contains a cubic term in $k$ similar to that obtained in (1.4.10) for a fluid confined by a system of elastic rings.

To proceed with the nonlinear equations, we first introduce dimensionless independent variables by setting $x = lx'$ and $t = (l/c_0)t'$. Similarly, we write the dependent variables as $\eta = a\eta'$, $u_1 = \epsilon c_0 u_1'$, $v_1 = \epsilon \delta c_0 v_1'$, and, from the first of (6.1.8), $f = \epsilon c_0 f'$. We also set $y_1 = h(1 + \epsilon\eta')$. To the first nonvanishing order in $\epsilon$ and $\delta$, the dimensionless velocities are

$$u_1' = f' - \tfrac{1}{2}\delta^2 \frac{\partial^2 f'}{\partial x'^2}$$
$$v_1' = -(1 + \epsilon\eta')\frac{\partial f'}{\partial x'} + \frac{1}{6}\delta^2 \frac{\partial^3 f'}{\partial x'^3} \qquad (6.1.14)$$

The first and second boundary conditions (6.1.2a) and (6.1.5) become

$$\frac{\partial \eta'}{\partial t'} + \frac{\partial f'}{\partial x'} + \epsilon\eta'\frac{\partial f'}{\partial x'} + \epsilon f'\frac{\partial \eta'}{\partial x'} - \frac{1}{6}\delta^2 \frac{\partial^3 f'}{\partial x'^3} = 0$$
$$\frac{\partial f'}{\partial t'} + \frac{\partial \eta'}{\partial x'} + \epsilon f'\frac{\partial f'}{\partial x'} - \frac{1}{2}\delta^2 \frac{\partial^3 f'}{\partial x'^2 \partial t'} = 0 \qquad (6.1.15)$$

These equations are now easily solved by using a perturbation technique described by Whitham (1974 p. 466). We first note that when $\epsilon$ and $\delta$ are neglected, we may write $f' = \eta'$ with $f_{t'} + f_{x'} = 0$. We now introduce into (6.1.15) the perturbation expansion

$$f' = \eta' + \epsilon f^{(1)} + \delta^2 f^{(2)} \qquad (6.1.16)$$

and retain terms to first order in $\epsilon$ and $\delta^2$. When one equation in (6.1.15) is then subtracted from the other and we note that $f_{t'}^{(1)} = -f_{x'}^{(1)}$ and $f_{t'}^{(2)} = -f_{x'}^{(2)}$ to the order of accuracy being considered, we obtain

$$\epsilon\left(2\frac{\partial f^{(1)}}{\partial x'} + \eta'\frac{\partial \eta'}{\partial x'}\right) + \delta^2\left(2\frac{\partial f^{(2)}}{\partial x'} - \frac{2}{3}\frac{\partial^3\eta'}{\partial x'^3}\right) = 0 \qquad (6.1.17)$$

Since $\epsilon$ and $\delta$ are independent, the coefficients of $\epsilon$ and $\delta^2$ must vanish separately. We then have $f^{(1)} = -\frac{1}{4}(\eta')^2$ and $f^{(2)} = \frac{1}{3}\partial^2\eta'/\partial x'^2$. With $f'$ now expressed in terms of $\eta'$, we find that the first of (6.1.15) yields

$$\frac{\partial \eta'}{\partial t'} + \frac{\partial \eta'}{\partial x'} + \frac{3}{2}\epsilon\eta'\frac{\partial \eta'}{\partial x'} + \frac{1}{6}\delta^2\frac{\partial^3\eta'}{\partial x'^3} = 0 \qquad (6.1.18)$$

The term $\partial\eta'/\partial x'$ can be eliminated by the transformation $\xi = x' - t'$, $\tau = t'$. The resulting equation has the form of the Korteweg–deVries equation. If it is multiplied by $9\epsilon/\delta^4$ and then $3\epsilon\eta'/2\delta^2$ is set equal to $w$, we obtain $w_\tau + 6ww_\xi + w_{\xi\xi\xi} = 0$, where $\tau' = \delta^2\tau/6$. When the single-soliton solution (1.4.18) is expressed in dimensional variables, we find that

$$\eta = \eta_0 \operatorname{sech}^2\left(\frac{x - Vt}{L}\right) \qquad (6.1.19)$$

where

$$\eta_0 = \frac{ch^3}{3l^2} \qquad (6.1.20a)$$

$$L = \left(\frac{4h^3}{3\eta_0}\right)^{1/2} \qquad (6.1.20b)$$

$$V = c_0\left(1 + \frac{1}{2}\frac{h_0}{h}\right), \qquad c_0^2 = gh \qquad (6.1.20c)$$

The constant $c$ in (6.1.20a) is the arbitrary constant appearing in (1.4.18). In dimensional variables (6.1.18) becomes

$$\frac{\partial \eta}{\partial t} + c_0\frac{\partial \eta}{\partial x} + \frac{3}{2}\frac{c_0}{h}\eta\frac{\partial \eta}{\partial x} + \frac{1}{6}c_0h^2\frac{\partial^3\eta}{\partial x^3} = 0 \qquad (6.1.21)$$

A somewhat more systematic derivation of the Korteweg–deVries equation may be obtained by following the procedure to be introduced in Section 6.5 (Su and Gardner, 1969, Appendix).

Surface-wave experiments confirming the soliton theory of the Korteweg-deVries equation have been carried out (see, for example, Hammack and Segur, 1974). As a simple example of the application of the foregoing results to water wave experiments, note that the $n$-soliton solution corresponding to (6.1.19) would be $\eta(x,0)=\frac{1}{2}n(n+1)\eta_0 \text{sech}^2(x/L)$, where $L$ is given by (6.1.20b). This may be rewritten $\eta(x,0)=A\,\text{sech}^2(x/L)$, where $L^2 A = \frac{2}{3}n(n+1)h^3$. Once the initial size of the wave has been specified through $L$ and $A$, the number of solitons expected to emerge is seen to be determined by the depth $h$. If $n=1$ for $h=h_1$, a decrease in depth to $h_2=rh_1$ yields a number of solitons $n_2=\frac{1}{2}(-1+\sqrt{1+8/r^3}\,)$, which follows by merely solving $2h_1^3=n_2(n_2+1)h_2^3$ (Tappert and Zabusky, 1971).

## 6.2 SHALLOW WATER WAVES AND THE CUBIC SCHRÖDINGER EQUATION

There is a relation between the cubic Schrödinger equation and the Korteweg-deVries equation that arises when we look for solutions to the latter equation in the form of a slowly modulated wavetrain. The cubic Schrödinger equation can also be obtained directly from the fundamental equations for water waves (Hasimoto and Ono, 1972; Davey, 1972) without the restriction to shallow water theory that is implied by the Korteweg-deVries equation and extensive experimental and numerical investigations of this case have been carried out (Lake et al., 1977). However, this more general application will not be considered here.

To understand how the cubic Schrödinger equation might be expected to arise in a description of surface waves, let us first recall that the simplest *linear* description of surface-wave phenomena was shown to be governed by the nondispersive wave equation $(gh)\eta_{xx} - \eta_{tt} = 0$. The more accurate linear equation (6.1.13) was found to introduce dispersion. For waves in one direction these linear results are all obtained from the Korteweg-deVries equation (6.1.21) when the nonlinear term $\eta\,\partial\eta/\partial x$ is neglected. The simplest *nonlinear* corrections are obtained by considering small-amplitude solutions of the full Korteweg-deVries equation (6.1.21). This is found to lead to an amplitude-dependent dispersion relation. It is not surprising that the cubic Schrödinger equation could arise at this point, since this equation introduces an amplitude-dependent dispersion relation in a very natural way. For instance, the equation $if_t + a_1|f|^2 + a_2 f_{xx} = 0$ has solutions in the form $f = A\exp[i(kx - \omega t)]$ provided that the amplitude-dependent dispersion relation $\omega = a_2 k^2 - a_1|A|^2$ is satisfied.

In more quantitative terms, the cubic Schrödinger equation arises when we look for a solution of the Korteweg-deVries equation in the form of the Fourier series

$$\eta(x,t)= \sum_{n=1}^{\infty} \epsilon^n \sum_{l=-\infty}^{\infty} \eta^{(n,l)}(x,t)e^{il(kx-\omega t)} \qquad (6.2.1)$$

The expansion parameter $\epsilon$ is assumed to be much less than unity and the coefficients $\eta^{(n,l)}$ are assumed to vary slowly over a wavelength $\sim k^{-1}$ and also during one cycle $\sim \omega^{-1}$. We will follow the analysis of Shimizu and Ichikawa (1972) and show that $\eta^{(1,1)}$ satisfies a cubic Schrödinger equation. For $\eta(x,t)$ to be real we must have $\eta^{(1,-1)} = \eta^{(1,1)*}$. It will be found that we need retain only terms with $n \leqslant 3$ and $l \leqslant 2$ to obtain this result.

We expect $\eta^{(1,1)}$ to represent a disturbance that is much longer than a wavelength moving at some velocity $V$. This may be accomplished by assuming that $\eta^{(1,1)}$ is a function of a variable $\xi = \epsilon(x - Vt)$. In addition, slow temporal variations in this profile may be introduced by allowing a higher-order time-dependence $\tau = \epsilon^2 t$. We thus have

$$\eta(x,t) = \sum_{n=1}^{\infty} \epsilon^n \bar{\eta}^{(n)}(\xi, \tau, x, t) \tag{6.2.2}$$

where

$$\bar{\eta}^{(n)} = \sum_{l=-\infty}^{\infty} \eta^{(n,l)}(\xi, \tau) e^{il(kx - \omega t)} \tag{6.2.3}$$

Time and space derivatives in the Korteweg–deVries equation must now be replaced by

$$\frac{\partial}{\partial t} \to \frac{\partial}{\partial t} - \epsilon V \frac{\partial}{\partial \xi} + \epsilon^2 \frac{\partial}{\partial \tau}$$

$$\frac{\partial}{\partial x} \to \frac{\partial}{\partial x} + \epsilon \frac{\partial}{\partial \xi} \tag{6.2.4}$$

where the $x,t$ derivatives refer to the phase and the $\xi,\tau$ derivatives to the amplitude.

Substitution of (6.2.2) into the Korteweg–deVries equation (6.1.19) and separation of terms according to the first three powers of $\epsilon$ yields

$$\bar{\eta}_t^{(1)} + c_0 \bar{\eta}_x^{(1)} + \gamma \bar{\eta}_{xxx}^{(1)} = 0 \tag{6.2.5a}$$

$$\bar{\eta}_t^{(2)} + c_0 \bar{\eta}_x^{(2)} + \gamma \bar{\eta}_{xxx}^{(2)} + (c_0 - V)\bar{\eta}_\xi^{(1)} + \beta \bar{\eta}^{(1)} \bar{\eta}_x^{(1)} + 3\gamma \bar{\eta}_{xx\xi}^{(1)} = 0 \tag{6.2.5b}$$

$$\bar{\eta}_t^{(3)} + c_0 \bar{\eta}_x^{(3)} + \gamma \bar{\eta}_{xxx}^{(3)} + (c_0 - V)\bar{\eta}_\xi^{(2)} + 3\gamma \bar{\eta}_{xx\xi}^{(2)} + \beta \{ [\bar{\eta}^{(1)}\bar{\eta}^{(2)}]_x$$

$$+ \bar{\eta}^{(1)}\bar{\eta}_x^{(1)} \} + \bar{\eta}_\tau^{(1)} + 3\gamma \bar{\eta}_{x\xi\xi}^{(1)} = 0 \tag{6.2.5c}$$

where $\beta = 3c_0/2h$ and $\gamma = c_0 h^2/6$.

We now separate terms according to powers of $l$. For terms of first order in $\epsilon$ we find that

$$-il(\omega - kc_0 + \gamma l^2 k^3)\eta^{(1,l)} = 0 \tag{6.2.6}$$

For $l=0$ no information is obtained. For $l=\pm 1$ we obtain the dispersion relation for the linear theory, namely

$$\omega = kc_0 - \gamma k^3 \qquad (6.2.7)$$

With $\omega$ determined by (6.2.7), we see from (6.2.6) that $\eta^{(1,l)}=0$ for $|l|>1$.
For arbitrary $l$, the second-order equation (6.2.5b) yields

$$-i l \eta^{(2,l)}(\omega - kc_0 + \gamma k^3 l) + \eta_\xi^{(1,l)}(c_0 - V - 3k^2 l^2 \gamma)$$
$$+ ik\beta \sum_p p\eta^{(1,l-p)}\eta^{(1,p)}=0 \qquad (6.2.8)$$

Since $\eta^{(1,l)}=0$ for $|l|>1$, only a few terms will remain in the summation.
For $l=0$ we find that

$$(V-c_0)\eta_\xi^{(1,0)}=0 \qquad (6.2.9)$$

As long as $V\neq c_0$ we see that $\eta^{(1,0)}$ has no $\xi$ dependence. Since a term $\eta^{(1,0)}(\tau)$ will not be of interest for propagation problems, we set $\eta^{(1,0)}=0$.
For $l=1$ we find that

$$\eta_\xi^{(1,1)}(V-c_0+3\gamma k^2)=0 \qquad (6.2.10)$$

or

$$V = c_0 - 3\gamma k^2 \qquad (6.2.11)$$

From $l=2$ we obtain

$$\eta^{(2,2)} = \frac{\frac{3}{2}\left[\eta^{(1,1)}\right]^2}{k^2 h^3} \qquad (6.2.12)$$

In third order we have

$$i l \eta^{(3,l)}(-\omega + kc_0 - \gamma k^3 l^2) + \eta_\xi^{(2,l)}(-V + c_0 - 3\gamma k^2 l^2)$$

$$+ \eta_\tau + \beta\left[ikl\sum_p \eta^{(1,l-p)}\eta^{(2,p)} + \sum \eta^{(1,l-p)}\eta_\xi^{(1,p)}\right]$$

$$+ 3i\gamma kl\eta_{\xi\xi}^{(1,l)}=0 \qquad (6.2.13)$$

For $l=0$ we obtain

$$\frac{\partial}{\partial \xi}\left[(c_0 - V)\eta^{(2,0)} + \beta|\eta^{(1,1)}|^2\right]=0 \qquad (6.2.14)$$

which may be integrated to yield

$$(c_0 - V)\left[\eta^{(2,0)}(\xi,\tau) - \eta^{(2,0)}(-\infty,\tau)\right] + \beta\left[|\eta^{(1,1)}|^2 - F(\tau)\right] = 0 \quad (6.2.15)$$

where $F(\tau) = |\eta^{(1,1)}(-\infty,\tau)|^2$. Assuming that $\eta^{(2,0)}(-\infty,\tau) = 0$ and using (6.2.11), we find that

$$\eta^{(2,0)} = -\frac{3}{k^2 h^3}\left[|\eta^{(1,1)}|^2 - F\right] \quad (6.2.16)$$

For $l = 1$ we obtain

$$\eta_\tau^{(1,1)} + ik\beta\left[\eta^{(1,1)}\eta^{(2,0)} + \eta^{(1,-1)}\eta^{(2,2)}\right] + 3ik\gamma\eta_{\xi\xi}^{(1,1)} = 0 \quad (6.2.17)$$

which becomes

$$\eta_\tau^{(1,1)} - \left(\frac{9ic_0}{4kh^3}\right)\left[|\eta^{(1,1)}|^2 - 2F\right]\eta^{(1,1)} + 3ik\gamma\eta_{\xi\xi}^{(1,1)} = 0 \quad (6.2.18)$$

For $F = 0$, which is the case for localized solutions, this is the cubic Schrödinger equation. Extensive experimental confirmation of multisoliton interaction of surface waves (on deep water) in conformity with the cubic Schrödinger equation has been obtained by Lake et al. (1977).

Although we are primarily concerned with soliton solutions, the steady-state waveform is also worth brief consideration, since it readily yields the amplitude-dependent dispersion relation mentioned above. Consider a solution of (6.2.18) in the form $\eta^{(1,1)} = A \exp[i(K\xi - \Omega\tau)]$, so that $F = |A|^2$. The assumed form of $\eta^{(1,1)}$ is a solution of (6.2.18) provided that the dispersion relation

$$\Omega = -3\gamma kK^2 + \frac{9c_0|A|^2}{4kh^2} \quad (6.2.19)$$

is satisfied. Returning to (6.2.1), we see that to first order in $\epsilon$ the displacement can be written

$$\eta = 2\epsilon A \cos(\kappa x - \tilde{\omega}t) \quad (6.2.20)$$

where

$$\kappa = k + \epsilon K$$

$$\tilde{\omega} = \omega + \epsilon KV + \epsilon^2\Omega \quad (6.2.21)$$

It is now seen to be convenient to set $A = \frac{1}{2}$. Using (6.2.7) and (6.2.11), we now find that

$$\frac{\tilde{\omega}}{c_0 \kappa} = 1 - \frac{1}{6}(h\kappa)^2 + \frac{9}{16}\frac{\epsilon^2}{(\kappa h)^2} \tag{6.2.22}$$

which contains the first amplitude correction to the dispersion relation for water waves (Whitham, 1974, p. 473).

## 6.3  ION PLASMA WAVES AND THE KORTEWEG–deVRIES EQUATION

We shall consider a description of ion acoustic waves, that is, fluctuations in the ion density of a two-component plasma. In simple physical terms, the ions carry out low-frequency density and velocity fluctuations near the ion plasma frequency, while the electrons preserve an approximate local charge neutrality by following the ion motion. Higher-frequency fluctuations near the electron plasma frequency are ignored. The basic equations are as follows (Spitzer, 1956):

Conservation of ion density:

$$\frac{\partial \bar{n}_i}{\partial t} + \nabla \cdot (\bar{n}_i \mathbf{v}_i) = 0 \tag{6.3.1}$$

where $\bar{n}_i$ and $\mathbf{v}_i$ are the number density and average velocity of the ions, respectively.

Conservation of ion and electron momentum:

$$\bar{n}_i m_i \frac{d\mathbf{v}_i}{dt} = Ze\bar{n}_i \bar{\mathbf{E}} - \nabla p_i$$

$$\bar{n}_e m_e \frac{d\mathbf{v}_e}{dt} = -e\bar{n}_e \bar{\mathbf{E}} - \nabla p_e \tag{6.3.2}$$

in which $m_i$ and $m_e$ are the ion and electron masses, respectively; $-e$ the electron charge; and $Ze$ the ion charge. The electron pressure is assumed to satisfy the equation of state $p_e = n_e k T_e$. We shall consider the case of low ion temperature ($T_i \ll T_e$) and set $T_i = 0$. The corresponding equation of state for the ions thus implies that $p_i = 0$. The electric field $\bar{E}$ that is set up by lack of complete localized charge neutrality satisfies

$$\nabla \cdot \bar{\mathbf{E}} = 4\pi e (Z\bar{n}_i - \bar{n}_e) \tag{6.3.3}$$

The plasma current is defined by

$$\mathbf{j} = e(Z\bar{n}_i \mathbf{v}_i - \bar{n}_e \mathbf{v}_e) \tag{6.3.4}$$

We shall consider only the case in which there is no plasma current, hence $j = 0$. The time derivative of (6.3.4) with the help of (6.3.2) yields

$$\bar{E} + \frac{1}{e\bar{n}_e} \nabla p_e = 0 \tag{6.3.5}$$

when we ignore the term $(m_e/m_i)Z^2\bar{n}_i/\bar{n}_e$, which is much less than unity $(m_e/m_i \approx 10^{-3})$.

We now introduce the dimensionless variables $x' = x/L$, $t' = \omega_0 t$, $\bar{E} = kT_e E/eL$, $n = \bar{n}_i/n_0$, $n_e = \bar{n}_e/Zn_0$, and $u = v/\omega_0 L$, where $L = (kT_e/4\pi Ze^2 n_0)^{1/2}$ is the electron Debye shielding length and $\omega_0 = (4\pi n_0 Z^2 e^2/m_i)^{1/2}$ is the ion plasma frequency (Spitzer, 1956). For the case of spatial variation in the $x$ direction only, (6.3.1) to (6.3.3) reduce to

$$\frac{\partial n}{\partial t'} + \frac{\partial}{\partial x'}(nu) = 0 \tag{6.3.6a}$$

$$\frac{\partial u}{\partial t'} + u\frac{\partial u}{\partial x'} = E \tag{6.3.6b}$$

$$E + \frac{1}{n_e}\frac{\partial n_e}{\partial x'} = 0 \tag{6.3.6c}$$

$$\frac{\partial E}{\partial x'} = n - n_e \tag{6.3.6d}$$

Before proceeding with a perturbation analysis of (6.3.6), we first note that in the limit of complete charge neutrality (i.e., $n = n_e$), (6.3.6b) and (6.3.6c) yield

$$\frac{\partial u}{\partial t'} + u\frac{\partial u}{\partial x'} + \frac{1}{n}\frac{\partial n}{\partial x'} = 0 \tag{6.3.7}$$

It is now this equation and (6.3.6a) that provide a complete description of the ion motion. Solutions of this pair of equations are known to exhibit pulse steepening and ultimately shock formation (Tidman and Krall, 1971, Sec. 6.1). Also, from the linearized form of (6.3.6a) and (6.3.7) we find that Fourier modes of the density and velocity in the form

$$n = 1 + \delta n e^{i(k'x' - \omega't')}$$
$$u = \delta u e^{i(k'x' - \omega't')} \tag{6.3.8}$$

can only exist when $\omega' = k'$. Thus the ratio $\omega'/k'$, which is the phase velocity of the waves in (6.3.8), is independent of frequency in the linear limit. Hence, in this limit the waves are nondispersive.

We now return to the full set of four equations given in (6.3.6). As shown below, the linear version of these equations now leads to dispersive propagation. Also, instead of shock formation, a symmetric steady density profile (the single-soliton solution of the Korteweg–deVries equation) can arise. Hence we again find that soliton formation may be considered as a balance between pulse sharpening due to nonlinear effects and pulse spreading due to dispersion.

When the Fourier modes given in (6.3.8) as well as

$$n_e = 1 + \delta n_e e^{i(k'x' - \omega't')}$$

$$E = \delta E e^{i(k'x' - \omega't')}$$

(6.3.9)

are substituted into the linearized version of (6.3.6), we obtain the dispersion relation

$$V^2 \equiv \left(\frac{\omega'}{k'}\right)^2 = 1 - \omega'^2 \tag{6.3.10}$$

For $\omega' \ll 1$ we may use the approximate relation $k' = \omega' + \frac{1}{2}\omega'^3$. The phase term in each of the exponentials then becomes $i[\omega'(x' - t') + \frac{1}{2}\omega'^3 x']$. As in Section 1.4, it is convenient to introduce new independent variables that incorporate this dispersive effect. This topic will be considered in more detail in Section 6.5. Setting $\xi = \omega'(x' - t')$ and $\eta = \omega'^3 x'$, we find that (6.3.6a) to (6.3.6d) become

$$-\frac{\partial \eta}{\partial \xi} + \frac{\partial}{\partial \xi}(\eta u) + \omega'\frac{\partial}{\partial \eta}(nu) = 0 \tag{6.3.11a}$$

$$-\omega'\frac{\partial u}{\partial \xi} + \omega'u\left[\frac{\partial u}{\partial \xi} + \omega'^2\frac{\partial u}{\partial \eta}\right] = E \tag{6.3.11b}$$

$$\omega'\frac{\partial \eta_e}{\partial \xi} + \omega'^3\frac{\partial n_e}{\partial \eta} = -n_e E \tag{6.3.11c}$$

$$\omega'\frac{\partial E}{\partial \xi} + \omega'^3\frac{\partial E}{\partial \eta} = \eta - n_e \tag{6.3.11d}$$

We now introduce the perturbation expansions

$$\eta_e = 1 + \omega'^2 \eta_e^{(1)} + \omega'^4 n_e^{(2)} + \cdots$$

$$n = 1 + \omega'^2 n^{(1)} + \omega'^4 n^{(2)} + \cdots$$

$$u = \omega'^2 u^{(1)} + \omega'^4 u^{(2)} + \cdots$$

$$E = \omega'^3 E^{(1)} + \omega'^5 E^{(2)} + \cdots$$

(6.3.12)

That the leading terms in the expansions of $E$ and $u$ should have the given $\omega'$ dependence is evident from (6.3.11c) and (6.3.11b), respectively. We now determine the two lowest-order contributions in (6.3.11). The lowest-order equations are

$$-\frac{\partial n^{(1)}}{\partial \xi} + \frac{\partial u^{(1)}}{\partial \xi} = 0$$

$$-\frac{\partial u^{(1)}}{\partial \xi} = E^{(1)}$$

$$\frac{\partial \eta_e^{(1)}}{\partial \xi} = -E^{(1)}$$

$$\eta^{(1)} - \eta_e^{(1)} = 0$$

(6.3.13)

from which it follows that $n_i^{(1)} = n_e^{(1)} = u_i^{(1)}$. Since we shall only consider the steady-state solution below, an arbitrary function of $\eta$ has been neglected. In the next order we find

$$-\frac{\partial n^{(2)}}{\partial \xi} + \frac{\partial}{\partial \xi} \left[ u^{(2)} + u^{(1)} n^{(1)} \right] + \frac{\partial u^{(1)}}{\partial \eta} = 0$$

$$-\frac{\partial u^{(2)}}{\partial \xi} + u^{(1)} \frac{\partial u^{(1)}}{\partial \xi} = E^{(2)}$$

$$\frac{\partial n^{(2)}}{\partial \xi} + \frac{\partial n_e^{(1)}}{\partial \eta} = -n_e^{(1)} E^{(1)} - E^{(2)}$$

(6.3.14)

$$\frac{\partial E^{(1)}}{\partial \xi} = n^{(2)} - n_e^{(2)}$$

These equations are now readily combined to yield an equation for $u^{(1)}$. We arrive at the Korteweg–deVries equation in the form

$$\frac{\partial u^{(1)}}{\partial \eta} + u^{(1)} \frac{\partial u^{(1)}}{\partial \xi} + \frac{1}{2} \frac{\partial^3 u^{(1)}}{\partial \xi^3} = 0$$

(6.3.15)

Recalling that $u^{(1)} = n^{(1)}$ and returning to dimensional variables, we obtain the steady-state solution

$$\bar{n}_i - n_0 = \omega'^2 n_0 n^{(1)}$$

$$= \delta n \, \mathrm{sech}^2\left( \frac{x - vt}{D} \right)$$

(6.3.16)

where $\delta n = \frac{3}{2} c \omega'^2 n_0$ and $V = \omega_0 L / (1 - \delta n / 3 n_0) \approx \omega_0 L (1 + \delta n / 3 n_0)$. We also have $(L/D)^2 = (\delta n / 6 n_0)(1 - \delta n / 3 n_0)^2 \approx \delta n / 6 n_0$. From these relations we again obtain the standard results of soliton theory that the width of the soliton ($\approx D$) decreases with increasing $\delta n$ while $V$, the velocity of the soliton, increases with increasing $\delta n$. Experimental verification of these results as well as pulse breakup of the sort exhibited in Figure 1.2 has been observed (Ikezi, 1973).

## 6.4 CLASSICAL MODEL OF ONE-DIMENSIONAL DISLOCATION THEORY—SINE–GORDON EQUATION

Some of the phenomena displayed by a layer of atoms in a solid are exhibited by the classical model of a row of particles attached to each other by springs. The effect of the adjacent layers of atoms (the substrate) is represented by a periodic potential. The simplest equilibrium situation is one in which there is a particle in each trough of the potential. However, the balance of potential energy between that in the springs and that in the periodically varying potential may lead to other equilibrium configurations. In particular, each successive particle in a sequence may be progressively farther from the corresponding trough of the potential. Over some distance the chain may be expanded (or contracted) so that the number of particles is one less (or more) than the number of troughs. Such a configuration is referred to as a negative (positive) dislocation.

In order to describe the dynamics of the chain within a completely classical framework, we shall consider a row of particles, each of mass $m$, attached to one another by linear springs with equal spring constants $k$. The particles slide over a sinusoidally corrugated surface so that the periodic potential is provided by gravity. A summary of the statics and dynamics of such a mechanical system has been presented in a series of papers by Frank and van der Merwe (1949). Their approach is followed here. A discussion of the relation of the model to the macroscopic properties of a solid may be found in Indenbom (1958) and Seeger and Schiller (1966).

The geometry of the system is shown in Figure 6.2. The various quantities involved in setting up the governing equations are listed below:

$a$ = period of substrate
$b$ = spacing between particles of chain when connecting spring is unstrained
$X_n$ = location of $n$th particle
$\bar{x}_n$ = location of $n$th trough = $a(n + \frac{3}{4})$
$x_n = X_n - \bar{x}_n$ = displacement of $n$th particle from $n$th trough

$$V_g = \text{gravitational potential energy of particles}$$

$$= mgh\Sigma\left(1 + sin\ \frac{2\pi X_n}{a}\right) = mgh\Sigma\left(1 - cos\ \frac{2\pi x_n}{a}\right)$$

$V_k$ = potential energy due to strain of spring

$$= \tfrac{1}{2}k\Sigma\left[ X_{n+1} + 1 - (X_n + b)\right]^2 = \tfrac{1}{2}k\Sigma(X_{n+1} + 1 - X_n - a + b)^2$$

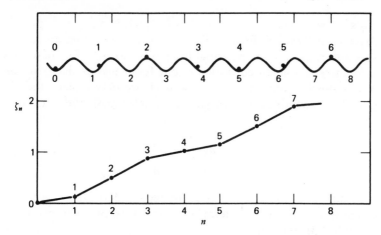

**Figure 6.2** Dislocation on chain of masses located over sinusoidal troughs.

In the definitions of $V_g$ and $V_k$, the sum is over all particles in the chain.

Introducing the abbreviations $\zeta_n = x_n/a$, $P_0 = a/(b-a)$, $W = 2mgh$, and $l_0^2 = ka^2/2W$, we find that the total potential energy $V = V_k + V_g$ may be written

$$V = Wl_0^2 \sum (\zeta_{n+1} - \zeta_n - P_0^{-1})^2 + \frac{1}{2} W \sum (1 - \cos 2\pi\zeta_n) \qquad (6.4.1)$$

Various possible equilibrium configurations of the masses are obtained by solving the system of equations that arise when we set $\partial V/\partial \zeta_p = 0$. Noting that each $\zeta_n$ occurs in two adjacent terms in the first sum in (6.4.1), we obtain

$$\frac{\partial V}{\partial \zeta_p} = -2l_0^2 W(\zeta_{p+1} - 2\zeta_p + \zeta_{p-1}) + \pi W \sin 2\pi\zeta_p = 0 \qquad (6.4.2)$$

Admissible equilibrium configurations are therefore the solutions of the equations

$$\zeta_{p+1} - 2\zeta_p + \zeta_{p-1} = \frac{\pi}{2l_0^2} \sin 2\pi\zeta_p \qquad (6.4.3)$$

It is instructive to consider a simple solution of these equations in some detail. If we choose two adjacent values of $\zeta_p$ as well as $l_0$, the other displacements are readily obtained from (6.4.3). As an example, we set $\zeta_0 = 0$ and choose $\zeta_1$ and $l_0$ so that $\zeta_2 = \frac{1}{2}$. Setting $p = 1$ in (6.4.3), we find that

$$\frac{1}{l_0^2} = \frac{1 - 4\zeta_1}{\pi \sin 2\pi\zeta_1}, \qquad \zeta_1 < \frac{1}{4} \qquad (6.4.4)$$

If we now set $\zeta_1 = \frac{1}{8}$, then (6.4.3) yields $\zeta_3 = \frac{7}{8}$, $\zeta_4 = 1$, $\zeta_5 = \frac{9}{8}$, $\zeta_6 = \frac{3}{2}$, $\zeta_7 = \frac{15}{8}$, and so on. A graph of these results as well as a diagram indicating the location of the masses with respect to the potential troughs is shown in Figure 6.2. Over a distance $5a$ there is a misfit consisting of one fewer masses than troughs.

If $\zeta$ varies slowly from one site to the next, we may replace the discrete labeling $\zeta_n$ by a continuously varying parameter $\zeta(n)$ and introduce the expansion

$$\zeta_{n+1} - \zeta_n = \frac{d\zeta}{dn} + \frac{1}{2!}\frac{d^2\zeta}{dn^2} + \cdots$$

We may terminate this expansion after the second derivative since this term provides the first nonvanishing term when the difference expression is substituted into (6.4.3). We obtain

$$\frac{d^2\zeta}{dn^2} = \frac{\pi}{2l_0^2}\sin 2\pi\zeta \tag{6.4.5}$$

In integrating this equation it is helpful to understand the significance of $d\zeta/dn$. Since

$$\frac{d\zeta}{dn} \approx \frac{x_{n+1} - x_n}{a} = \frac{X_{n+1} - X_n - a}{a}$$

$$= \frac{X_{n+1} - X_n}{a} - 1 \tag{6.4.6}$$

we see that $d\zeta/dn$ is the amount by which the distance between successive particles, measured in units of $a$, exceeds unity. If we set $d\zeta/dn = \epsilon$ where $\zeta = 0$, that is, where successive masses are in the troughs of the substrate potential, the first integral of (6.4.5) is

$$\left(\frac{d\zeta}{dn}\right)^2 = \epsilon + \frac{1-\cos 2\pi\zeta}{2l_0^2} \tag{6.4.7}$$

To consider a simple example, assume that $\epsilon = 0$, so that far from the dislocation the particles are in the troughs. Then

$$\frac{d\zeta}{dn} = \pm\frac{\sin\pi\zeta}{l_0} \tag{6.4.8}$$

An expansion of the chain (decrease in density) corresponds to the positive sign and we shall consider only this case. Integrating and arbitrarily setting $\zeta = \frac{1}{2}$ at $n = 0$, we obtain $\pi n/l_0 = \ln(\tan\pi\zeta/2)$, or

$$\zeta = \frac{2}{\pi}\tan^{-1}(e^{n\pi/l_0}) \tag{6.4.9}$$

Therefore, in a region of magnitude $l_0$, the displacement increases from 0 to 1. There is thus one less particle than there are troughs in the substrate potential over this region. We see that $l_0$ is a measure of the size of the dislocation, measured in units of $a$.

The potential energy associated with this dislocation is readily obtained from (6.4.1) by replacing the sum over discrete $\zeta_n$ by an integral over $\zeta(n)$ and using (6.4.9). Assuming the chain to be of infinite extent, we then write

$$V = l_0^2 W \int_{-\infty}^{\infty} dn \left( \frac{d\zeta}{dn} \right)^2 + \frac{1}{2} W \int_{-\infty}^{\infty} dn \, (1 - \cos 2\pi\zeta)$$

$$= 2W \int_{-\infty}^{\infty} dn \, \sin^2 \pi\zeta$$

$$= \frac{4 l_0 W}{\pi} \tag{6.4.10}$$

We now turn to a consideration of the *motion* of the particles on the chain. The kinetic energy of the chain is

$$T = \frac{1}{2} m \sum (\dot{x}_n)^2 = \frac{1}{2} ma^2 \sum (\dot{\zeta}_n)^2 \tag{6.4.11}$$

The Lagrangian $L = T - V$ then yields the equation of motion for the $p$th mass in the form

$$\frac{d}{dt} \left( \frac{\partial L}{\partial \dot{\zeta}_p} \right) - \frac{\partial L}{\partial \zeta_p} = 0 \tag{6.4.12}$$

Using (6.4.1) for the potential energy, we find that

$$ma^2\ddot{\zeta}_p + 2l_0^2 W(-\zeta_{p+1} + 2\zeta_p - \zeta_{p-1}) + \pi W \sin 2\pi\zeta_p = 0 \tag{6.4.13}$$

In the continuum limit introduced above, this equation becomes

$$\frac{a^2}{c^2} \frac{\partial^2 \zeta}{\partial t^2} - \frac{\partial^2 \zeta}{\partial n^2} + \frac{\pi}{2l_0^2} \sin 2\pi\zeta = 0 \tag{6.4.14}$$

where $c^2 = ka^2/m$. If we now introduce the new dependent variable $\sigma = 2\pi\zeta$ as well as the independent variables $u = \pi n/l_0$ and $v = \pi ct/l_0 a$, we obtain the equation

$$\frac{\partial^2 \sigma}{\partial u^2} - \frac{\partial^2 \sigma}{\partial v^2} = \sin \sigma \tag{6.4.15}$$

which is the sine–Gordon equation considered in Section 5.2. We now see that the single-soliton solution

$$\sigma = 4\tan^{-1} e^{m(u-\beta v)}, \qquad m = (1-\beta^2)^{-1/2} \tag{6.4.16}$$

given in (5.2.11) describes the uniform translation of the dislocation given in (6.4.9).

The total energy $E$ associated with a disturbance moving on the chain is $E = T + V$. In the continuum limit this becomes

$$E = \frac{l_0^2 W^2 a}{c^2} \int dn \left( \frac{\partial \zeta}{\partial t} \right)^2 + l_0^2 W \int dn \left( \frac{\partial \zeta}{\partial n} \right)^2 + \tfrac{1}{2} W \int dn \, (1 - \cos 2\pi\zeta)$$

$$= \frac{l_0 W}{2\pi} \int du \, \mathcal{K}(u, v) \tag{6.4.17}$$

where

$$\mathcal{K} = \tfrac{1}{2}\left[ (\sigma_u)^2 + (\sigma_v)^2 \right] + (1 - \cos\sigma) \tag{6.4.18}$$

We have already seen in Section 5.2 that $\int du \, \mathcal{K} = 8m = 8(1-\beta^2)^{-1/2}$ when $\sigma$ is given by the single-soliton solution (6.4.16). Hence the energy associated with the motion of a single soliton is $E_s = E_0(1-\beta^2)^{-1/2}$, where $E_0$ is the rest energy $V$ given in (6.4.10). The multisoliton results discussed in Section 5.2 may be taken over in their entirety.

## 6.5 CHOICE OF EXPANSION PARAMETERS

We have a encountered a number of examples in which an evolution equation has been obtained after the introduction of a change of variables of the form

$$\xi = \epsilon^\alpha (x - ct)$$

$$\tau = \epsilon^\beta t \tag{6.5.1}$$

where $\epsilon \ll 1$ and certain values have been chosen for the constants $\alpha$ and $\beta$. We now consider a simple example that should help to clarify the relation between the constants $\alpha$ and $\beta$ and the resulting evolution equation. We follow the approach presented in Su and Gardner (1969).

As a simple example that can lead to either of two nonlinear evolution equations in appropriate limits, we consider the ion–acoustic wave calculations of Section 6.3 when viscosity is included. If we include a (bulk) viscosity term in the description of the ion motion, the pressure term in the first of (6.3.3) will be replaced by $p - \frac{4}{3}\mu \, \partial v_i / \partial x$ (Whitham, 1974, p. 150). We again assume vanishing ion temperature, so that $p_i = 0$. Introducing the dimensionless variables used in Section 6.3, we find that the conservation of ion momentum (6.3.6b) becomes

$$n\left(\frac{\partial u}{\partial t'} + u\frac{\partial u}{\partial x'}\right) = nE + \delta\frac{\partial^2 u}{\partial x'^2} \tag{6.5.2}$$

where $\delta = 4\mu/8m_i n_0\omega_0 L^2$. To understand the significance of the parameter $\delta$, we may write $m_i n_0$ as a mass density $\rho_i$ and replace $\omega_0 L$ by a characteristic velocity $U_0$, we then have

$$\delta = \frac{4}{3}\frac{\mu}{\rho V_0 L} \tag{6.5.3}$$

In fluid dynamics the reciprocal of this ratio of parameters is known as a Reynolds number (Batchelor, 1967).

In order to recast the four equations (6.3.6) into the general form considered by Su and Gardner, we first introduce the potential function $E = -\varphi_x$. Taking $\varphi = 0$ at equilibrium where $n = 1$, we may integrate (6.3.6c) to yield $n_e = e^\varphi$. We may now write (6.3.6d) in the form

$$F(n,\varphi,\varphi_{xx}) \equiv n - e^\varphi - \varphi_{xx} = 0 \tag{6.5.4}$$

We may also combine the conservation of mass and momentum relations (6.3.6a) and (6.5.2) into

$$(nu)_t + (nu^2 + P)_x = 0 \tag{6.5.5}$$

where

$$P \equiv e^\varphi - \tfrac{1}{2}(\varphi_x)^2 - \delta u_x \tag{6.5.6}$$

The four equations to be analyzed are (6.3.6a) and (6.5.5) together with (6.5.4) and (6.5.6).

We now introduce the perturbation expansions

$$n = 1 + \epsilon n^{(1)} + \epsilon^2 n^{(2)} + \cdots$$
$$u = \epsilon u^{(1)} + \epsilon^2 u^{(2)} + \cdots \tag{6.5.7}$$
$$\varphi = \epsilon \varphi^{(1)} + \epsilon^2 \varphi^{(2)} + \cdots$$

and first consider only the linearized equations. The conservation of mass and momentum are then given respectively by

$$n_t^{(1)} + u_x^{(1)} = 0$$
$$u_t^{(1)} + \varphi_x^{(1)} = 0 \tag{6.5.8}$$

Although the exact linearized relation between $n^{(1)}$ and $\varphi^{(1)}$ given by (6.3.6) is $n^{(1)} = \varphi^{(1)} - \varphi^{(1)}_{xx}$, we shall develop a perturbation procedure in which nonlinearity, dissipation, and dispersion [the latter two effects being represented by derivative terms such as $\varphi^{(1)}_{xx}$] are treated as perturbations. We thus assume a linear relation between $n^{(1)}$ and $\varphi^{(1)}$ that is simply $n^{(1)} = \varphi^{(1)}$. We shall see subsequently that after the coordinate transformation (6.5.1) is introduced, the term $\varphi^{(1)}_{xx}$ will in fact enter as a higher-order correction. We now find that $n^{(1)}$ satisfies the linear wave equation $n^{(1)}_{xx} - n^{(1)}_{tt} = 0$, and similarly for $u^{(1)}$ and $\varphi^{(1)}$.

Let us fix attention on one of the traveling-wave solutions for $n^{(1)}$ and set

$$n^{(1)} = n^{(1)}(x - ct) \tag{6.5.9}$$

A perturbation procedure is now developed to account for the modification that takes place in this solution because of nonlinearities, dissipation, and dispersion. We assume that the solution varies slowly in a coordinate system that moves with the wave velocity $c$ (which equals unity in this example) and assume that $n^{(1)}$ is a function of the variables $\xi$ and $\tau$ introduced in (6.5.1) with $c = 1$. The derivatives are then related by

$$\frac{\partial}{\partial x} \to \epsilon^\alpha \frac{\partial}{\partial \xi}, \qquad \frac{\partial}{\partial t} \to \epsilon^\alpha \frac{\partial}{\partial \tau} - \epsilon^\beta \frac{\partial}{\partial \xi} \tag{6.5.10}$$

and the conservation of mass and momentum equations, respectively, are transformed to

$$\epsilon^{\beta - \alpha} n_\tau - n_\xi + (nu)_\xi = 0 \tag{6.5.11a}$$

$$\epsilon^{\beta - \alpha}(nu)_\tau - (nu)_\xi + (nu^2 + P)_\xi = 0 \tag{6.5.11b}$$

The perturbation expansion will thus proceed in integral powers of $\epsilon$ if we set $\beta - \alpha = 1$. The four equations to be analyzed may now be written

$$\epsilon n_\tau - n_\xi + nu_\xi + un_\xi = 0$$

$$\epsilon nu_\tau - nu_\xi + nuu_\xi + P_\xi = 0$$

$$P = e^\varphi - \tfrac{1}{2}\epsilon^{2\alpha}(\varphi_\xi)^2 - \delta\epsilon^\alpha u_\xi \tag{6.5.12}$$

$$F = n - e^\varphi + \epsilon^{2\alpha}\varphi_{\xi\xi} = 0$$

The first-order equations are

$$-n_\xi^{(1)} + u_\xi^{(1)} = 0$$

$$-u_\xi^{(1)} + P_\xi^{(1)} = 0 \tag{6.5.13}$$

$$P^{(1)} = \varphi^{(1)}, \qquad n^{(1)} - \varphi^{(1)} = 0$$

After integration of the first two relations, we have $n^{(1)} = u^{(1)} = \varphi^{(1)} = p^{(1)}$. Here an arbitrary function of $t$ has been discarded since we will only be interested in pulse solutions that vanish as $\xi \to \pm \infty$. When the equality of the first-order quantities is used in the equations for the second-order quantities, we obtain

$$n_\tau^{(1)} - n_\xi^{(2)} + 2n^{(1)}n_\xi^{(1)} + u_\xi^{(2)} = 0$$

$$n_\tau^{(1)} - u_\xi^{(2)} + P_\xi^{(2)} = 0$$

$$P^{(2)} = \varphi^{(2)} + \tfrac{1}{2}(n^{(1)})^2 - \delta \epsilon^{\alpha-1} n_\xi^{(1)} \tag{6.5.14}$$

$$F^{(2)} = n^{(2)} - \varphi^{(2)} - \tfrac{1}{2}(n^{(1)})^2 + \epsilon^{2\alpha-1} n_{\xi\xi}^{(1)} = 0$$

We can now easily combine these four equations to obtain an equation for $n^{(1)}$ alone. The result is

$$n_\tau^{(1)} + n^{(1)}n_\xi^{(1)} - \tfrac{1}{2}\delta \epsilon^{\alpha-1} n_{\xi\xi}^{(1)} + \tfrac{1}{2}\epsilon^{2\alpha-1} n_{\xi\xi\xi}^{(1)} = 0 \tag{6.5.15}$$

Therefore, if viscous effects are included, so that $\delta \neq 0$, the lowest-order contribution is obtained by setting $\alpha = 1$. The third derivative term is then of higher order in $\epsilon$ and is neglected. The density then satisfies

$$n_\tau^{(1)} + n^{(1)}n_\xi^{(1)} - \tfrac{1}{2}\delta n_{\xi\xi}^{(1)} = 0 \tag{6.5.16}$$

which is known as the Burgers equation (Whitham, 1974, p. 96). Although this equation does not possess soliton solutions, it is a nonlinear evolution equation with an interesting history and will be considered again briefly in Chapter 8.

On the other hand, if viscous effects are neglected, so that $\delta = 0$, the lowest-order correction term in (6.5.15) is obtained by setting $\alpha = \tfrac{1}{2}$. We then obtain

$$n_\tau^{(1)} + n^{(1)}n_\xi^{(1)} + \tfrac{1}{2} n_{\xi\xi\xi}^{(1)} = 0 \tag{6.5.17}$$

which is the Korteweg–deVries equation. In both cases we see that the unperturbed solution is recovered when nonlinear and dissipative or dispersive effects are neglected, for then we obtain $n_\tau^{(1)} = 0$, so that $n^{(1)} = n^{(1)}(\xi)$ as in (6.5.9).

A more general matrix version of the procedure employed here may be found in Taniuti and Wei (1968) and Leibovich and Seebass (1974, p. 203).

CHAPTER 7

# Applications II

In Chapter 6 we considered some of the physical contexts in which the more common soliton equations have arisen. These equations could then be solved by using the methods described in earlier chapters. The linear equations required for solution by these techniques would have to be introduced in an ad hoc manner, however, since they seem to play no apparent role in the physical consideration that leads to the various nonlinear evolution equations. We now consider two examples in which physical or geometric considerations yield not only the soliton equations but also the attendant linear equations.

The first example is concerned with the motion along a vortex filament of a helical loop in the shape of the filament. The shape of the filament may be likened to that of a twisted curve in space and thus can be described by the Serret–Frenet equations. These latter equations are the set of linear differential equations upon which the theory of twisted curves is based (Eisenhart, 1909; Struik, 1961). The Serret–Frenet equations, in turn, can be recast into a form identical with that of the equations for the two-component inverse method. In this example the evolution equation is found to be the cubic Schrödinger equation. The relation of certain other soliton equations to helical curves is also briefly considered.

The second example is concerned with the propagation of a pulse of light energy through an atomic medium. The Schrödinger equation for the idealized atoms in the medium is shown to lead again to the linear equations of the two-component inverse method. In the limit that the motion of the atoms is neglected, this example reduces to the sine–Gordon equation. When atomic motion is retained, however, we obtain an example of soliton propagation that, in some respects, is somewhat more general than that usually encountered.

## A Soliton on a Vortex Filament

We consider the motion of an isolated vortex filament in three dimensions. When a vortex is not perfectly straight, it feels the fluid motion due to its own vorticity and is thus swept about in some selfconsistent manner. We shall begin with a brief discussion of the approximations used in specializing the general hydrodynamic equations to treat this problem. The approach follows that of Batchelor (1967, p. 509).

## 7.1 SELF-INDUCTION OF A VORTEX

The vorticity vector $\omega$ is defined by $\omega = \nabla \times \mathbf{u}$, where $\mathbf{u}$ is the vector specifying the velocity field of the fluid. For a incompressible fluid, as noted in Section 6.1, the conservation of mass reduces to $\nabla \cdot \mathbf{u} = 0$. The equations governing $\mathbf{u}$ are thus of the same form as those governing the magnetic field vector $\mathbf{B}$ in magnetostatics and may be solved by the same techniques. If we introduce a vector potential by the definition $\mathbf{u} = \nabla \times \mathbf{A}$, then with $\nabla \cdot \mathbf{A} = 0$ (as will be seen below to be the case), we have

$$\omega = \nabla \times \nabla \times \mathbf{A} = -\nabla^2 \mathbf{A} \tag{7.1.1}$$

In an infinite volume of fluid, where surface integrals do not contribute, the solution of this equation is (Jackson, 1962, p. 141)

$$\mathbf{A} = \frac{1}{4\pi} \int d^3\mathbf{r}' \frac{\omega(\mathbf{r}')}{|\mathbf{r} - \mathbf{r}'|} \tag{7.1.2}$$

and therefore

$$\mathbf{u} = \nabla \times \mathbf{A} = -\frac{1}{4\pi} \int d^3\mathbf{r}' \frac{(\mathbf{r} - \mathbf{r}') \times \omega(\mathbf{r}')}{|\mathbf{r} - \mathbf{r}'|^3} \tag{7.1.3}$$

To see that this solution implies $\nabla \cdot A = 0$, we first note that $\nabla \cdot \omega = 0$, since $\omega = \nabla \times \mathbf{u}$ and the divergence of a curl is identically zero. We thus have

$$\nabla \cdot \mathbf{A} = \frac{1}{4\pi} \int d^3\mathbf{r}' \omega \cdot \nabla \left( \frac{1}{|\mathbf{r} - \mathbf{r}'|} \right) = -\frac{1}{4\pi} \int d^3\mathbf{r}' \omega \cdot \nabla' \left( \frac{1}{|\mathbf{r} - \mathbf{r}'|} \right)$$

$$= -\frac{1}{4\pi} \int d^3\mathbf{r}' \nabla' \cdot \left( \frac{\omega}{|\mathbf{r} - \mathbf{r}'|} \right) = -\frac{1}{4\pi} \int d^2\mathbf{S}' \cdot \frac{\omega}{|\mathbf{r} - \mathbf{r}'|} = 0 \tag{7.1.4}$$

since there are no surfaces present.

If the vorticity is assumed to be confined to a tube of cross-sectional area $dA$, and to be of constant magnitude across the tube, then (7.1.3) becomes

$$\mathbf{u} = -\frac{1}{4\pi} \int dA \frac{\delta l (\mathbf{r} - \mathbf{r}') \times \omega}{|\mathbf{r} - \mathbf{r}'|^3}$$

$$= -\frac{1}{4\pi} \int dA |\omega| \int \frac{(\mathbf{r} - \mathbf{r}') \times \delta l}{|\mathbf{r} - \mathbf{r}'|^3} \tag{7.1.5}$$

where $\delta l$ is a length increment along the direction of the filament.

The circulation of a vortex is defined as

$$\Gamma \equiv \int_c \mathbf{v} \cdot ds = \int_S \nabla \times \mathbf{v} \cdot \mathbf{n} \, dS = \int_S |\omega| \, dA \tag{7.1.6}$$

where $ds$ is an increment of length along a curve that encircles the vortex tube and $S$ is the area of the tube enclosed by $c$. Thus (7.1.5) becomes

$$\mathbf{u}(\mathbf{r}) = -\frac{\Gamma}{4\pi} \int \frac{(\mathbf{r}-\mathbf{r}')\times\delta\mathbf{l}}{|\mathbf{r}-\mathbf{r}'|^3} \qquad (7.1.7)$$

where the integral is taken along the vortex line.

To determine the self-induced motion of the vortex, we must evaluate the integral in (7.1.7) at field points near the vortex. We shall find that the result diverges logarithmically if we attempt to treat the vortex filament as a curve of vanishing thickness. We also encounter a logarithmic divergence if the integral along the filament is taken to infinity.

We consider a curved portion of the filament and approximate it by a circular arc. The vectors $\mathbf{r}$ and $\mathbf{r}'$ are as shown in Figure 7.1. The three unit vectors are the tangent vector $\mathbf{t}$, the normal vector $\mathbf{n}$, and the binormal vector $\mathbf{b}$ to the arc. The field vector $\mathbf{r}$ will be assumed small but nonvanishing. If the filament is in a 1–2 plane as shown in the figure, then

$$\mathbf{r}' = x_1'\mathbf{t} + x_2'\mathbf{n}$$
$$= R\mathbf{t}\sin\theta + R(1-\cos\theta)\mathbf{n} \qquad (7.1.8)$$

where $R$ is the radius of curvature of the filament at 0. For small $\theta$ this is approximately

$$\mathbf{r}' \simeq l\mathbf{t} + \tfrac{1}{2}\kappa l^2\mathbf{n} \qquad (7.1.9)$$

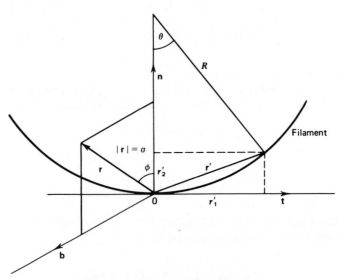

**Figure 7.1**  Section of vortex filament.

where $l = R\theta$ and $\kappa = 1/R$, the curvature of the filament. Then, to first order in $\kappa l$,

$$d\mathbf{r}' = \delta \mathbf{l} = (\mathbf{t} + \kappa l \mathbf{n})\, dl \qquad (7.1.10)$$

We then find that

$$(\mathbf{r} - \mathbf{r}') \times \delta \mathbf{l} = \left[ x_3 \mathbf{n} + \kappa l x_3 \mathbf{t} - \left( x_2 + \tfrac{1}{2}\kappa l^2 \right)\mathbf{b} \right] dl \qquad (7.1.11)$$

Near the filament we may ignore $x_2$ and $x_3$ and obtain

$$(\mathbf{r} - \mathbf{r}') \times \delta \mathbf{l} \simeq -\tfrac{1}{2}\kappa l^2 \mathbf{b}\, dl \qquad (7.1.12)$$

We may also write

$$|\mathbf{r} - \mathbf{r}'|^2 \simeq r^2 + l^2 \qquad (7.1.13)$$

where $r^2 = x_2^2 + x_3^2$. A term $\kappa x_2$ has been neglected in comparison with unity and we have assumed that $\kappa l < < 1$. The integral in (7.1.7) is now

$$\mathbf{u} = -\frac{\Gamma}{4\pi} \int_{-L}^{L} \frac{-\tfrac{1}{2}\kappa l^2 \mathbf{b}\, dl}{(r^2 + l^2)^{3/2}}$$

$$= \frac{\Gamma}{4\pi} \kappa \mathbf{b} \int_{0}^{L} \frac{l^2\, dl}{(r^2 + l^2)^{3/2}} \qquad (7.1.14)$$

As noted earlier, this result diverges logarithmically if we attempt to set $r = 0$. We also obtain a logarithmic divergence at the upper limit $L$. For $L/r >> 1$ we find that the velocity near the vortex is

$$\mathbf{u} \simeq \frac{\Gamma}{4\pi} \kappa \mathbf{b}\left( \ln \frac{2L}{r} - 1 \right) \qquad (7.1.15)$$

The velocity at the vortex is thus in the direction of the binormal $\mathbf{b}$ and proportional to the curvature of the filament. This velocity is now identified with that of the vortex filament. If $s$ represents length measured along the filament and $\mathbf{X}(s,t)$ represents a position vector to the filament, then

$$\mathbf{u} = \frac{\partial \mathbf{X}}{\partial t} \qquad (7.1.16)$$

Introducing a new time coordinate

$$t' = t \frac{\Gamma}{4\pi}\left( \ln \frac{2L}{\sigma} - 1 \right)$$

we have $\partial \mathbf{X}/\partial t' = \kappa \mathbf{b}$. Henceforth we shall measure time in this manner but omit the prime. Thus the equation governing the motion of the filament is

$$\frac{\partial \mathbf{X}}{\partial t} = \kappa \mathbf{b} \qquad (7.1.17)$$

We now derive the equation that determines the shape of the filament when it is governed by this equation of motion.

## 7.2  MOTION OF THE FILAMENT

Displacement of the position vector along the filament yields a unit vector tangent to the curve; thus

$$\frac{\partial \mathbf{X}}{\partial s} = \mathbf{t} \qquad (7.2.1)$$

As shown in elementary presentations of curve theory (Eisenhart, 1909, p.16; Struik, 1961, p. 29; Hildebrand, 1976, p. 278), the tangent vector and the vectors $\mathbf{b}$ and $\mathbf{n}(\mathbf{n} \times \mathbf{b} = \mathbf{t})$ satisfy the Serret–Frenet equations

$$\begin{aligned}
\mathbf{t}_s &= \kappa \mathbf{n} \\
\mathbf{n}_s &= \tau \mathbf{b} - \kappa \mathbf{t} \\
\mathbf{b}_s &= -\tau \mathbf{n}
\end{aligned} \qquad (7.2.2)$$

where $\kappa$ is the curvature, $\tau$ the torsion of the curve, and the subscript $s$ indicates a partial derivative with respect to distance along the curve. When $\kappa$ and $\tau$ are specified at each point of the curve, the shape of the curve, except for its location in space, is uniquely determined.

We will now follow a presentation due to Hasimoto (1972) and show that the complex quantity $\psi(s,t) = \kappa(s,t) \exp[i \int_0^s ds' \tau(s',t)]$, which contains both $\kappa$ and $\tau$ and thus completely determines the shape of the filament, satisfies the equation

$$i\psi_t + \psi_{ss} + \tfrac{1}{2}\left[|\psi|^2 + A(t)\right]\psi = 0 \qquad (7.2.3)$$

which may be transformed into the cubic Schrödinger equation (5.3.1) by the substitution $\Psi = \psi \exp[-i/2 \int_0^t dt' A(t')]$.

It is the association of the Serret–Frenet equations (7.2.2) with the cubic Schrödinger equation that is of particular interest, since the Serret–Frenet equations are known to be equivalent to a Riccati equation (Eisenhart, 1909, p. 25). We have already seen in Section 2.12 how a Riccati equation may be transformed to linear equations of the type used to solve nonlinear evolution equations by inverse scattering methods. In the present problem these linear equations therefore arise in a natural way.

To see how the complex quantity satisfying the cubic Schrödinger equation arises, we first combine the last two of the Serret–Frenet equations (7.2.2) into the complex form

$$(\mathbf{n}+i\mathbf{b})_s + i\tau(\mathbf{n}+i\mathbf{b}) = -\kappa\mathbf{t} \tag{7.2.4}$$

On introducing

$$\mathbf{N} \equiv (\mathbf{n}+i\mathbf{b})e^{i\int_0^s ds'\tau} \tag{7.2.5}$$

and

$$\psi = \kappa e^{i\int_0^s ds'\tau} \tag{7.2.6}$$

we have

$$\mathbf{N}_s = -\psi\mathbf{t} \tag{7.2.7}$$

in place of (7.2.4), while the first of (7.2.2) may be written

$$\mathbf{t}_s = \tfrac{1}{2}(\psi^*\mathbf{N} + \psi\mathbf{N}^*) \tag{7.2.8}$$

The asterisk indicates a complex conjugate.

We now develop equations for the time derivatives of $\mathbf{N}$ and $\mathbf{t}$. From the definition of $\mathbf{t}$ we have

$$\begin{aligned} \mathbf{t}_t = (\mathbf{X}_s)_t = \mathbf{X}_{ts} &= (\kappa\mathbf{b})_s \\ &= \kappa_s\mathbf{b} - \kappa\tau\mathbf{n} \end{aligned} \tag{7.2.9}$$

Since

$$\begin{aligned} \kappa_s\mathbf{b} - \kappa\tau\mathbf{n} &= \mathrm{Re}\big[\kappa_s(\mathbf{b}+i\mathbf{n}) + i\kappa\tau(\mathbf{b}+i\mathbf{n})\big] \\ &= \mathrm{Re}\big[i(\kappa_s + i\kappa\tau)(\mathbf{n}-i\mathbf{b})\big] \\ &= \mathrm{Re}(i\psi_s\mathbf{N}^*) \\ &= \frac{i}{2}(\psi_s N^* - \psi_s^* N) \end{aligned} \tag{7.2.10}$$

the time derivative of the tangent vector is

$$\mathbf{t}_t = \frac{i}{2}(\psi_s\mathbf{N}^* - \psi_s^*\mathbf{N}) \tag{7.2.11}$$

The time derivative of $\mathbf{N}$ must in general be a linear combination of $\mathbf{N}$, $\mathbf{N}^*$, and $\mathbf{t}$. We may therefore write

$$\mathbf{N}_t = \alpha\mathbf{N} + \beta\mathbf{N}^* + \gamma\mathbf{t} \tag{7.2.12}$$

where $\alpha$, $\beta$, and $\gamma$ are to be determined. From the definition of N and the fact that **t**, **n**, and **b** are orthogonal, we readily find

$$\mathbf{N} \cdot \mathbf{N} = \mathbf{t} \cdot \mathbf{N} = 0, \qquad \mathbf{N} \cdot \mathbf{N}^* = 2 \tag{7.2.13}$$

Taking the dot product of (7.2.12) with each of N, N*, and t, we obtain

$$2\beta = \mathbf{N} \cdot \mathbf{N}_t = \tfrac{1}{2} (\mathbf{N} \cdot \mathbf{N}) \, \mathbf{t} = 0 \tag{7.2.14}$$

as well as

$$2\alpha = \mathbf{N}^* \cdot \mathbf{N}_t \tag{7.2.15}$$

We thus have

$$2(\alpha + \alpha^*) = (\mathbf{N} \cdot \mathbf{N}^*)_t = 0 \tag{7.2.16}$$

and hence find that $\alpha$ is purely imaginary. We will set $\alpha = i\mathsf{R}$, where $\mathsf{R}$ is real. Finally, since $(N \cdot t)_t = 0$,

$$\gamma = \mathbf{t} \cdot \mathbf{N}_t = -\mathbf{N} \cdot \mathbf{t}_t = -\frac{i}{2} \mathbf{N} \cdot (\psi_s \mathbf{N}^* - \psi_s^* \mathbf{N}) \tag{7.2.17}$$

or $\gamma = -i\psi_s$. Thus, we finally obtain

$$\mathbf{N}_t = i\mathsf{R}\mathbf{N} - i\psi_s \mathbf{t} \tag{7.2.18}$$

The time derivative of (7.2.7) and the $s$ derivative of (7.2.18) yield

$$\mathbf{N}_{st} = \frac{i}{2} \psi \psi_s^* \mathbf{N} - \frac{i}{2} \psi \psi_s \mathbf{N}^* - \psi_t \mathbf{t} \tag{7.2.19}$$

$$\mathbf{N}_{ts} = \mathbf{N} \left( i\mathsf{R}_s - \frac{i}{2} \psi_s \psi^* \right) - \frac{i}{2} \psi \psi_s \mathbf{N}^* - it(\mathsf{R}\psi + \psi_{ss}) \tag{7.2.20}$$

Equating components of these two expressions we find that $\mathsf{R}_s = \tfrac{1}{2}(\psi^* \psi_s + \psi \psi_s^*)$, or

$$\mathsf{R} = \tfrac{1}{2} \big[ |\psi|^2 + A(t) \big] \tag{7.2.21}$$

where $A(t)$ arises upon integration, and

$$\psi_t - i\psi_{ss} - i\mathsf{R}\psi = 0 \tag{7.2.22}$$

Combining these results, we obtain

$$i\psi_t + \psi_{ss} + \tfrac{1}{2} \big[ |\psi|^2 + A(t) \big] \psi = 0 \tag{7.2.23}$$

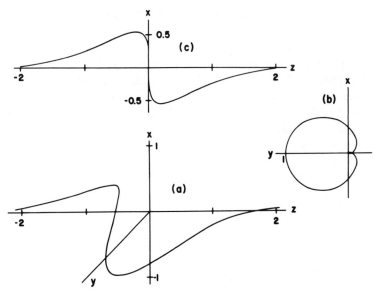

**Figure 7.2** (*a*) Shape of vortex filament corresponding to single-soliton solution of cubic Schrödinger equation, $\tau = 1$, $\kappa = 2\,\text{sech}\,s$. (*b*) Projection of curve on $xy$ plane. (*c*) Projection of curve on $xz$ plane. (With permission of American Institute of Physics.)

which may be put in the form of the cubic Schrödinger equation (5.3.1) by the transformation

$$u = \tfrac{1}{2}\psi \exp\left[ -\frac{i}{2}\int_0^t dt'\, A(t') \right] \tag{7.2.24}$$

The single-soliton solution of this equation was given in (5.3.13). Let us now consider the physical significance of this solution in the present context. A comparison of that solution with (7.2.6) and (7.2.24) shows that

$$\kappa = 4\beta\,\text{sech}\left[ 2\beta(s + 4\alpha t) \right]$$

$$\tau = -2\alpha \tag{7.2.25}$$

$$A(t) = 8(\alpha^2 - \beta^2)t$$

The solution thus represents a helical curve of constant torsion $-2\alpha$ with a curvature $\kappa$ that decreases from a maximum value $4\beta$ at the point $s = -4\alpha t$, to zero as $s \to \pm\infty$. This single loop of helical motion moves along the vortex line with a velocity $-4\alpha = 2\tau$. Also, the length of the disturbance is on the order if $(2\beta)^{-1} = 2\,/\,_{Kmax}$. An example with $\alpha = -\tfrac{1}{2}$ and $\beta = \tfrac{1}{2}$ is shown in Figure 7.2. Additional figures have been given by Hasimoto (1972).

## 7.3  SHAPE OF THE SINGLE-SOLITON FILAMENT

The shape of the curve may be considered to be completely determined when we know the coordinates of the position vector $\mathbf{x}(s,t)$ with respect to some coordinate system. Determination of these coordinates when the curvature and torsion are specified as a function of arc length $s$ (and time $t$ in the present problem) is, in principle, a standard calculation in elementary differential geometry although it requires the solution to a Riccati equation. We shall only outline the method here. A fuller discussion may be found in Struik (1961) and Eisenhart (1909).

Let us refer the curve to a rectangular coordinate system that is oriented in such a way that undisturbed portions of the filament coincide with the $z$ axis. The three unit vectors may be decomposed as

$$\begin{aligned}
\mathbf{t} &= \mathbf{i}t_1 + \mathbf{j}t_2 + \mathbf{k}t_3 \\
\mathbf{n} &= \mathbf{i}n_1 + \mathbf{j}n_2 + \mathbf{k}n_3 \\
\mathbf{b} &= \mathbf{i}b_1 + \mathbf{j}b_2 + \mathbf{k}b_3
\end{aligned} \tag{7.3.1}$$

Since $\mathbf{X}_s = \mathbf{t}$, we have

$$x = \int^s t_1\, ds, \qquad y = \int^s t_2\, ds, \qquad z = \int^s t_3\, ds \tag{7.3.2}$$

The origin of coordinates is arbitrary. The components of the tangent vector are determined by solving the Serret–Frenet equations (7.2.2) with the conditions that $t_1 \rightarrow 1$ as $s \rightarrow \pm \infty$. These vector differential equations are equivalent to the three sets of scalar equations

$$\begin{aligned}
t_{is} &= \kappa n_i \\
n_{is} &= \tau b_i - \kappa t_i \qquad i = 1, 2, 3 \\
b_{is} &= -\tau n_i
\end{aligned} \tag{7.3.3}$$

Each set of these equations has a first integral $t_i^2 + n_i^2 + b_i^2 = 1$. Reduction of the Serret–Frenet equations to a Riccati equation has been summarized by Eisenhart (1909, p. 25) and is based upon the observation that these first integrals may be written in the factored form $(t + in)(t - in) = (1 - b)(1 + b)$. (The subscript $i$ is superfluous in the following transformations and will be omitted until the final results have been obtained.) If we introduce the functions $\varphi$ and $\chi$ through the definitions

$$\varphi = \frac{t + in}{1 - b} = \frac{1 + b}{t - in} \tag{7.3.4a}$$

$$-\frac{1}{\chi} = \varphi^* = \frac{t - in}{1 - b} = \frac{1 + b}{t + in} \tag{7.3.4b}$$

then it is a simple matter to express $t$, $n$, and $b$ in terms of $\varphi$ and $\chi$. We find

$$t = \frac{1 - \varphi\chi}{\varphi - \chi}, \qquad n = i\frac{1 + \varphi\chi}{\varphi - \chi}, \qquad b = \frac{\varphi + \chi}{\varphi - \chi} \qquad (7.3.5)$$

We now determine the differential equations satisfied by $\varphi$ and $\chi$. They will be equivalent to the Serret–Frenet equations (7.3.3). Differentiation of (7.3.4a) and use of the Serret–Frenet equations yields

$$\varphi_s = \frac{t_s + in_s}{1 - b} + \frac{t + in}{(1 - b)^2} b_s$$

$$= \frac{\kappa n - i\kappa t + i\tau b}{1 - b} - \frac{\tau n}{1 - b}\varphi$$

$$= -i\kappa\varphi + \frac{\tau(ib - \varphi n)}{1 - b} \qquad (7.3.6)$$

From (7.3.4a) we also have

$$n\varphi = \tfrac{1}{2}i\left[1 + b - \varphi^2(1 - b)\right] \qquad (7.3.7)$$

When $n$ is eliminated from the last form of (7.3.6), we find that $b$ is also eliminated and we obtain the Riccati equation

$$\varphi_s + i\kappa\varphi + \tfrac{1}{2}i\tau(1 - \varphi^2) = 0 \qquad (7.3.8)$$

The same equation is found to be satisfied by $\chi$.

Some of the more common properties of the Riccati equation were summarized in Section 2.12. In particular, the general solution may always be put in the form

$$\varphi_i = \frac{c_i P + Q}{c_i R + S} \qquad (7.3.9)$$

where we have now reinstated the subscript $i$. Each $c_i$ is a constant of integration and the functions $P$, $Q$, $R$, and $S$, which are the same for each of the three components, are functions of $s$ (and $t$). It is a somewhat lengthy matter to express the components of the tangent vector $t_i$ in terms of $P$, $Q$, $R$, and $S$. The procedure is explained in detail in Struik (1961). There is, of course, a nonuniqueness in the final result that allows for various orientations of the coordinate system. As shown in texts on differential geometry, one possibility for the orientation leads to

$$t_1 = \frac{P^2 - R^2 - (Q^2 - S^2)}{2T}$$

$$t_2 = \frac{i[P^2 - R^2 + (Q^2 - S^2)]}{2T} \qquad (7.3.10)$$

$$t_3 = \frac{RS - PQ}{T}$$

in which

$$T = PS - QR \qquad (7.3.11)$$

To obtain the functions $P$, $Q$, $R$, and $S$ for the single-soliton curve, we must determine the general solution of the Riccati equation (7.3.8) when $\kappa = 4\beta \operatorname{sech}[2\beta(s + 4\alpha t)]$ and $\tau = -2\alpha$. On setting $\xi = 2\beta(s + 4\alpha t)$ and $\nu = -\alpha/\beta$, (7.3.8) reduces to the equation considered in Ex. 26 of Chapter 2. The functions $P$, $Q$, $R$, and $S$ that occur in the general solution were listed there.

When the expression for the components of the tangent vector are formed from these results by using (7.3.10), we find that the integrals in (7.3.2) become

$$x = \int_0^s t_1 \, ds = -2h \int_0^s d\xi \frac{\partial}{\partial \xi} (\operatorname{sech} \xi \sin \nu \xi) = -2h \operatorname{sech} s \sin \nu s$$

$$y - 2h = \int_0^s t_2 \, ds = 2h \int_0^s d\xi \frac{\partial}{\partial \xi} (\operatorname{sech} \xi \cos \nu \xi) = 2h \operatorname{sech} s \cos \nu s - 1 \qquad (7.3.12)$$

$$z = \int_0^s t_3 \, ds = s - h \tanh s$$

where $h = [\beta(\nu^2 + 1)]^{-1}$. This parametric representation has been used in constructing the curves shown in Figure 7.2.

## 7.4 OTHER SOLITON EQUATIONS

The previous example led to the cubic Schrödinger equation because of the particular expression for the velocity of the position vector, namely $\mathbf{X}_t = \kappa \mathbf{b}$. This particular motion led, in turn, to the relation $\mathbf{t}_t = \kappa_s \mathbf{b} - \kappa \tau \mathbf{n}$ given in (7.2.9). If a specific choice for the velocity $\mathbf{X}_t$ is not introduced and we merely use a general expression of the form

$$\mathbf{t}_t = \lambda^* \mathbf{N} + \lambda \mathbf{N}^* + \mu \mathbf{t} \qquad (7.4.1)$$

then a repetition of the calculation leading to (7.2.18) shows that we must require $\mu = 0$ and $\lambda = -\frac{1}{2}\gamma$ where $\gamma$ is the coefficient of $\mathbf{t}$ in (7.2.12). The other results obtained previously, namely $\beta = 0$ and $\alpha = i$R, still obtain. When the two mixed second derivatives of N are again equated, we find

$$\psi_t + \gamma_s - i\text{R}\psi = 0 \tag{7.4.2a}$$

$$\text{R}_s = \tfrac{1}{2}i(\gamma\psi^* - \gamma^*\psi) \tag{7.4.2b}$$

In the example that led to the cubic Schrödinger equation, we obtained $\gamma = -i\psi_s$. This relation gave an integrable expression for $\text{R}_s$. Other evolution equations may be associated with the motion of twisted curves by assuming other integrable forms for $\gamma$. We shall consider the equation that follows from the choice

$$\gamma = f\psi + ik\psi_s + a\psi_{ss} \tag{7.4.3}$$

where $f$ is a real function while $k$ and $a$ are real constants. This more general expression for $\text{R}_s$ is again readily integrated and we obtain

$$\text{R} = -\tfrac{1}{2}k|\psi|^2 + \tfrac{1}{2}ia(\psi^*\psi_s - \psi\psi_s^*) + \Gamma(t) \tag{7.4.4}$$

The evolution equation (7.4.2a) then reads

$$\psi_t + (f\psi)_s + ik\psi_{ss} + a\psi_{sss} + \tfrac{1}{2}ik|\psi|^2\psi$$
$$+ \tfrac{1}{2}a|\psi|^2\psi_s - \tfrac{1}{2}a\,\psi^2\psi_s^* - i\Gamma\psi = 0 \tag{7.4.5}$$

Let us now confine attention to evolution equations that contain only $\psi$, $|\psi|$, and derivatives of $\psi$. We must therefore eliminate the term $\psi^2\psi_s^*$ in (7.4.5). This is readily accomplished by setting $f = f(\psi, \psi^*)$ and requiring $\psi\,\partial f/\partial\psi^* = \tfrac{1}{2}a\psi^2$. Then, since $f$ is real, integration of this relation yields

$$f = \tfrac{1}{2}a|\psi|^2 + c \tag{7.4.6}$$

where $c$ is a real constant. The equation for $\psi$ now takes the form

$$\psi_t + \tfrac{3}{2}a|\psi|^2\psi_s + c\psi_s + ik\psi_{ss} + a\psi_{sss} + \tfrac{1}{2}ik|\psi|^2\psi - i\Gamma\psi = 0 \tag{7.4.7}$$

This evolution equation and the associated linear equations are brought into conformity with standard forms by introducing $u = \tfrac{1}{2}\psi e^{-i\eta s}$. The resulting equation for $u$ contains terms in $u$ and $u_s$. The coefficients of these terms are $i(c\eta - k\eta^2 - a\eta^3 - \Gamma)$ and $(c - 2k\eta - 3a\eta^2)$, respectively. The evolution equation is thus simplified if we choose $c$ and $\Gamma$ so as to make these terms vanish. This requires that

$$c = 2k\eta + 3a\eta^2 \tag{7.4.8}$$

$$\Gamma = k\eta^2 + 2a\eta^3 \tag{7.4.9}$$

The evolution equation is then of the form

$$u_t + 3A|u|^2 u_s + iB|u|^2 u + iCu_{ss} + Du_{sss} = 0 \qquad (7.4.10)$$

where $A = 2a$, $B = 2(3a\eta + k)$, $C = k + 3a\eta$, and $D = a$. The coefficients are seen to satisfy the relation $AC = BD$. The evolution equation (7.4.10) with this restriction on the coefficients, is known as the Hirota equation (Hirota, 1973). It reduces to the cubic Schrödinger equation when $A = D = 0$ (i.e., for $a = 0$), and reduces to the modified Korteweg–deVries equation if $k + 3a\eta = 0$, so that $B = C = 0$.

The linear equations associated with the Hirota equation can be obtained by following the procedure used in Section 7.3. First, the linear vector equations are converted to a Riccati equation as in obtaining (7.3.8), and then this Riccati equation is converted to the standard linear equations of two component inverse scattering theory as in Section 2.12. The linear vector equations are

$$\mathbf{N}_s = -\psi \mathbf{t}$$
$$\mathbf{t}_s = \tfrac{1}{2}(\psi^* \mathbf{N} + \psi \mathbf{N}^*) \qquad (7.4.11)$$

and

$$\mathbf{N}_t = i R \mathbf{N} + \gamma \mathbf{t}$$
$$\mathbf{t}_t = -\tfrac{1}{2}(\gamma^* N + \gamma N^*) \qquad (7.4.12)$$

Each of the three components of (7.4.11) and (7.4.12) has a first integral of the form $|N|^2 + t^2 = 1$, where $N$ is now some one of the three components of $\mathbf{N}$ and similarly for $t$. If we set $N = u + iv$ and write the first integral as $(u + iv)(u - iv) = (1 + t)(1 - t)$, the functions

$$\varphi = \frac{u + iv}{1 - t} \qquad (7.4.13)$$

and

$$-\frac{1}{\chi} = \frac{u - iv}{1 - t} = \varphi^* \qquad (7.4.14)$$

are found to satisfy the Riccati equations

$$\varphi_s - \tfrac{1}{2}\psi^* \varphi^2 - \tfrac{1}{2}\psi = 0$$
$$\varphi_t - iR\varphi + \tfrac{1}{2}\gamma^* \varphi^2 + \tfrac{1}{2}\gamma = 0 \qquad (7.4.15)$$

The procedure is the same as that used to obtain the Riccati equation (7.3.8).

When we set $\varphi = w_1/w_2$ and follow the procedure employed in Section 2.12, we obtain the linear equations

$$w_{1s} = \tfrac{1}{2}\psi w_2$$
$$w_{2s} = -\tfrac{1}{2}\psi^* w_1 \tag{7.4.16}$$

and

$$w_{1t} = \tfrac{1}{2}iR w_1 - \tfrac{1}{2}\gamma w_2$$
$$w_{2t} = \tfrac{1}{2}\gamma^* w_1 - \tfrac{1}{2}iR w_2 \tag{7.4.17}$$

The substitutions $\psi = 2ue^{i\eta s}$ used in obtaining (7.4.10) as well as $v_1 = w_1 e^{-i\eta s/2}$ and $v_2 = w_2 e^{i\eta s/2}$ yield the linear equations in standard form. In particular, with $\eta = 2\zeta$, we find that

$$v_{1s} + i\zeta v_1 = u v_2$$
$$v_{2s} - i\zeta v_2 = -u^* v_1 \tag{7.4.18}$$

and

$$v_{1t} = \tfrac{1}{2}iR v_1 - \tfrac{1}{2}\tilde{\gamma} v_2$$
$$v_{2t} = \tfrac{1}{2}\tilde{\gamma}^* v_1 - \tfrac{1}{2}iR v_2 \tag{7.4.19}$$

where

$$\tilde{\gamma} = \gamma e^{-2i\zeta s} = 4(a|u|^2 + k\zeta + 4a\zeta^2) + 2(k + 4a\zeta)u_s + 2au_{ss}$$
$$R = -2(k + 4a\zeta)|u|^2 + 2ia(u^* u_s - u u_s^*) + 4k\zeta + 16a\zeta^3 \tag{7.4.20}$$

With $a=0$ and $k=-1$, we recover the cubic Schrödinger equation (5.3.1) and the associated linear equations (5.3.5) while for $u$ real and $a=1, k=-6\zeta$, we recover the modified Korteweg–deVries equation (5.1.7) and the corresponding linear equations (5.1.9).

The sine–Gordon equation may also be obtained by the procedure outlined here. If we consider curves of constant curvature $\kappa_0$, then $\psi = \kappa_0 \exp(i\sigma)$, where $\sigma(s,t) = \int_0^s \tau(s',t)\,ds'$. If we also use $\gamma = \text{constant} = \gamma_0$ in place of (7.4.3), then (7.4.2b) yields $R_s = \gamma_0 \kappa_0 \sin\sigma$. In addition, (7.4.2a) reduces to $R = \sigma_t$ and therefore satisfies the sine–Gordon equation in the form $\sigma_{st} = \gamma_0 \kappa_0 \sin\sigma$.

It is also instructive to consider this example without the use of complex notation. Since the trihedral of unit vectors $\mathbf{n}$, $\mathbf{b}$, and $\mathbf{t}$ moves as a rigid body, the time dependence of these vectors may be written (Goldstein, 1950)

$$\mathbf{t}_t = \mathbf{\Omega} \times \mathbf{t}$$
$$\mathbf{b}_t = \mathbf{\Omega} \times \mathbf{b}$$
$$\mathbf{n}_t = \mathbf{\Omega} \times \mathbf{n} \tag{7.4.21}$$

where $\mathbf{\Omega} = \omega_1 \mathbf{t} + \omega_2 \mathbf{b} + \omega_3 \mathbf{n}$. A corresponding vector form of the Serret–Frenet equations is also possible. We may write $\mathbf{t}_s = \mathbf{d} \times \mathbf{t}$, and so on, where $\mathbf{d}$ is the Darboux vector $\mathbf{d} = \tau \mathbf{t} + \kappa \mathbf{b}$.

Equating the mixed derivatives $t_{st} = t_{ts}$, $b_{st} = b_{ts}$, and $n_{st} = n_{ts}$, we obtain

$$\frac{\partial \kappa}{\partial t} - \frac{\partial \omega_2}{\partial s} = \tau \omega_3$$

$$\frac{\partial \tau}{\partial t} - \frac{\partial \omega_1}{\partial s} = -\kappa \omega_3 \qquad (7.4.22)$$

$$\frac{\partial \omega_3}{\partial s} = \tau \omega_2 - \kappa \omega_1$$

If we consider curves of constant curvature and also set $\omega_1 = 0$, we find that the first and third of (7.4.22) may be integrated to yield $\omega_2 = \cos \sigma$, $\omega_3 = \sin \sigma$, and $\tau = \sigma_s$. The second of (7.4.22) now results in

$$\sigma_{st} = -\kappa_0 \sin \sigma \qquad (7.4.23)$$

Extensions of these geometric considerations have been carried out by Lakshmanan (1979) and Reiter (1980).

## Coherent Optical Pulse Propagation

The interaction of intense light radiation with various forms of matter is a fruitful source of problems in nonlinear wave propagation. Different models of an atomic medium have been devised to isolate the numerous types of phenomena that can occur. In particular, if the frequency of the light wave is almost exactly equal to a transition frequency between two populated energy levels of the atoms that comprise the material, then strongly resonant interactions between the light and matter can take place.

In the theoretical description of this strongly resonant situation it is frequently possible to ignore all other energy levels of the atoms and treat the interaction of light with a so-called two-level atom. The quantum theory of the two-level atom is relatively simple. An additional simplification follows from the treatment of the light wave by classical electrodynamics. This assumption is justified in examining nearly all manifestations of the interaction of intense light waves with atoms. Within the framework of these simplifications, the resonant interaction of intense light with matter can be treated quite thoroughly. As we shall see, the governing equations in this theoretical model lead to yet another context in which solitons may occur.

Extensive investigation of this strongly resonant situation led to the observation of soliton behavior both in experiments and in numerical solutions of the governing equations (McCall and Hahn, 1969). Additional results have been obtained by Patel and Slusher (1968) and Gibbs and Slusher (1972). The effect is known as self induced transparency.

A schematic outline of a typical experimental arrangement is shown in Figure 7.3. A light pulse of predetermined shape enters the region containing the two-level atoms. We shall assume that the medium is a gas, although some

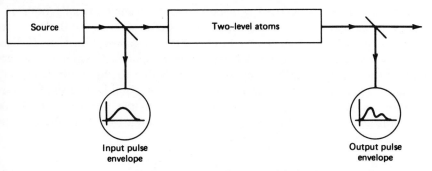

**Figure 7.3**  Schematic diagram of experiment in optical pulse propagation.

xperiments have been performed in solids. We inquire as to the shape of this ulse after it has passed through the medium and emerged at the other end. We shall find that the envelope of the light wave can exhibit soliton behavior.

The theoretical model of this process is especially interesting because it rovides not only the nonlinear evolution equations that possess soliton roperties but also the linear equations for obtaining solutions by inverse cattering techniques. There is thus no need to introduce these linear equa- ions independently, although this may be done and it leads to linear quations that are somewhat more general than those introduced through hysical considerations.

All theoretical models have their limitations, of course, and in the present nstance one of the important deficiencies in the model is the absence of level degeneracy. Although numerical solutions of the equations that arise when evel degeneracy is included have been found to exhibit interesting soliton ike phenomena (Dodd et al., 1975) these equations have not yet been solved y any analytical methods. They are, however, briefly considered in Section .13.

## .5  DESCRIPTION OF THE ELECTROMAGNETIC FIELD

The electromagnetic field vectors that characterize the light wave are overned by the Maxwell equations which we shall write in the form (see for xample, Jackson, 1962)

$$\nabla \times \mathbf{E} = -\frac{1}{c}\frac{\partial \mathbf{B}}{\partial t} \tag{7.5.1a}$$

$$\nabla \times \mathbf{B} = \frac{4\pi}{c}\mathbf{j} + \frac{1}{c}\frac{\partial \mathbf{D}}{\partial t} \tag{7.5.1b}$$

$$\nabla \cdot \mathbf{E} = 4\pi\rho \tag{7.5.1c}$$

$$\nabla \cdot \mathbf{B} = 0 \tag{7.5.1d}$$

n which $\mathbf{D} = \mathbf{E} + 4\pi\mathbf{P}$, where $\mathbf{P}$ is the polarization of the medium.

Since we shall be concerned with propagation through spatially homogeneous systems that possess charge neutrality, we may set $\rho = j = 0$. The only source for the light wave in the medium is then the polarization term $P$. The polarization is due to departures from complete spherical symmetry in the shape of the atoms while the departure itself is due to the electromagnetic field of the light wave in the medium. It is this interaction between light wave and medium that introduces the nonlinearity into the problem. The light wave satisfies a wave equation that may be obtained by taking the curl of (7.5.1a) and using (7.5.1b). We obtain

$$\nabla^2 \mathbf{E} - \frac{1}{c^2} \frac{\partial^2 \mathbf{E}}{\partial t^2} = \frac{4\pi}{c^2} \frac{\partial^2 \mathbf{P}}{\partial t^2} \qquad (7.5.2)$$

This equation is found to be nonlinear when the dependence of $\mathbf{P}$ upon $\mathbf{E}$ that is described above is used.

We now confine attention to the propagation of plane-wave fronts and a plane-polarized wave. Reasonable correspondence between theory and experiment has been achieved with this simplification and at the present time soliton propagation has only been treated satisfactorily in one dimension. Since the duration of the light pulse tends to be in the nanosecond ($10^{-9}$ sec) to picosecond ($10^{-12}$ sec) regime while an optical cycle is on the order of $10^{-15}$ sec, even the shortest pulses contain many optical cycles. It is thus appropriate to write the magnitude of the field $E(x,t)$ as a rapidly oscillating traveling wave with a more slowly varying envelope. We may also allow for a slow variation in the phase of the carrier wave and write

$$E(x,t) = \mathcal{E}(x,t) \cos\left[ kx - \omega t + \varphi(x,t) \right] \qquad (7.5.3)$$

Slow variations of $\mathcal{E}$ and $\varphi$ on the length and time scales of the carrier wave imply that

$$\frac{\partial \mathcal{E}}{\partial x} << k\mathcal{E}, \qquad \frac{\partial \mathcal{E}}{\partial t} << \omega\mathcal{E} \qquad (7.5.4)$$

and similarly for $\varphi$. We will find that the approximate equations governing $\mathcal{E}$ and $\varphi$ are simpler than the wave equation (7.5.2). Let us first consider the response of the medium to this electric field.

## 7.6 THE TWO-LEVEL ATOM

As mentioned previously, we shall be concerned with an idealized medium composed of atoms in which there are only two energy levels. The energy difference between the upper level $a$ and the lower level $b$ is assumed to be nearly equal to the frequency of the incident light $\omega_0 = 2\pi\nu_0$. The resonance condition $E_a - E_b \equiv \hbar\omega_{ab} \approx \hbar\omega_0$, where $\hbar$ is Planck's constant divided by $2\pi$, is thus assumed to be satisfied. The wave function for the atom may then be written as a time-dependent linear combination of normalized wave functions for the two levels, that is,

$$\psi(\mathbf{r}, t) = a(t)\psi_a(\mathbf{r}) + b(t)\psi_b(\mathbf{r}) \tag{7.6.1}$$

where $\int |\psi_\alpha|^2 d\tau = 1, \alpha = a, b$. Normalization of $\psi$ implies that

$$\int \psi^* \psi \, d\tau = \int |\psi|^2 d\tau = \int \left( |a\psi_a|^2 + |b\psi_b|^2 \right) d\tau = |a|^2 + |b|^2 = 1 \tag{7.6.2}$$

The population difference for $n_0$ such atoms per unit volume is

$$n = n_0 \left( \int \left( |a\psi_a|^2 - |b\psi_b|^2 \right) d\tau \right)$$
$$= n_0(|a|^2 - |b|^2) \tag{7.6.3}$$

The wave function $\psi$ satisfies the time dependent Schrödinger equation

$$H\psi = i\hbar \frac{\partial \psi}{\partial t} \tag{7.6.4}$$

where the Hamiltonian $H$ is assumed to contain a term $-(\hbar^2/2m)\nabla^2$ that describes the external translation of the atom, a term $H_0$ which is the Hamiltonian for the free atom, and the term $H_I = -\mathbf{d} \cdot \mathbf{E}$ due to the interaction of the atom with the incident light wave. Here $\mathbf{d}$ is the dipole moment operator for the atom $\mathbf{d} = -e\mathbf{r}$, where $e$ is the electron charge and $\mathbf{r}$ is the internal atomic coordinate. We shall assume that $\mathbf{d}$ is parallel to $\mathbf{E}$ so that $H_I = -dE$. The polarization of an atom is then

$$p = \int \psi^*(d)\psi \, d\tau \tag{7.6.5}$$

If we assume that the atom has no permanent dipole moment so that $\int d\tau \psi_\alpha r \psi_\alpha = 0, \alpha = a, b$, then (7.6.5) reduces to

$$p = p_0(a^*b + b^*a) \tag{7.6.6}$$

where

$$p_0 \equiv -e \int d\tau \psi_a^* r \psi_b = -e \int d\tau \psi_b^* r \psi_a.$$

The Schrödinger equation may be transformed into a pair of simultaneous equations for the amplitudes $a(t)$ and $b(t)$ by using the orthogonality properties customarily exhibited by wave functions, that is, $\int \psi_a^* \psi_b \, d\tau = \delta_{ab}$. Multiplying (7.6.4) by $\psi_a$, integrating over all space, and then repeating the procedure with $\psi_b$, we obtain

$$a_t + i\omega_a a = -iVb \tag{7.6.7a}$$
$$b_t + i\omega_b b = -iVa \tag{7.6.7b}$$

where $\omega_a = E_a/\hbar$, $\omega_b = E_b/\hbar$, and $V = -p_0 E(x,t)/\hbar$. The field $E(x,t)$ is th field at the location of the atom. Since atomic velocities are much less tha the light velocity, this distinction is negligible except in rapidly varying phas terms. A full discussion of this topic would be somewhat lengthy (McCall an Hahn, 1969; Icsevgi and Lamb, 1969) and will not be taken up in its entiret here. For an atom moving with a velocity $v$, we introduce a Galilea coordinate $x_g$ and a laboratory coordinate $x_l$ related by $x_l = x_g + vt$. The fiel at the atom is then

$$E(x_l, t) = \mathcal{E}(x,t) \cos\left[ k(x_g + vt) - \omega t + \varphi(x,t) \right] \tag{7.6.8}$$

The distinction between coordinates is ignored in the slowly varying ampli tude $\mathcal{E}$ and phase $\varphi$ but in the rapidly varying part of the phase it introduce the Doppler shift $\Delta\omega = kv$. If we set

$$a = iv_1 \exp\left[ -i\omega_a\left(t - \frac{x_g}{c}\right) + \frac{i}{2} t\Delta\omega \right]$$

$$b = v_2 \exp\left[ -i\omega_b\left(t - \frac{x_g}{c}\right) - \frac{i}{2} t\Delta\omega \right] \tag{7.6.9}$$

in (7.6.7) and assume that the fields generated at the second harmonic $2\omega_0$ ar of small enough intensity that they may be neglected, we obtain

$$v_{1t} + \frac{1}{2} i\Delta\omega v_1 = \frac{p_0 \mathcal{E}}{2\hbar} e^{i\varphi} v_2$$

$$v_{2t} - \frac{1}{2} i\Delta\omega v_2 = -\frac{p_0 \mathcal{E}}{2\hbar} e^{-i\varphi} v_1 \tag{7.6.10}$$

We thus find that $v_1$ and $v_2$ satisfy equations of the type used in th two-component inverse method *except that the space and time variables ar interchanged.*

The normalization condition (7.6.2) becomes $|v_1|^2 + |v_2|^2 = 1$ and the nor malized population density difference $\mathfrak{N} = n/n_0$ is

$$\mathfrak{N} = |v_1|^2 - |v_2|^2 \tag{7.6.11}$$

From (7.6.6) the polarization for an atom with a Doppler shift $\Delta\omega$ become

$$p = p_0\left\{ -iv_1^* v_2 e^{i\omega(t - x_g/c) - i\Delta\omega t} + \text{c.c.} \right\}$$

Writing $\Phi = kx_l - \omega t + \varphi$, we find that

$$p = p_0(\mathcal{C} \cos\Phi + \mathcal{S} \sin\Phi) \tag{7.6.12}$$

where the polarization envelope functions $\mathcal{C}$ and $\mathcal{S}$ are given by

$$\mathcal{C}(\Delta\omega, x, t) = i\left(v_1 v_2^* e^{-i\varphi} - v_1^* v_2 e^{i\varphi}\right)$$
$$\mathcal{S}(\Delta\omega, x, t) = -\left(v_1 v_2^* e^{-i\varphi} + v_1^* v_2 e^{i\varphi}\right)$$

(7.6.13)

The polarization induced in the atom by the light wave is seen to have a component with amplitude $p_0\mathcal{C}$ that is in phase with the electric field of the light wave and a component with amplitude $p_0\mathcal{S}$ that is 90 degrees out of phase with the light wave. The polarization components $\mathcal{C}$ and $\mathcal{S}$ are comparable in duration with the slowly varying electric field envelope $\mathcal{E}(x, t)$.

## 7.7  EQUATIONS OF THE MODEL

Since the atoms have a distribution of velocities, there is a corresponding distribution in frequency shifts $\Delta\omega$. If $n_0$ is the number of atoms per unit volume and the distribution in frequency shifts (velocity) is given by $g(\Delta\omega)$, then the total polarization per unit volume is

$$P(x, t) = n_0 \int_{-\infty}^{\infty} d(\Delta\omega) g(\Delta\omega) p(\Delta\omega, x, t)$$

$$\equiv n_0 \langle p(\Delta\omega, x, t) \rangle \qquad (7.7.1)$$

The angular brackets $\langle \rangle$ are used to indicate the average of a quantity over the distribution $g(\Delta\omega)$. The function $g(\Delta\omega)$ is frequently assumed to have a Gaussian form. An example will be considered in (7.11.9). The source term in the wave equation (7.5.2) is the second derivative of this expression for the polarization. Neglecting the slower time dependence in the envelope functions $\mathcal{C}$ and $\mathcal{S}$, we have

$$\frac{\partial^2 P}{\partial t^2} \approx -\omega_0^2 P = -n_0 \omega_0^2 \int d(\Delta\omega) g(\Delta\omega)(\mathcal{C} \cos\Phi + \mathcal{S} \sin\Phi) \qquad (7.7.2)$$

When the form of the field $E(x, t)$ given in (7.5.3) is inserted in (7.5.2) and the second derivatives of the slowly varying quantities $\mathcal{E}$ and $\varphi$ as well as product terms such as $\mathcal{E}_t \varphi_t$ are neglected, we find that the coefficients of the terms $\cos\Phi$ and $\sin\Phi$ satisfy, respectively, the equations

$$\frac{\partial \mathcal{E}}{\partial t} + c\frac{\partial \mathcal{E}}{\partial x} = 2\pi n_0 \omega_0 p_0 \langle \mathcal{S}(\Delta\omega, x, t) \rangle \qquad (7.7.3a)$$

$$\mathcal{E}\left(\frac{\partial \varphi}{\partial t} + c\frac{\partial \varphi}{\partial x}\right) = 2\pi n_0 \omega_0 p_0 \langle \mathcal{C}(\Delta\omega, x, t) \rangle \qquad (7.7.3b)$$

These equations may be combined into the complex form

$$\left(\frac{\partial \mathcal{E}}{\partial t} + c\frac{\partial \mathcal{E}}{\partial x}\right)(\mathcal{E}e^{i\varphi}) = 2\pi n_0 \omega_0 p_0 \langle \mathcal{S} + i\mathcal{C} \rangle e^{i\varphi} \qquad (7.7.4)$$

Equations 7.7.3 or 7.7.4 must be solved simultaneously with the equations for $v_1$ and $v_2$ given in (7.6.10). These latter equations yield $\mathcal{S}$ and $\mathcal{C}$ by use of the definitions in (7.6.13). A more convenient procedure for some purposes is to obtain equations for the time dependence of $\mathcal{S}$ and $\mathcal{C}$ directly. Differentiating (7.6.13) and (7.6.11), we find that

$$\mathcal{S}_t = \frac{p_0 \mathcal{E}}{\hbar} \mathcal{N} + (\Delta\omega + \varphi_t)\mathcal{C} \qquad (7.7.5a)$$

$$\mathcal{N}_t = -\frac{p_0 \mathcal{E}}{\hbar} \mathcal{S} \qquad (7.7.5b)$$

$$\mathcal{C}_t = -(\Delta\omega + \varphi_t)\mathcal{S} \qquad (7.7.5c)$$

which are frequently referred to as the Bloch equations (Bloch, 1946).* The similarity of the Bloch equations to a single component of the Serret–Frenet equations (7.2.2) should be noted.

A special case of (7.7.3) and (7.7.5) is of some interest. If $g(\Delta\omega)$ is an even function and $\mathcal{C}(\Delta\omega, x, t)$ is an odd function of $\Delta\omega$, the source of $\varphi$ in (7.7.3b) is zero. We may thus consider propagation problems in which there is no phase variation. The simplest example of this case is obtained when atomic motion is also neglected. The governing equations are then reducible to the sine–Gordon equation. We now consider this case.

## 7.8 STATIONARY ATOMS—SINE–GORDON LIMIT

Translational motion of the atoms is eliminated by introducing a Dirac delta function and writing $g(\Delta\omega) = \delta(\Delta\omega)$. From (7.7.5c) we see that if $\mathcal{C}(0, x, 0)$ is set equal to zero, it will remain zero. Introducing the new dependent variable

$$\tilde{\mathcal{E}} = \frac{p_0 \mathcal{E}}{\hbar} \qquad (7.8.1)$$

---

*In obtaining (7.7.5), we have neglected the effects of collisions that produce a decay in the polarization and population terms. When collisions are included, the left-hand side of the equations in (7.7.5) are replaced by $\mathcal{S}_t + \mathcal{S}/T_2$, $\mathcal{N}_t + (\mathcal{N} - 1)/T_1$, and $\mathcal{C}_t + \mathcal{C}/T_2$, respectively, where $T_1$ is the collision time for collisional deexcitation and $T_2$ (usually $T_2 << T_1$) is the collision time for dephasing collisions (Allen and Eberly, 1975).

which has dimensions of frequency, we find that the governing equations are

$$\frac{\partial \tilde{\mathcal{E}}}{\partial t} + c\frac{\partial \tilde{\mathcal{E}}}{\partial x} = \Omega^2 \mathcal{S}(0,x,t) \tag{7.8.2a}$$

$$\frac{\partial \mathcal{S}}{\partial t} = \tilde{\mathcal{E}}\,\mathcal{N} \tag{7.8.2b}$$

$$\frac{\partial \mathcal{N}}{\partial t} = -\tilde{\mathcal{E}}\,\mathcal{S} \tag{7.8.2c}$$

where

$$\Omega^2 = \frac{2\pi n_0 \omega_0 p_0^2}{\hbar} \tag{7.8.3}$$

and the functions $\mathcal{N}$ and $\mathcal{S}$ are evaluated at zero frequency shift. The latter two equations in (7.8.2) have the integral $\mathcal{N}^2 + \mathcal{S}^2 = 1$, so that we may write

$$\mathcal{S} = \pm \sin\sigma$$
$$\mathcal{N} = \pm \cos\sigma \tag{7.8.4}$$

as well as

$$\tilde{\mathcal{E}} = \frac{\partial\sigma}{\partial t} \tag{7.8.5}$$

Writing $\sigma(x,t) = \int_{-\infty}^{t} dt'\,\tilde{\mathcal{E}}(x,t')$, so that $\sigma(x,-\infty)$ vanishes, we see that $\mathcal{N}(0,x,-\infty) = \pm 1$. This limit refers to the state of the system before the pulse arrives. The upper sign corresponds to $\mathcal{N} = (n_a - n_b)/n_0 = +1$, which refers to the case in which the atomic population is initially inverted. The medium then acts as an amplifier for the light wave passing through it. The lower sign in (7.8.5) applies when the atomic population is in the lower level. As we shall see below, it is only in this latter instance that stable soliton propagation can take place.

Introducing the coordinate transformation

$$\xi = \frac{\Omega x}{c}$$
$$\tau = \Omega\left(t - \frac{x}{c}\right) \tag{7.8.6}$$

which implies that

$$\frac{\partial}{\partial t} = \Omega\frac{\partial}{\partial\tau}, \qquad \frac{\partial}{\partial x} = \frac{\Omega}{c}\left(\frac{\partial}{\partial\xi} - \frac{\partial}{\partial\tau}\right) \tag{7.8.7}$$

we find that the first of (7.8.2) becomes the sine–Gordon equation in the form

$$\frac{\partial^2 \sigma}{\partial \xi \partial \tau} = \pm \sin \sigma \qquad (7.8.8)$$

where the upper sign applies to an amplifier and the lower sign to an attenuator if $\sigma(x, -\infty) = 0$. As we shall see below, the multisoliton solutions of the sine–Gordon equation obtained in Chapter 5 have a simple physical interpretation in the present context.

Let us first consider the stability of the single-soliton solution $\sigma = 4\tan^{-1}[\exp(a\tau \pm \xi/a)]$, where the upper (lower) sign is used with the upper (lower) sign in the sine–Gordon equation (7.8.8). In terms of a dimensionless electric field variable

$$u = \frac{\tilde{\mathscr{E}}}{\Omega} \qquad (7.8.9)$$

this latter equation may be written

$$\frac{\partial u}{\partial \xi} = \pm \sin \sigma \qquad (7.8.10)$$

Near the trailing edge of the pulse (i.e., as $t \to +\infty$), $\sigma$ becomes $\int_{-\infty}^{\infty} d\tau\, u(\xi, \tau)$, which may be thought of as an area under the pulse profile. For the single-soliton solution this integral has the value $2\pi$. The single-soliton solution is thus frequently referred to as a $2\pi$ pulse. If the area under the pulse is slightly larger than $2\pi$ so the $\sigma = 2\pi + \epsilon$, then

$$\frac{\partial u}{\partial \xi} = \pm \sin(2\pi + \epsilon) = \pm \epsilon \qquad (7.8.11)$$

For the upper sign, which corresponds to the amplifier, $\partial u/\partial \xi > 0$, which leads to still further increase in the area, while for the attenuator (lower sign) $\partial u/\partial \xi < 0$ and $\sigma$ decreases toward $2\pi$. Similarly, for $\sigma = 2\pi - \epsilon$, the area increases to $2\pi$ for the attenuator but diverges away from $2\pi$ for the amplifier. It thus follows that the $2\pi$ pulse only has a stable area in an attenuator, so that the soliton can only be expected to occur in the attenuator. Further insight into this area stability will be obtained when we consider moving atoms in the next section.

We can readily understand the occurrence of soliton phenomena when a coherent light wave interacts with two-level atoms. If the atoms are initially in the lower atomic state, the leading edge of the light pulse inverts the atomic population and is thus attenuated, while the trailing edge of the pulse returns the population to the initial state by means of stimulated emission. This process is realizable only if both inversions in population take place before the atom can experience a collision that causes the atomic wave functions to lose their phase coherence with the incident light wave. The pulse must also have sufficient intensity to invert the population; otherwise, it is merely attenuated in the manner expected for small-signal propagation in an attenuator. When conditions for the process are met, it is found (as shown at the end of Section 7.10) that a steady-state pulse profile is established, and that this pulse envelope then propagates without attenuation at a velocity that may be a few orders of magnitude less than the phase velocity of light in the medium.

The steady-state electric field profile that effects this inversion is that of the simple-soliton solution of the sine–Gordon equation for the attenuator, namely, $\sigma = 4 \tan^{-1}[\exp(a\tau - \xi/a)]$. The field profile is

$$\tilde{\mathscr{E}} = \frac{p_0}{\hbar} \mathscr{E} = \Omega \frac{\partial \sigma}{\partial \tau} = \frac{2}{\tau_p} \operatorname{sech}\left[ \frac{t - x/v}{\tau_p} \right] \tag{7.8.12}$$

where $\tau_p = (a\Omega)^{-1}$ and

$$\frac{1}{v} = \frac{1}{c}\left(1 + \frac{1}{a^2}\right) \tag{7.8.13}$$

If the pulse area is much larger than $2\pi$, the population can be inverted repeatedly. Each region of the pulse that encounters the medium in the ground state will be attenuated while each region that interacts with an inverted population will be amplified. As this process of alternating amplification and decay continues, the pulse can be expected to split up. Decomposition into two solitons leads to field profiles similar to those in Figure 1.2. The total area under the pulse remains $4\pi$. The analytical expression for this solution is obtained from the two soliton solution given in (5.2.37).

Oscillatory solutions of the form shown in Figure 5.2 also play a role in the present problem. Pulses with negative values of $\mathscr{E}$ are admissible since a negative value of $\mathscr{E}$ can be interpreted as a phase change of $\pi$ in the carrier wave of the pulse. This interpretation is consistent with (7.7.3b), which shows that when $\mathscr{E}$ goes to zero, the phase may be expected to change quite rapidly. The total area under oscillatory pulses such as that given by (5.2.19) is $\int \tilde{\mathscr{E}} \, dt = (p_0/\hbar)[\sigma(x, \infty) - \sigma(x, -\infty)] = 0$ and they may thus be considered to be zero area pulses.

Finally, we mention the amplifier. As noted above, the soliton is unstable in this case. An altogether different type of solution has been found that describes some important features of the amplification of coherent optical pulses. The appropriate solution is the similarity solution described in Section 5.2. The pulse profile was shown in Figure 5.5b. From Figure 5.5a and c we see that $\sigma(\infty) - \sigma(-\infty) = \pi$ and the solution is frequently referred to as a $\pi$ pulse. This solution will be considered again in Section 7.11.

## 7.9   MOVING ATOMS AND THE AREA THEOREM

The results of Section 7.8 display the essential nonlinearities of coherent optical pulse propagation in their simplest form. Since the governing equation is the sine–Gordon equation, the problem may be extensively analyzed by using the known solutions to that equation that were described in Section 5.2. We may also introduce the associated linear equations as was done in Section 5.2 and proceed with inverse scattering considerations.

Extensions of the theory to include moving atoms and a variable phase $\varphi(x,t)$ provide further refinements that lead to more satisfactory comparisons between theory and experiment. Also, with the appearance of the detuning parameter $\Delta\omega$ in (7.6.9) and (7.7.5), the equations that describe the physics of the problem are themselves already in a form that can be analyzed by inverse scattering methods. This situation is analogous to that described in Section 7.4 for helical curves. In this regard the similarity between the Bloch equations (7.7.5) and a scalar component of the Serret–Frenet equations (7.3.3) should be noted.

In this section we include the detuning parameter $\Delta\omega$ but continue to ignore the slow phase variations $\varphi(x,t)$. This latter term will be considered in the next section when a more general inverse scattering procedure is introduced.

Before considering pulse profiles, we shall obtain a very simple relation that describes how the total area under the pulse $\theta(x)$ defined by

$$\theta(x) = \frac{p_0}{\hbar} \int_{-\infty}^{\infty} dt\, \mathcal{E}(x,t) \tag{7.9.1}$$

is affected by propagation in both attenuators and amplifiers. The result, known as the area theorem, provides an overall view of many features of the subject (McCall and Hahn, 1969).

The area theorem may be obtained from (7.7.3a). Integrating over all time, we obtain

$$\frac{d\theta}{dx} = \frac{2\pi n_0 \omega_0 p_0^2}{\hbar c} \int_{-\infty}^{\infty} dt \int_{-\infty}^{\infty} d(\Delta\omega) g(\Delta\omega) \mathcal{S}(\Delta\omega, x, t) \tag{7.9.2}$$

The integrand is rewritten by first noting that (7.7.5a) and (7.7.5c) can be combined to yield

$$\frac{\partial}{\partial t} \frac{d\theta}{dx} e^{i\Delta\omega t} (\mathcal{S} + i\mathcal{C}) = e^{i\Delta\omega t} \mathcal{N}(\Delta\omega, x, t) \tilde{\mathcal{E}}(x,t) \tag{7.9.3}$$

Since phase variations are being ignored, as noted previously, the term $\varphi(x,t)$ has again been set equal to zero. Integration of (7.9.3) and separation of real and imaginary terms yields

$$\mathcal{S}(\Delta\omega, x, t) = \int_{-\infty}^{t} dt'\, \tilde{\mathcal{E}}(x,t') \mathcal{N}(\Delta\omega, x, t') \cos \Delta\omega(t - t') \tag{7.9.4}$$

When this form for $\mathfrak{S}$ is introduced into (7.9.2) and the order of the $t$ and $t'$ integrations is interchanged, which requires a change in the limits of integration as indicated, we have

$$\frac{d\theta}{dx} = \frac{2\pi n_0 \omega_0 p_0^2}{\hbar c} \int_{-\infty}^{\infty} dt' \, u(x,t') \int_{-\infty}^{\infty} d(\Delta\omega) g(\Delta\omega) \tilde{\mathfrak{E}}(x,t')$$

$$\times \mathfrak{N}(\Delta\omega, x, t) \int_{t'}^{\infty} dt \cos \Delta\omega(t - t') \qquad (7.9.5)$$

Substituting $\tau = t - t'$ in the last integral and recalling that $\pi\delta(\Delta\omega) = \int_0^\infty d\tau \cos \Delta\omega\tau$, we obtain

$$\frac{d\theta}{dx} = \frac{2\pi^2 n_0 \omega_0 p_0^2 g(0)}{\hbar c} \int_{-\infty}^{\infty} dt' \, \tilde{\mathfrak{E}}(x,t') \mathfrak{N}(0, x, t') \qquad (7.9.6)$$

That is, the area is only changed by the atoms that are on resonance. Expressions for $\mathfrak{N}(0, x, t)$ and $u(x, t)$ were obtained in Section 7.8, where we found that $\tilde{\mathfrak{E}} = \partial\sigma/\partial t$ as well as $\mathfrak{N}(0, x, t) = \pm \cos\sigma$. With these representations the integral in (7.9.6) can be performed and we have

$$\frac{d\theta}{dx} = \pm \frac{1}{2}\alpha \sin\theta \qquad \begin{array}{l} + \text{ amplifier} \\ - \text{ attenuator} \end{array} \qquad (7.9.7)$$

where

$$\alpha = \frac{4\pi^2 n_0 \omega_0 p_0^2 g(0)}{\hbar c} \qquad (7.9.8)$$

The physical significance of the constant $\alpha$ follows directly from the linearized version of (7.9.7), in which $\sin\theta$ is replaced by $\theta$. The area is then seen to amplify or decay in the characteristic length $\alpha^{-1}$. In the limit that $g(\Delta\omega)$ becomes a delta function, so that the sine–Gordon limit is recovered, we find that $g(0)$ becomes infinite and the characteristic length tends to zero.

The solution of (7.9.7) that satisfies $\theta = \theta_0$ at $x = x_0$ is

$$\tan\left(\frac{\theta}{2}\right) = \tan\left(\frac{\theta_0}{2}\right) \exp\left[\pm\left(\frac{\alpha}{2}\right)(x - x_0)\right] \qquad (7.9.9)$$

and is depicted schematically in Figure 7.5.

Since (7.9.8) contains a choice of signs, it is actually two distinct differential equations. The two solutions are obtained from Figure 7.4 by reading the diagram from right to left for the plus sign (amplifier) and from left to right for the minus sign (attenuator). Hence we see that an infinitesimal area will grow to $\pi$ in an amplifier while any area less than $\pi$ will evolve to zero in an attenuator. This second result allows for not only the well-known decay of a

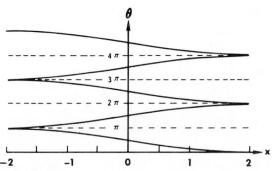

**Figure 7.4**   Area theorem—graph of (7.9.9). (With permission of American Institute of Physics.)

pulse as it propagates in an attenuator, but also for evolution into a non-vanishing zero area pulse, that is, one in which the total area under the pulse envelope is zero but the area under the pulse energy ($\propto \mathcal{E}^2$) is not zero. This is possible if the positive portions of a pulse envelope are equal in area to the negative portions. As noted above, the regions of positive and negative envelope are merely regions in which there is a relative difference of 180 degrees in the phase of the carrier wave. In an attenuator, initial pulse areas between $\pi$ and $3\pi$ will evolve into the steady state $2\pi$ pulse of self-induced transparency. The figure also shows that the $2\pi$ pulse is unstable in an amplifier and will evolve into either a $\pi$ or $3\pi$ pulse.

Figure 7.4 refers only to the total area of a pulse and gives no information at all about either the possible breakup of a pulse into two or more pulses with the same total area or whether a continually amplifying $\pi$ pulse will retain an area of $\pi$ by virtue of pulse narrowing or by developing negative regions in the pulse envelope.

## 7.10   SOLUTION BY AN INVERSE METHOD

The equations in (7.6.9) for the two-level atom can be solved by the two-component inverse method. This approach, in a somewhat more general setting, will be summarized in Section 7.12. Here we shall use the one-component inverse method for a Schrödinger equation with a complex potential.

Employing the transformation mentioned in Chapter 1, we first introduce new dependent variables $y$ and $z$ according to

$$y = v_1 + v_2$$
$$iz = v_1 - iv_2 \qquad (7.10.1)$$

and the dimensionless time variable $\tau = \Omega(t - x/c)$, where $\Omega$ is given by (7.8.3). From (7.6.10), the new variables $y$ and $z$ satisfy the equations

$$y_\tau + \tfrac{1}{2} iuy = \eta z$$
$$z_\tau - \tfrac{1}{2} iuz = -\eta y \qquad (7.10.2)$$

where

$$\eta = \frac{\Delta\omega}{2\Omega} \tag{7.10.3}$$

and $u = \tilde{\xi}/\Omega$ as given in (7.8.9). Following the procedure in Section 1.5, we find that the second-order equations are

$$y_{\tau\tau} + (\eta^2 - \mathcal{V})y = 0$$
$$z_{\tau\tau} + (\eta^2 - \mathcal{V}^*)z = 0 \tag{7.10.4}$$

where $\mathcal{V}$ is the complex potential

$$\mathcal{V} = -\tfrac{1}{4}(u^2 + 2iu_\tau) \tag{7.10.5}$$

We must also determine the boundary conditions to be imposed upon the functions $y$ and $z$. For the *attenuator* we require $a \to 0$ and $|b| \to 1$ as $\tau \to -\infty$ (i.e., before the arrival of the pulse). According to (7.6.9) and (7.6.10), this in turn implies that $v_1 \to 0$ and $v_2 \sim e^{i\eta\tau}$. Thus, $y$ and $z$ vary as $e^{i\eta\tau}$ as $t \to -\infty$. From the definition of the fundamental solutions for the Schrödinger equation given in (2.8.4) and (2.10.7), we see that $y$ and $z$ are therefore proportional to $f_2(-\eta,\xi,\tau)$. Ignoring a space-dependent phase factor that does not enter the calculation, we may set

$$y(\eta,\xi,\tau) = \alpha f_2(-\eta,\xi,\tau) \tag{7.10.6}$$

where $\alpha$ is a constant. Since the initial condition as $\tau \to -\infty$ and the defining equations (7.10.4) satisfy $z^*(\eta,\xi,\tau) = y(-\eta,\xi,\tau)$, we may set $z^*(\eta,\xi,\tau) = \alpha f_2(\eta,\xi,\tau)$. When $v_1$ and $v_2$ are expressed in terms of $y$ and $z$, we now find that (7.6.10) and (7.6.12) are equivalent to

$$\mathcal{R} - i\mathcal{S} = yz^* = -f_2(\eta,\xi,\tau)f_2(-\eta,\xi,\tau) \tag{7.10.7}$$

when we set $\alpha = e^{i\pi/4}$. This expression reduces to $\mathcal{R} \to -1$ as $\tau \to -\infty$.

For the *amplifier* we require that $|a| \to 1$ and $b \to 0$ as $\tau \to -\infty$. We then find $y \sim e^{-i\eta\tau}$, $z \sim e^{-i\eta\tau}$, which shows that $y$ and $z$ are proportional to $f_2(\eta,\xi,\tau)$ and again leads to $z^*(\eta,\xi,\tau) = y(-\eta,\xi,\tau)$.

Returning to the case of the attenuator, we note that as $\tau \to +\infty$, that is, after the pulse has passed by, we may use

$$f_2(-\eta,\xi,\tau) = c_{11}(-\eta,\xi)f_1(-\eta,\xi,\tau) + c_{12}(-\eta,\xi)f_1(\eta,\xi,\tau) \tag{7.10.8}$$

Dividing by $c_{12}$ to obtain standard expressions for scattering theory (with $x \to -\tau$), we have

$$t(-\eta,\xi)f_2(-\eta,\xi,\tau) = R_R(-\eta,\xi)f_1(-\eta,\xi,\tau) + f_1(\eta,\xi,\tau) \underset{\tau\to+\infty}{\to} R_R e^{-i\eta\tau} + e^{i\eta\tau} \tag{7.10.9}$$

If we wish to consider lossless propagation (i.e., solitons), the solution must again approach $a \to 0$, $|b| \to 1$ after passage of the pulse. As noted above, this implies that $y \sim e^{+i\eta\tau}$ and hence we must set $r_R(-\eta,\xi)=0$ in (7.10.9). According to (2.8.16), this implies that $r_L=0$.

The physical considerations of the present problem are thus seen to provide a natural motivation for the association of vanishing reflection coefficients with lossless (soliton) propagation. For nonsoliton propagation $r_R(-\eta,\xi)$ does not vanish and the spatial variation of the reflection coefficient must be determined. This topic will be considered after we have summarized the equations required for solution by inverse scattering.

We found in Section 3.1 that the potential may be obtained from the expression

$$\mathcal{V}(\xi,\tau)=2\frac{\partial}{\partial\tau}A_L(\tau,\tau;\xi) \tag{7.10.10}$$

which is the adaptation of (3.1.4) to the notation of the present problem. The function $A_L(\tau,t;\xi)$ satisfies (3.2.8) in the form

$$A_L(\tau,t;\xi)+\Omega_L(\tau+t;\xi)+\int_{-\infty}^{\tau} dt'\, A_L(\tau,t';\xi)\Omega_L(t'+t;\xi)=0 \tag{7.10.11}$$

Using (7.10.5) and separating real and imaginary parts, we have

$$u^2(\xi,\tau)=-8\,\mathrm{Re}\,\frac{\partial}{\partial\tau}A_L(\tau,\tau;\xi) \tag{7.10.12a}$$

$$u(\xi,\tau)=-4\,\mathrm{Im}\,A_L(\tau,\tau;\xi) \tag{7.10.12b}$$

The spatial dependence of the reflection coefficient is obtained from an examination of the governing equation in regions where the pulse is small. In the linear region as the pulse just begins to arrive at some location $\xi$, we may assume that $\mathcal{N}(\eta,\xi,\tau)=+1$ for an amplifier and $-1$ for an attenuator. In dimensionless form the Bloch equations (7.7.5) with $\varphi=0$ then reduce to

$$\mathcal{S}_\tau(\eta,\xi,\tau)=\pm u(\xi,\tau)+2\eta\mathcal{C}(\eta,\xi,\tau)$$
$$\mathcal{C}_\tau(\eta,\xi,\tau)=-2\eta\mathcal{S}(\eta,\xi,\tau) \tag{7.10.13}$$

We may also ignore the integral in the Marchenko equation (7.10.11) since it is a nonlinear term and thus small when $\mathcal{E}$ is small. As $\tau \to -\infty$ we then have

$$u(\xi,\tau)=-4\,\mathrm{Im}\,A_L(\tau,\tau;\xi)=-4i\Omega_L(2\tau;\xi)$$
$$=-4i\left[r_L(2\tau,\xi)-\sum m_n e^{-2ik_l\tau}\right] \tag{7.10.14}$$

where

$$r_L(2\tau,\xi)=\int_{-\infty}^{\infty}\frac{d\eta}{2\pi}R_L(\eta,\xi)e^{2i\eta\tau} \tag{7.10.15}$$

and $m_n=ic_{22}/c_{12}$.

Since $u(\xi,\tau)$ is real, we see from (7.10.14) that, for the complex potential $\mathcal{V}$ considered in this problem, $r_L$ must be purely imaginary and the $m_n$ must be either purely imaginary or occur in pairs located symmetrically about the imaginary axis.

We may now solve (7.10.13) for $\mathcal{S}$ and $\mathcal{C}$ in terms of $u$. For the contribution to the continuous spectrum we introduce a traveling wave representation of the form

$$\mathcal{S}(\eta,\xi,\tau)=\int_{-\infty}^{\infty}\frac{d\zeta}{2\pi}\overline{\mathcal{S}}(\zeta,\eta)e^{ik(\zeta)\xi-i\zeta\tau} \tag{7.10.16a}$$

$$\mathcal{C}(\eta,\xi,\tau)=\int_{-\infty}^{\infty}\frac{d\zeta}{2\pi}\overline{\mathcal{C}}(\zeta,\eta)e^{ik(\zeta)\xi-i\zeta\tau} \tag{7.10.16b}$$

$$u(\xi,\tau)=\int_{-\infty}^{\infty}\frac{d\zeta}{2\pi}\bar{u}(\zeta)e^{ik(\zeta)\xi-i\zeta\tau} \tag{7.10.16c}$$

into (7.10.13) and find that

$$\overline{\mathcal{S}}(\zeta,\eta)=\pm\frac{i\zeta\bar{u}(\zeta)}{\zeta^2-4\eta^2}$$

Substituting into the wave equation

$$\frac{\partial u}{\partial\xi}=\langle\mathcal{S}\rangle \tag{7.10.17}$$

which is the dimensionless version of (7.7.3a), we obtain the dispersion relation

$$k(\zeta)=\pm\zeta\left\langle\frac{1}{\zeta^2-4\eta^2}\right\rangle \tag{7.10.18}$$

In carrying out the integral in the averaging operation, the singularities are avoided by noting that finite relaxation times in (7.10.13) would require replacement of $\zeta$ by $\zeta+i\epsilon$. We may now use (7.10.16c) and (7.10.14) to write $\bar{u}(-2\zeta)e^{ik(-2\zeta)}=-2iR_L(\zeta,\xi)$. Consequently, we find that the space dependence of the reflection coefficient is given by

$$R_L(\zeta,\xi)=R_L(\zeta,0)e^{ik(-2\zeta)\xi} \tag{7.10.19}$$

where $k(-2\zeta)$ is obtained from (7.10.18).

A similar consideration of the bound-state contributions in the attenuator case yields

$$m_n(\zeta_n, \xi) = m_n(\zeta_n, 0) \exp\left( -\frac{1}{2} i\zeta_n \xi \left\langle \frac{1}{\zeta_n^2 - \eta^2} \right\rangle \right) \tag{7.10.20}$$

We now consider the application of these results to some simple problems in coherent pulse propagation.

The pure multisoliton solutions can be obtained by using the expression for $A_L$ given in (3.3.15). In the notation of the present section, we have

$$u = 4\frac{\partial}{\partial \tau} \{ \text{Im}[\ln(\det V)] \}$$
$$= 4\frac{\partial}{\partial \tau} \tan^{-1}\left[ \frac{\text{Im}(\det V)}{\text{Re}(\det V)} \right] \tag{7.10.21}$$

where $V$ is the $N \times N$ matrix $V = I + M$ given in (3.3.7).

For the single-soliton solution with a pole at $\zeta = ia/2$, where $a$ is real, we find

$$V = 1 + \int_{-\infty}^{\tau} d\tau' \, m_1(\xi) e^{a\tau'} \tag{7.10.22}$$

Setting $m_1(\xi) = ic(\xi)$, where $c(\xi)$ is real, and using (7.10.20), we obtain

$$V = 1 + i\frac{c(0)}{a} \exp\left[ a\left( \tau - \xi \left\langle \frac{1}{a^2 + 4\eta^2} \right\rangle \right) \right] \tag{7.10.23}$$

The field envelope is then

$$\tilde{\mathcal{E}} = \frac{p_0}{\hbar}\mathcal{E} = \Omega \frac{\partial \sigma}{\partial \tau} = \frac{2}{\tau_p} \text{sech}\left[ \frac{1}{\tau_p}\left( t - \frac{x}{v} \right) \right] \tag{7.10.24}$$

where $\tau_p = (a\Omega)^{-1}$ and the pulse velocity $v$ is given by

$$\frac{c}{v} = 1 + \left\langle \frac{1}{a^2 + 4\eta^2} \right\rangle \tag{7.10.25}$$

The first form for the velocity given in (7.10.25) reduces to that obtained in (7.8.12) in the sine–Gordon limit when we use $g(\Delta\omega) = \delta(\Delta\omega)$ for the averaging. This same prescription may be used in incorporating inhomogeneous broadening into the various multisoliton solutions of the sine–Gordon equation.

The velocity $v$ obtained from (7.10.25) may be two or three orders of magnitude lower than $c$, the phase velocity of the light wave in the medium.

Numerical values indicative of experimental results (Gibbs and Slusher, 1972) are $\omega_0 \sim 10^{15}$ sec$^{-1}$, $n_0 \sim 10^{12}$ cm$^{-3}$, $p_0 \sim 6 \times 10^{-16}$ (cgs), $\tau_p = 7 \times 10^{-9}$ sec. Then $\Omega^2 \sim 2 \times 10^{20}$ sec$^{-2}$. We thus have $(\Omega \tau_p)^2 \sim 3000$, so that the envelope velocity is reduced by three orders of magnitude, since the function averaged over the frequency distribution leads to a factor of order unity.

We have seen that the properties of solitons can explain a number of important features of coherent optical pulse propagation in an attenuator. The complete solution of the Marchenko equation that is possible for pure multisoliton propagation as well as the decay of the nonsoliton contribution to the solution leads to a very satisfactory description of nonlinear propagation in an attenuator. In addition, a pulse below a certain threshold (a $\pi$ pulse) merely attenuates in a manner to be expected of small-amplitude pulses. Nonlinear effects do not become important at all in this case. In addition, the concept of the zero area pulse arises naturally in the theory and has been observed experimentally (Grieneisen et al., 1972).

As in Section 4.4 for the Korteweg–deVries equation, once the poles associated with the reflection coefficient for an initial pulse profile have been determined, the final pulse amplitudes may be obtained. Pairs of poles that are located symmetrically about the imaginary axis determine the structure of the breather solutions that emerge.

## 7.11 PROPAGATION IN AN AMPLIFIER

Since the soliton is not a stable mode for propagation of a light pulse into an inverted population, as we saw in Section 7.8, propagation in an amplifier can be expected to be quite different from that in an attenuator. The analysis is also much more difficult since the nonsoliton part of the solution is now amplified rather than attenuated and must therefore be retained. A pulse that is initially small will be amplified until nonlinear effects become dominant.

There is one case that can be analyzed quite simply, however. We saw in Section 5.2 that the similarity solution for the sine–Gordon equation yields a pulse profile that has the properties that we require of a $\pi$ pulse in an amplifier. The results shown in Figure 5.5 may be applied directly to the case of pulse amplification in the limit of no inhomogeneous broadening [i.e., $g(\Delta\omega) = \delta(\Delta\omega)$]. We need merely interpret Figure 5.5$b$ as a graph of $\Omega^{-1}\tilde{\mathcal{E}}/\xi$ versus $\xi\tau$. Scaling laws for $\pi$ pulse propagation in a lossless amplifier may be inferred from these results. Since the abscissa for the pulse envelope is $\xi\tau$, the duration of the pulse narrows linearly with increasing distance of propagation. Also, since $\Omega^{-1}\tilde{\mathcal{E}} = \xi\sigma'$, the amplitude of the pulse envelope increases linearly with distance.

In Section 3.4 we saw that a truncated potential with a reflection coefficient that is a rational function could be analyzed almost as easily as a pure multisoliton solution. The Marchenko equation could be solved by a method similar to that used with the reflectionless potentials of pure soliton propagation. Hence truncated initial pulse profiles seem to be a preferable choice for constructing initial pulse profiles for analyzing propagation in an amplifier. Because of the analytical complexities associated with the time-dependent phase term in the reflection coefficient, however, the *propagation* of even these initial pulse profiles has yet to be treated satisfactorily. We now consider an example that shows how some of the features expected of pulse propagation in an amplifier may be investigated with these truncated potentials. Extensive approximation is required in carrying out the analysis, however, and the example is only considered in order to show how reasonable results can be obtained with such approximate procedures.

The initial pulse profile

$$\mathcal{E}(0,\tau) = 2u(\tau - \tau_0) \operatorname{sech} \tau \tag{7.11.1}$$

where $u(\tau - \tau_0)$ is a unit step function, is one of the easiest to treat. As $\tau_0$ varies from $-\infty$ to $+\infty$, the area under the pulse decreases from $2\pi$ to zero. From (7.10.5), the associated complex potential is

$$\mathcal{V}(0,\tau) = -\tfrac{1}{2}u(t - t_0)\operatorname{sech}^2\left(\tfrac{1}{2}\tau + i\frac{\pi}{4}\right) - i\delta(\tau - \tau_0)\operatorname{sech}\tau_0 \tag{7.11.2}$$

We now consider the solution of the Schrödinger equation (7.10.4) with this complex potential. The fundamental solution that reduces to $e^{-i\eta\tau}$ as $\tau$ approaches $-\infty$, and thus, according to the results of Section 7.10, describes an inverted population before the pulse arrives, is the solution $f_2(\eta,\tau;0)$. Of more interest are the coefficients $c_{ij}(\eta,0)$ that relate the various fundamental solutions. Since the potential $\mathcal{V}$ is truncated and also contains a delta function, these coefficients may be obtained immediately by using the results of Ex. 23 in Chapter 2. For $\tau > \tau_0$ the potential reduces to $-\tfrac{1}{2}\operatorname{sech}^2\theta$, where $\theta = \tfrac{1}{2}\tau + i\tfrac{1}{4}\pi$. The fundamental solutions are then given by Ex. 15 of Chapter 2. In particular, we have

$$f_1(\eta,\tau;0) = \frac{e^{i\eta\tau}(2i\eta - \tanh\theta)}{2i\eta - 1} \tag{7.11.3}$$

This, then, is a fundamental solution for the untruncated potential. Using the results of Ex. 22 in Chapter 2, we have

$$\tilde{c}_{12}(\eta,0) = e^{-i\eta\tau_0}\frac{\dot{f}_1(\eta,\tau_0;0) + (i\eta - a)f_1(\eta,\tau_0;0)}{2i\eta} \tag{7.11.4}$$

where now $a = -i\operatorname{sech}\tau_0$. Setting $\theta_0 = \tfrac{1}{2}\tau_0 + i\tfrac{1}{4}\pi$ and noting that $2a\tanh\theta_0 = \operatorname{sech}^2\theta_0$, we find that

$$\tilde{c}_{12}(\eta,0) = \frac{2i\eta - \tanh\tau_0}{2i\eta - 1} \tag{7.11.5}$$

Similarly, we obtain

$$\tilde{c}_{11}(\eta,0) = -\tilde{c}_{22}(-\eta,0) = \frac{ae^{-2i\eta\tau_0}}{2i\eta - 1} \qquad (7.11.6)$$

In arriving at this result, we have used the relation $a + \tanh\theta_0 = \tanh\tau_0$. The reflection coefficient $R_L(\eta,0) = \tilde{c}_{22}(\eta,0)/\tilde{c}_{12}(\eta,0)$ is then

$$R_L(\eta,0) = -\frac{1}{2}\frac{e^{2i\eta\tau_0}\operatorname{sech}\tau_0}{\eta + (i/2)\tanh\tau_0} \qquad (7.11.7)$$

Using (2.8.32) for the Fourier transform and (7.10.19) for the spatial dependence, we find that at an arbitrary distance $\xi$ into the medium,

$$r_L(\xi,\tau) = -\frac{1}{2}\int_{-\infty}^{\infty}\frac{d\eta}{2\pi}\frac{e^{-i\eta(\tau - 2\tau_0) + ik(-2\eta)\xi}}{\eta + (i/2)\tanh\tau_0} \qquad (7.11.8)$$

For $\tau_0 > 0$, the initial pulse is less than a $\pi$ pulse and the pole in the integrand is in the lower half-plane. For $\tau_0 < 0$, a contribution from the pole that is then in the upper half-plane would have to be included. This more-complicated case will not be treated here. The occurrence of poles in the upper half-plane will lead to the multipeaking of amplified pulses that has been observed in numerical solutions of this problem (Hopf and Scully, 1969, Fig. 4).

The frequency distribution of the inhomogeneous broadening is frequently assumed to be Gaussian. In dimensionless form, we then have

$$\bar{g}(\eta) = \frac{\Omega T_2^*}{\sqrt{\pi}}e^{-(\Omega T_2^*\eta)^2} \qquad (7.11.9)$$

where $T_2^*$ is known as the inhomogeneous relaxation time and $\int_{-\infty}^{\infty}d\eta\,\bar{g}(\eta) = 1$. The time $T_2^*$ is a measure of the decay of a pulse profile due to dispersive (bandwidth) effects and is a reversible phenomenon (Hahn, 1950; Abella et al., 1966).

The propagation constant in (7.10.14) now takes the form

$$k(\zeta) = -\frac{\zeta(\Omega T_2^*)^2}{4\sqrt{\pi}}\int_{-\infty}^{\infty}\frac{dz\,w(z)}{z^2 - \left(\frac{1}{2}\Omega T_2^* + i\epsilon\right)^2} \qquad (7.11.10)$$

where an infinitesimal amount of damping has been introduced to move the singularities off of the real axis. The function $w(z)$ is the complex error function (Abramowitz and Stegun, 1964, p. 297)

$$w(z) = e^{-z^2}\left(1 + \frac{2i}{\sqrt{\pi}}\int_0^z dt\,e^{t^2}\right), \qquad \operatorname{Im}z > 0 \qquad (7.11.11)$$

Since the imaginary part of $w(z)$ is an odd function of $z$, it may be added to the integrand in (7.10.14) to yield the integral given in (7.11.10). The function $w(z)$ has the limiting forms

$$
\begin{aligned}
w(z) &= 1 + \frac{2iz}{\sqrt{\pi}} + \cdots, && |z| \ll 1 \\
w(z) &= \frac{i}{\sqrt{\pi}}\left(\frac{1}{z} + \frac{1}{2z^3} + \cdots\right), && |z| \gg 1
\end{aligned}
\tag{7.11.12}
$$

Consequently, the function $w(z)$ vanishes as $z^{-1}$ on the arc at infinity in the upper half-plane and the integral in (7.11.10) may be evaluated by the theory of residues. Since $w(z)$ is analytic in the upper half-plane, we obtain

$$
k(\zeta) = -\tfrac{1}{2}i\sqrt{\pi}\,\Omega T_2^* w\left(\tfrac{1}{2}\zeta\Omega T_2^*\right)
\tag{7.11.13}
$$

In the limit $\Omega T_2^* \gg 1$, the Gaussian distribution (7.11.9) becomes a delta function and the problem reduces to the sine–Gordon equation, as shown in Section 7.8. Here we shall only consider the opposite limit of extremely large inhomogeneous broadening $\Omega T_2^* \ll 1$. The small-argument form of $w(z)$ given in (7.11.12) is then appropriate. It will be found that even though the function $k(\zeta)$ is integrated over all $\zeta$ in (7.11.8), use of only the small-argument approximation for $w(z)$ leads to the expected results. Setting $w(z) = 1$, we then find that $k(-2\zeta) = -\tfrac{1}{2}i\sqrt{\pi}\,\Omega T_2^* = -i\pi\bar{g}(0)$ and

$$
r_L(\xi, \tau) = e^{\gamma\xi} r_L(0, \tau)
\tag{7.11.14}
$$

where $\gamma = 2\pi\bar{g}(0) = 4\pi\Omega g(0)$. When expressed in dimensional form, the spatial factor in (7.11.14) is $e^{\alpha x/2}$, where $\alpha$ is the linear amplification factor given in (7.9.8).

The evaluation of the integral in (7.11.8) is now quite simple and we obtain

$$
r_L(\xi, \tau) = \tfrac{1}{2}ie^{\gamma\xi}u(\tau - 2\tau_0)\operatorname{sech}\tau_0 \exp\left[-\tfrac{1}{2}(\tau - 2\tau_0)\tanh\tau_0\right]
\tag{7.11.15}
$$

Suppressing the $\xi$ dependence, since it is only parametric, we find that the Marchenko equation (7.10.11) now takes the form

$$
\begin{aligned}
A_L(v, z) &+ \tfrac{1}{2}i\beta e^{-(1/2)(v+z)\tanh\tau_0} \\
&+ \tfrac{1}{2}i\beta \int_{-z}^{u} ds\, A_L(v, s)e^{-(1/2)(s+z)\tanh\tau_0} = 0
\end{aligned}
\tag{7.11.16}
$$

where

$$
v = \tau - \tau_0, \qquad z = t - \tau_0, \qquad s = y - \tau_0
\tag{7.11.17}
$$

and

$$
\beta = e^{\gamma\xi}\operatorname{sech}\tau_0
\tag{7.11.18}
$$

In (7.11.17) the lower limit of $-z$ rather than $-\infty$ is a result of the truncation of the pulse profile. Since the spatial dependence of the transmission coefficient is not readily available by the procedure being employed here, although it can be obtained with the more general technique to be described in the next section, we are unable to proceed as in Section 3.4 to construct $r_R$ and then obtain the solution of the integral equation (3.2.7), in which the range of integration is from $z$ to $+\infty$.

In solving (7.11.15), an obvious initial choice for simplification of the equation is the factored form

$$A_L(v,z)=F(v,z)e^{-(1/2)z\tanh\tau_0}\tag{7.11.19}$$

Differentiating the resulting equation for $F(v,z)$ with respect to $z$, we obtain

$$\frac{\partial F(v,z)}{\partial z}+\tfrac{1}{2}i\beta F(v,-z)e^{z\tanh\tau_0}=0\tag{7.11.20}$$

A subsequent differentiation of this equation and elimination of $\partial F(v,-z)/\partial z$ by use of the equation that follows from (7.11.20) by changing $z$ to $-z$ yields

$$\left(\frac{\partial^2}{\partial z^2}-\tanh\tau_0\frac{\partial}{\partial z}-\frac{1}{4}\beta^2\right)F(v,z)=0\tag{7.11.21}$$

This equation has solutions of the form $e^{\mu z}$, where

$$\mu=\tfrac{1}{2}(\tanh\tau_0\pm\delta)\tag{7.11.22}$$

in which

$$\delta=(\tanh^2\tau_0+\beta^2)^{1/2}\tag{7.11.23}$$

From (7.11.18) we see that both $\beta$ and $\delta$ depend upon $\xi$. According to (7.11.19) and (7.11.22), an appropriate form of $A_L(v,z)$ for solving the Marchenko equation (7.11.15) is thus

$$A_L(v,t)=a(v)e^{\delta z/2}+b(v)e^{-\delta z/2}\tag{7.11.24}$$

Introduction of this form of $A_L$ into the Marchenko equation leads to

$$a(v)=-\frac{1}{2}\left(\frac{e^{\delta v/2}}{\delta-\tanh\tau_0}-i\frac{e^{-\delta v/2}}{\beta}\right)^{-1}\tag{7.11.25a}$$

and

$$b(v)=-\frac{2i\beta}{\delta-\tanh\tau_0}a(v)\tag{7.11.25b}$$

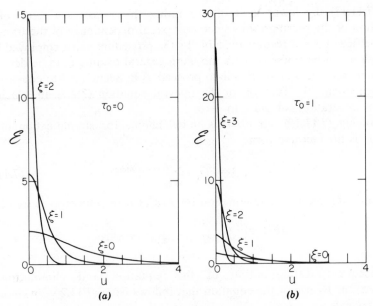

**Figure 7.5** Amplification of initial pulse profile given by (7.11.1) for (a) $\tau_0 = 0$, a $\pi$ pulse, and (b) $\tau_0 = 1$, initial area approximately $0.45\pi$. The analytical expression for these curves is given in (7.11.26). The dimensionless linear gain $\gamma$ is equal to unity for both figures. (With permission of American Institute of Physics.)

Recalling from (7.10.15) that $u = -4\operatorname{Im}A_L(\tau,\tau;\xi)$, we find that (7.11.24) yields the electric field profile in the form

$$u(\xi,\tau) = \frac{4\delta}{\beta(\delta - \tanh\tau_0)}\left[\frac{e^{\delta v}}{(\delta - \tanh\tau_0)^2} + \frac{e^{-\delta v}}{\beta^2}\right]^{-1} \qquad (7.11.26)$$

Representative graphs of this result are given in Figure 7.5.

The total area under the pulse is*

$$\theta(\xi) = \int_0^\infty dv\, u(v,\xi)$$

$$= 2\pi - 4\tan^{-1}\left(\frac{\beta}{\delta - \tanh\tau_0}\right) \qquad (7.11.27)$$

*In computing the area, we have used the relation

$$\int dx(be^{ax} + ce^{-ax})^{-1} = \frac{1}{a\sqrt{bc}}\tan^{-1}\left(e^{ax}\sqrt{b/c}\,\right)$$

This result may be rewritten in the form $\tan(\frac{1}{2}\theta) = e^{a\xi}\operatorname{csch}\tau_0$, or equivalently, as dictated by the area theorem (7.9.9),

$$\tan\left[\tfrac{1}{2}\theta(\xi)\right] = e^{a(\xi-\xi_0)}\tan\left[\tfrac{1}{2}\theta(\xi_0)\right] \tag{7.11.28}$$

As $\xi$ approaches infinity, $\theta(\xi)$ approaches $\pi$ for $0 < \tau_0 < \infty$. Also, as $\tau_0$ approaches 0, $\theta(\xi)$ approaches $\pi$ for all $\xi$.


## 7.12   THE TWO-COMPONENT METHOD

Coherent optical pulse propagation provides an instructive example of the use of the two-component inverse method. It is the more convenient approach to use when the phase term $\varphi(x,t)$ in (7.5.3) is retained since the complex potential required in the one-component method then becomes unwieldy. When the phase term is present in the Bloch equations (7.7.5), we can easily show that they may be written in dimensionless form as

$$\lambda_\tau + 2i\eta\lambda = \bar{u}\mathfrak{N} \tag{7.12.1a}$$

$$\lambda_\tau^* - 2i\eta\lambda^* = \bar{u}^*\mathfrak{N} \tag{7.12.1b}$$

$$\mathfrak{N}_\tau = -\tfrac{1}{2}(\bar{u}^*\lambda + \bar{u}\lambda^*) \tag{7.12.1c}$$

where $\eta = \Delta\omega/2\Omega$, as before, and

$$\lambda = (\mathcal{S} + i\mathcal{C})e^{i\varphi} \tag{7.12.2a}$$

$$\bar{u} = ue^{i\varphi} = \left(\frac{p_0\mathcal{E}}{\hbar\Omega}\right)e^{i\varphi} \tag{7.12.2b}$$

In addition, the Maxwell equations, which previously reduced to (7.10.17), now become

$$\frac{\partial\bar{u}}{\partial\xi} = \langle\lambda\rangle \tag{7.12.3}$$

As we saw in Chapter 5, customary usage of the two-component inverse method involves the introduction of a pair of coupled eigenvalue equations. In the present application of this method, it is convenient to interchange the roles of space and time with respect to previous usage. We thus introduce two functions $v_1$ and $v_2$ that satisfy

$$\begin{aligned} v_{1\tau} + i\zeta v_1 &= \tfrac{1}{2}\bar{u}v_2 \\ v_{2\tau} - i\zeta v_2 &= -\tfrac{1}{2}\bar{u}^*v_1 \end{aligned} \tag{7.12.4}$$

where $\bar{u}$ is the complex field profile given in (7.12.2b). It is now the spatial derivatives $v_{1\xi}$ and $v_{2\xi}$ that are expressible as linear combinations of $v_1$ and

$v_2$, as was the case for the time derivative in Section 5.4. It must now be shown that the equating of mixed second derivatives $v_{1\xi\tau} = v_{1\tau\xi}$ and $v_{2\xi\tau} = v_{2\tau\xi}$ as well as the requirement that the Maxwell–Bloch equation be satisfied leads to $\zeta_\xi = 0$, that is, the spatial independence of the eigenvalue parameter. It should be noted that although the linear equations (7.12.4) are of the same form as those in (7.6.9) that provide the basis for the inverse method in Section 7.6, the equations in (7.12.4) have a fundamentally different i terpretation. The eigenvalue parameter $\zeta$ in (7.12.4) is not the same as $\eta = \Delta\omega/2\Omega$ in (7.6.9). Although it should be clear on physical grounds that $\eta$ is independent of $\xi$, this independence must be demonstrated for the parameter $\zeta$ that is being introduced here. The method of the present section has the advantage that it may still be used to solve the sine–Gordon equation which arises in the limit that $g(\Delta\omega) = \delta(\Delta\omega)$ so that $\Delta\omega \to 0$. The previous method could not be used in this limit (since $\eta \to 0$), although the various multisoliton solutions obtained in Section 7.6 reduce to the corresponding solutions of the sine–Gordon equation in the limit $g(\Delta\omega) \to \delta(\Delta\omega)$.

The linear equations for $v_{1\xi}$ and $v_{2\xi}$ with the appropriate coefficients $A$, $B$, and $C$ may be obtained quite directly from the Bloch equations in the following way. We first add the term $2i\zeta\lambda$ to both sides of (7.12.1a) and write it in the form $\lambda_\tau + 2i\zeta\lambda = 2i(\zeta - \eta)\lambda + \bar{u}\mathfrak{N}$. We now divide this equation by $4i(\eta - \zeta - i\epsilon)$ and average the resulting equation over the inhomogeneous broadening. The small imaginary term in the denominator provides a prescription for avoiding integrands with singular denominators. (The final results are independent of this artifice, however, and a principal-value integral could also be employed.) We finally obtain

$$B_\tau + 2i\zeta B = \tfrac{1}{2}\bar{u}_\xi - \bar{u}A \qquad (7.12.5)$$

where

$$A = -\frac{i}{4}\int_{-\infty}^{\infty} \frac{d\eta\, g(\eta)\,\mathfrak{N}(\eta,\xi,\tau)}{\eta - \zeta - i\epsilon} \qquad (7.12.6a)$$

$$B = \frac{i}{4}\int_{-\infty}^{\infty} \frac{d\eta\, g(\eta)\lambda(\eta,\xi,\tau)}{\eta - \zeta - i\epsilon} \qquad (7.12.6b)$$

In obtaining (7.12.5) we have employed the wave equation (7.12.3). The result in (7.12.5) should be compared with (5.1.36b). Applications of the same procedure to the latter two equations in (7.12.1) leads to

$$C_\tau - 2i\zeta C = -\tfrac{1}{2}\bar{u}_\xi^* - \bar{u}^* A \qquad (7.12.7a)$$

$$A_\tau = \tfrac{1}{2}(\bar{u}^* B + uC) \qquad (7.12.7b)$$

where

$$C = \frac{i}{4}\int_{-\infty}^{\infty} \frac{d\eta\, g(\eta)\lambda^*(\eta,\xi,\tau)}{\eta - \zeta - i\epsilon} \qquad (7.12.8)$$

The relations among $A$, $B$, and $C$ expressed by (7.12.5) and (7.12.7) are of the form given in (5.1.36).

Let us now proceed to obtain the equations that express $v_{1\xi}$ and $v_{2\xi}$ as linear combinations of $v_1$ and $v_2$. If we multiply (7.12.7b) and (7.12.5) by $v_1$ and $v_2$, respectively, and add, we obtain

$$\frac{\partial}{\partial \tau}(v_1 A + v_2 B) + i\zeta(v_1 A + v_2 B) = \tfrac{1}{2}\bar{u}_\xi + \tfrac{1}{2}\bar{u}(v_1 C - v_2 A) \qquad (7.12.9)$$

With the definitions

$$\psi_{1\xi} \equiv A v_1 + B v_2 \qquad (7.12.10a)$$

$$\psi_{2\xi} \equiv C v_1 - A v_2 \qquad (7.12.10b)$$

(7.12.9) takes the form

$$\frac{\partial}{\partial \tau}(\psi_{1\xi} - v_{1\xi}) + i\zeta(\psi_{1\xi} - v_{1\xi}) = \tfrac{1}{2}\bar{u}(\psi_{2\xi} - v_{2\xi}) \qquad (7.12.11)$$

where $\tfrac{1}{2}\bar{u}_\xi v_2$ has been eliminated by using the first of (7.12.4). A similar combination of (7.12.7a) and (7.12.7b) leads to

$$\frac{\partial}{\partial \tau}(\psi_{2\xi} - v_{2\xi}) - i\zeta(\psi_{2\xi} - v_{2\xi}) = -\tfrac{1}{2}\bar{u}^*(\psi_{1\xi} - v_{1\xi}) \qquad (7.12.12)$$

Since (7.12.11) and (7.12.12) are of the same form as (7.12.4), the solution must be proportional to $v_1$ and $v_2$. We thus may write

$$\psi_{1\xi} - v_{1\xi} = \mu(\xi, \zeta) v_1$$
$$\psi_{2\xi} - v_{2\xi} = \mu(\xi, \zeta) v_2 \qquad (7.12.13)$$

where the function $\mu$ is to be determined. Employing (7.12.10), we find that

$$v_{1\xi} = (A - \mu) v_1 + B v_2$$
$$v_{2\xi} = C v_1 - (A + \mu) v_2 \qquad (7.12.14)$$

The function $\mu(\xi, \zeta)$ may be determined by specifying a choice for the asymptotic form of the functions $v_1$ and $v_2$. If we associate $v_2$ with the lower atomic level (the $b$ state) as in (7.6.9), then for an attenuator we require a solution that approaches $v_1 \to 0, |v_2| \to 1$ as $\tau \to -\infty$. Such a solution is $\begin{pmatrix} 0 \\ -1 \end{pmatrix} e^{i\zeta\tau}$ given in (2.11.6). Since $v_1$ and $v_{2\xi}$, then vanish as $\tau \to -\infty$, (7.12.4) yields

$$\mu(\xi, \zeta) = -A(\zeta, \xi, -\infty) = \frac{i}{4}\left\langle \frac{\mathfrak{N}(\eta, \xi, -\infty)}{\eta - \zeta - i\epsilon} \right\rangle \qquad (7.12.15)$$

where the averaging notation introduced in (7.7.1) has been used. We shall only consider the spatially homogeneous case and therefore set $\mathfrak{N}(\eta,\xi,-\infty) = -1$.

If we now equate mixed second derivatives of $v_1$ and $v_2$, we find that $\zeta_\xi = 0$. We could, of course, have obtained the foregoing results in the more customary way by starting with (7.12.4) and (7.12.14) and showing that the requirements $v_{1\xi\tau} = v_{1\tau\xi}$ and $v_{2\xi\tau} = v_{2\tau\xi}$ imply that $\zeta_\xi = 0$ provided that the Maxwell–Bloch equations are satisfied and the coefficients $A$, $B$, and $C$ satisfy (7.12.5) and (7.12.7) with these functions specified as in (7.12.6) and (7.12.8) (Gibbon et al., 1973; Ablowitz et al., 1974b).

When $\zeta = \eta$, the functions $v_1$ and $v_2$ reduce to those used in (7.6.9). To obtain the proper initial conditions for an attenuator, we set

$$\begin{pmatrix} v_1 \\ v_2 \end{pmatrix} = \tilde{\varphi} = \begin{pmatrix} \tilde{\varphi}_1 \\ \tilde{\varphi}_2 \end{pmatrix} = \begin{pmatrix} \varphi_2^* \\ -\varphi_1^* \end{pmatrix} \tag{7.12.16}$$

where $\tilde{\varphi}$ is defined in (3.9.7). We then find that the normalization condition becomes

$$(|v_1|^2 + |v_2|^2)_{\zeta=\eta} = (\tilde{\varphi}_1\varphi_2 - \tilde{\varphi}_2\varphi_1)_{\zeta=\eta} = 1 \tag{7.12.17}$$

Also, the population inversion and polarization may be written

$$\mathfrak{N} = (|v_1|^2 - |v_2|^2)_{\zeta=\eta} = (\tilde{\varphi}_1\varphi_2 + \varphi_1\tilde{\varphi}_2)_{\zeta=\eta} \tag{7.12.18a}$$

$$\lambda = -2(v_1 v_2^*)_{\zeta=\eta} = 2(\varphi_1\tilde{\varphi}_1)_{\zeta=\eta} \tag{7.12.18b}$$

As in Section 7.10, we obtain the field profile from that Marchenko equation which involves integration over the prior evolution of the solution from $-\infty$ up to $\tau$. Adapting the results of Section 3.9 to the notation of the present section, we have, for $\tau > t$,

$$A_1^*(\tau, t; \xi) + \int_{-\infty}^{\tau} dt' A_2(\tau, t'; \xi)\Omega_L(t' + t; \xi) = 0$$

$$-A_2^*(\tau, t; \xi) + \Omega_L(\tau + t; \xi) \tag{7.12.19}$$

$$+ \int_{-\infty}^{\tau} dt' A_1(\tau, t'; \xi)\Omega_L(\tau' + t; \xi) = 0$$

with

$$\Omega_L(z; \xi) = r_L(z, \xi) - i \sum_l \frac{c_{22}(k_l, \xi)}{\dot{c}_{12}(k_l, \xi)} e^{-ik_l z} \tag{7.12.20}$$

These results follow from taking the Fourier transform of (3.9.8b). Using (3.9.a0), we have $R_L(k,\xi) = -c_{22}(k,\xi)/c_{21}(k,\xi) = c_{11}^*(k,\xi)/c_{12}(k,\xi)$ and

$$r_L(z,\xi) = \int_{-\infty}^{\infty} \frac{dk}{2\pi} e^{ikz} \frac{c_{11}^*(k,\xi)}{c_{12}(k,\xi)} \tag{7.12.21}$$

From (3.9.3), the complex field envelope is given by $q = \frac{1}{2}u(\xi,\tau) = -2A_2(\tau,\tau;\xi)$.

We now consider the spatial dependence of the reflection coefficient $r_L(z,\xi)$. Instead of using the method employed in Section 7.10 again, we follow a procedure employed by Ablowitz et al., (1974) and examine the linear equations (7.12.14) in the limit $\tau \to +\infty$. As noted above, we shall employ the fundamental solution $\tilde{\varphi}$. Thus we set

$$\begin{pmatrix} v_1 \\ v_2 \end{pmatrix} = \tilde{\varphi} = \begin{pmatrix} \tilde{\varphi}_1 \\ \tilde{\varphi}_2 \end{pmatrix} = \begin{pmatrix} \varphi_2^* \\ -\varphi_1^* \end{pmatrix} \tag{7.12.22}$$

From (3.9.9a) we have

$$\tilde{\varphi} = c_{11}^* \begin{pmatrix} \psi_2^* \\ -\psi_1^* \end{pmatrix} - c_{12}^* \begin{pmatrix} \psi_1 \\ \psi_2 \end{pmatrix}$$

which has the limiting form

$$\tilde{\varphi} \underset{\tau \to +\infty}{\to} \begin{pmatrix} c_{11}^* e^{-i\xi\tau} \\ -c_{12}^* e^{i\xi\tau} \end{pmatrix} \tag{7.12.23}$$

This also implies that

$$\varphi \underset{\tau \to +\infty}{\to} \begin{pmatrix} c_{12} e^{-i\xi\tau} \\ c_{11} e^{i\xi\tau} \end{pmatrix} \tag{7.12.24}$$

The linear equations (7.2.14) are now

$$\tilde{\varphi}_{1\xi} = (A + A_-)\tilde{\varphi}_1 + B\tilde{\varphi}_2$$
$$\tilde{\varphi}_{2\xi} = C\tilde{\varphi}_1 - (A - A_-)\tilde{\varphi}_2 \tag{7.12.25}$$

where $A_-$ is given by (7.12.15). In the limit $\tau \to +\infty$, this yields

$$c_{11\xi}^* = (A_+ + A_-)c_{11}^* - c_{12}^* \lim_{\tau \to +\infty} (Be^{2i\xi\tau})$$
$$c_{12\xi}^* = -c_{11}^* \lim_{\tau \to +\infty} (Ce^{-2i\xi\tau}) - (A_+ - A_-)c_{12}^* \tag{7.12.26}$$

With the limiting forms

$$\mathfrak{N}(\xi, \eta, +\infty) = |c_{11}|^2 - |c_{12}|^2$$
$$\lambda(\xi, \eta, +\infty) = \lim_{\tau \to +\infty} (2c_{12}c_{11}^* e^{-2i\eta\tau}) \qquad (7.12.27)$$

which follow from (7.12.18) as well as the normalization condition $|c_{11}|^2 + |c_{12}|^2 = 1$, we find that (7.12.6a) and (7.12.15) yield

$$A_+ + A_- = -\tfrac{1}{2} i \left\langle \frac{|c_{12}(\eta)|^2}{\eta - \zeta - i\epsilon} \right\rangle$$
$$A_+ - A_- = -\tfrac{1}{2} i \left\langle \frac{|c_{11}(\eta)|^2}{\eta - \zeta - i\epsilon} \right\rangle \qquad (7.12.28)$$

Using the two prescriptions

$$\frac{1}{\eta - \zeta - i\epsilon} = \frac{P}{\eta - \zeta} + i\pi\delta(\eta - \zeta)$$

and                                                                                                          (7.12.29)

$$\lim_{\tau \to +\infty} \frac{e^{i\zeta\tau} - 1}{\Omega} = i\pi\delta(\Omega) - \frac{P}{\Omega}$$

for a principal value we also obtain

$$\lim_{\tau \to +\infty} (Be^{2i\zeta\tau}) = \tfrac{1}{2} i \int_{-\infty}^{\infty} d\eta\, g(\eta) \frac{c_{12}(\eta)c_{11}^*(\eta)e^{2i(\zeta - \eta)\tau}}{\eta - \zeta - i\epsilon}$$

$$= -\frac{\pi}{2} g(\zeta)c_{12}(\zeta)c_{11}^*(\zeta) + \tfrac{1}{2} \lim_{\tau \to +\infty} P \int_{-\infty}^{\infty} d\eta\, g(\eta)c_{12}c_{11} \left[ \frac{e^{2i(\zeta - \eta)\tau} - 1 + 1}{\eta - \zeta} \right]$$

$$= -\frac{\pi}{2} g(\zeta)c_{12}(\zeta)c_{11}^*(\zeta) + \tfrac{1}{2} P \int_{-\infty}^{\infty} \frac{d\eta\, g(\eta)}{\eta - \zeta}$$

$$+ \tfrac{1}{2} \int_{-\infty}^{\infty} d\eta\, g(\eta)c_{12}c_{11} \lim_{\tau \to +\infty} \left[ \frac{e^{2i(\zeta - \eta)\tau} - 1}{\eta - \zeta} \right]$$

$$= 0 \qquad (7.12.30)$$

and

$$\lim_{\tau \to +\infty} (Ce^{-2i\zeta\tau}) = -\pi g(\zeta)c_{11}(\zeta)c_{12}^*(\zeta) \qquad (7.12.31)$$

The linear equations (7.12.26) then reduce to the differential equations

$$c_{12\xi}^* = \tfrac{1}{2} i c_{12}^* \left\langle \frac{|c_{11}(\eta)|^2}{\eta - \zeta - i\epsilon} \right\rangle$$
$$c_{11\xi}^* = \tfrac{1}{2} i c_{11}^* \left\langle \frac{|c_{12}(\eta)|^2}{\eta - \zeta - i\epsilon} \right\rangle \qquad (7.12.32)$$

which have solutions

$$c_{11}^*(\eta,\xi) = c_{11}^*(\eta,0)\exp\left(\frac{1}{2}i\xi\left\langle\frac{|c_{12}|^2}{\eta-\zeta-i\epsilon}\right\rangle\right)$$

(7.12.33)

$$c_{12}^*(\eta,\xi) = c_{12}^*(\eta,0)\exp\left(\frac{1}{2}i\xi\left\langle\frac{|c_{11}|^2}{\eta-\zeta-i\epsilon}\right\rangle\right)$$

The reflection coefficient is then

$$r_L(\eta,\xi) = \frac{c_{11}^*(\eta,\xi)}{c_{12}(\eta,\xi)} = r_L(\eta,0)\exp\left(\frac{1}{2}i\xi\left\langle\frac{1}{\eta-\zeta-i\epsilon}\right\rangle\right)$$

(7.12.34)

This result should be compared with that given in (7.10.18) and (7.10.19).


*Exercise 1*

Show that the expression for the reflection coefficient $r_L(\eta,\xi)$ given in (7.12.34) is also obtained when the term $-i\epsilon$ is ignored in (7.12.6) and (7.12.8) and the integrals are treated as principal-value integrals.


*Exercise 2*

The final values for the upper- and lower-level populations are given by $n_a(\eta,\xi,+\infty)$ $=|c_{11}(\eta,\xi,+\infty)|^2$ and $n_b(\eta,\xi,+\infty)=|c_{12}(\eta,\xi,+\infty)|^2$. Show that

$$n_a(\eta,\xi,+\infty) = n_a(\eta,0,+\infty)\left[\,n_b(\eta,0,+\infty)e^{\pi\xi g(\eta)}+n_a(\eta,0,+\infty)\right]^{-1}$$

$$n_b(\eta,\xi,+\infty) = n_b(\eta,0,+\infty)\left[\,n_a(\eta,0,+\infty)e^{-\pi\xi g(\eta)}+n_b(\eta,0,+\infty)\right]^{-1}$$

Also show that the final population difference $\mathcal{R}(\eta,\xi,+\infty) = n_a(\eta,\xi,+\infty)$ $- n_b(\eta,\xi,+\infty)$ is of the form

$$1+\mathcal{R}(\eta,\xi,+\infty) = 2\frac{1+\mathcal{R}_0 e^{-\pi\xi g(\eta)}}{1-\mathcal{R}_0+(1+\mathcal{R}_0)e^{-\pi\xi g(\eta)}}$$

where $\mathcal{R}_0 \equiv \mathcal{R}(\eta,0,+\infty)$ (Ablowitz et al., 1974b).

Finally, the spatial dependence of each term in the summation in (7.12.20) is given by

$$\frac{c_{22}(k_l,\xi)}{\dot{c}_{12}(k_l,\xi)} = \frac{c_{22}(k_l,0)}{\dot{c}_{12}(k_l,0)}\exp\left(\frac{1}{2}i\xi\left\langle\frac{1}{\eta-k_l-i\epsilon}\right\rangle\right)$$

(7.12.35)


The equations governing this problem have now been completely formulated and the techniques used previously may again be employed to obtain multisoliton solutions. To proceed beyond the multisoliton solutions is, as always, a formidable task. For results based on an initially rectangular pulse profile, see Kaup (1977).

*Exercise 3*

Show that the single-soliton solution of (7.12.19) yields

$$\bar{u} = 4\beta e^{-i(\mu\xi + 2\alpha\tau + \theta)}\text{sech}(\nu\xi - 2\beta\tau + \gamma)$$

where $k_1 = \alpha + i\beta$, $\kappa_1 = \frac{1}{2}\langle 1/(\eta - k_l - i\epsilon)\rangle = \mu + i\nu$, and $ic_{22}(0)/\dot{c}_{12}(0) = e^{-\gamma + i\theta}$.

## Conserved Quantities

We saw in Section 4.5 that soliton amplitudes resulting from smoothly varying initial pulse profiles could be estimated quite readily and accurately by using the first few conserved quantities associated with the Korteweg-deVries equation. A similar procedure may be applied to the other soliton equations by using the quantities $g_n$ for the two-component eigenvalue problems that are given in (2.11.57) and (3.9.24) as long as breather-type solutions do not appear in the final result. The appearance of conserved quantities was associated with the time independence of the coefficient $c_{12}$, which is the reciprocal of the transmission coefficient in the associated scattering problem. In the present example (in which the roles played by space and time are interchanged) the coefficient $c_{12}$ contains space dependence as given by (7.12.33). We shall see that this space dependence is related to the energy that the pulse leaves in the medium as it passes through it. Even though the coefficient $c_{12}$ is no longer constant, information concerning final soliton amplitudes is still obtainable from the quantities $g_n$ by using a minor extension of the method used previously.

Adapting (3.9.26) to the notation of the present example, we have

$$(2i)^n C_n(\xi) = \int_{-\infty}^{\infty} d\tau\, g_n(\xi, \tau) \tag{7.12.36}$$

Using (3.9.23), we have

$$2iC_1(\xi) = \int_{-\infty}^{\infty} d\tau\, |q|^2 = \frac{1}{4}\int_{-\infty}^{\infty} d\tau\, |\bar{u}|^2 \tag{7.12.37}$$

Differentiating with respect to $\xi$ and using the Maxwell–Bloch equations (7.12.3) and (7.12.1c) we have

$$2i\frac{\partial C_1}{\partial \xi} = -\frac{1}{2}\int_{-\infty}^{\infty} d\tau \frac{\partial}{\partial \tau}\langle \mathfrak{N}(\eta, \xi, \tau)\rangle$$

$$= -\frac{1}{2}\langle \mathfrak{N}(\eta, \xi, +\infty) + 1\rangle \tag{7.12.38}$$

where we have set $\mathfrak{N}(\eta, \xi, -\infty) = -1$, since we are considering an attenuator and by $\mathfrak{N}(\eta, \xi, +\infty)$ we mean the value of the population inversion at point $\xi$

after the pulse has passed that point. The pulse profile eventually becomes a sequence of solitons which leave $\mathfrak{N}(\eta, \xi + \infty)$ with the value $-1$ for large values of $\xi$. This is evident from the exact value of $1 + \mathfrak{N}(\eta, \xi, +\infty)$ that is given in Ex. 2. If we now integrate (7.12.38) with respect to $\xi$ and use (7.12.36) with $n = 1$, we have

$$\int_{-\infty}^{\infty} d\tau\, g_1(\infty, \tau) - \int_{-\infty}^{\infty} d\tau\, g_1(0, \tau) = -\tfrac{1}{2} \int_0^{\xi} d\xi \langle 1 + \mathfrak{N}(\eta, \xi, +\infty) \rangle$$

$$(7.12.39)$$

According to the result of Ex. 2,

$$\int_0^{\xi} d\xi (1 + \mathfrak{N}) = -\frac{2}{\pi g(\eta)} \ln\left[ \frac{1 - \mathfrak{N}_0 + (1 + \mathfrak{N}_0) e^{-\pi \xi g}}{2} \right] \qquad (7.12.40)$$

where $\mathfrak{N}_0 = \mathfrak{N}(\eta, 0, +\infty)$. This latter term, the final population inversion at the interface where the pulse enters the medium, is unknown until the problem is solved. However, if we assume that only the population that is detuned by a reciprocal of the initial pulse width is affected by the pulse, and that this inversion of the population is Gaussian, we may write

$$\mathfrak{N}_0 = -1 + 2\epsilon \left( \frac{\tau_0}{\sqrt{\pi}} \right) e^{-(\pi \tau_0)^2} \qquad (7.12.41)$$

where $\tau_0$ is a measure of the pulse half-width and $\epsilon$ is a parameter that will be determined subsequently. If we further assume that $\epsilon \ll 1$, then, for large $\xi$, we have, from (7.12.40),

$$\int_0^{\infty} d\xi (1 + \mathfrak{N}) = \frac{2\epsilon \tau_0}{\pi^{3/2} g(\eta)} e^{-(\pi \tau_0)^2} \qquad (7.12.42)$$

and (7.12.39) becomes

$$\int_{-\infty}^{\infty} d\tau\, g_1(\infty, \tau) - \int_{-\infty}^{\infty} d\tau\, g_1(0, \tau) = -\frac{\epsilon}{\pi} \qquad (7.12.43)$$

Similar calculations involving $g_2$ and $g_4$ may also be carried out. To simplify these considerations, let us consider a situation in which there is no phase variation in the field profile. From (3.9.24) with $q = \tfrac{1}{2}\bar{u} = \tfrac{1}{2}u$, where $u$ is real, we find that

$$g_3 = \tfrac{1}{16}\left[ u^4 - 4(u_\tau)^2 \right]$$

$$g_5 = \tfrac{1}{32}\left[ u^6 + 8(u_{\tau\tau})^2 - 20u^2(u_\tau)^2 \right] \qquad (7.12.44)$$

Using the Maxwell–Bloch equations, we can show that

$$(2i)^3 \frac{\partial}{\partial \xi} C_3(\xi) = \frac{1}{16} \int_{-\infty}^{\infty} d\tau \frac{\partial}{\partial \tau} \left[ 32\eta^2 \mathfrak{N} - 4u^2 \mathfrak{N} \right]$$

$$= 2\eta^2 \left[ 1 + \mathfrak{N}(\eta, \xi, +\infty) \right] \qquad (7.12.45)$$

where we have used the fact that the solution $u$ ultimately becomes a sequence of localized pulses so that $u$ vanishes as $\tau \to \pm \infty$. In addition, a similar but more tedious calculation shows that

$$(2i)^5 \frac{\partial C_5}{\partial \xi} = \frac{1}{32} \int_{-\infty}^{\infty} d\tau \left[ 64\eta \mathfrak{N} + 6u^4 \mathfrak{N} - 8(u_\tau)^2 \mathfrak{N} \right.$$

$$- 32\eta^2 u^2 \mathfrak{N} + 16u^2 u_\tau S + 64\eta^2 u^2 \mathfrak{N} + 16\eta u^3 C \Big]$$

$$= 2\eta^4 \left[ 1 + \mathfrak{N}(\eta, \xi, +\infty) \right] \qquad (7.12.46)$$

We thus obtain

$$\int_{-\infty}^{\infty} d\tau\, g_3(\infty, \tau) - \int_{-\infty}^{\infty} d\tau\, g_3(0, \tau) = \frac{4\epsilon}{\pi} \int_{-\infty}^{\infty} d\eta\, \eta^2 e^{-\eta^2 \tau_0^2}$$

$$= \frac{4\epsilon}{\sqrt{\pi}\, \tau_0^3} \qquad (7.12.47)$$

$$\int_{-\infty}^{\infty} d\tau\, g_5(\infty, \tau) - \int_{-\infty}^{\infty} d\tau\, g_5(0, \tau) = \frac{2\epsilon}{\pi} \int_{-\infty}^{\infty} d\eta\, \eta^4 e^{-\eta^2 \tau_0^2}$$

$$= \frac{3\epsilon}{2\sqrt{\pi}\, \tau_0^5} \qquad (7.12.48)$$

If we assume an initial pulse profile of the form

$$u(0, \tau) = \frac{\theta_0}{\pi \tau_0} \operatorname{sech}\left( \frac{\tau}{\tau_0} \right) \qquad (7.12.49)$$

and a final pulse that consists of a number of $2\pi$ pulses, we may write

$$u(\xi, \tau) \underset{\xi \to +\infty}{\longrightarrow} \sum_{n=1}^{N} (2a_n) \operatorname{sech}\left[ a_n \left( \tau - \tau_i - \frac{\xi}{v_i} \right) \right] \qquad (7.12.50)$$

where $N$ is the number of $2\pi$ pulses. Here $v_i$ is the pulse velocity and $t_i$ is the phase term for the $i$th pulse. We find that (7.12.43), (7.12.47), and (7.12.48) may be put in the form

$$\sum_{n=1}^{N} x_n = \frac{1}{2}r - \frac{1}{2}\left(\frac{\gamma}{r}\right)$$

$$\sum_{n=1}^{N} x_n^3 = \frac{r-2}{r} + \frac{12\gamma}{r^3} \qquad (7.12.51)$$

$$\sum_{n=1}^{N} x_n^5 = \frac{8(r^4 - 5r^2 + 7)}{3r^3} - \frac{480\gamma}{r^5}$$

where

$$x_n = \frac{2a_n}{u(0,\tau)_{\max}} = \frac{2a_n \pi \tau_0}{\theta_0}$$

$$\gamma = \frac{2\epsilon \tau_0}{\pi} \qquad (7.12.52)$$

$$r = \frac{\theta_0}{\pi}$$

For $3\pi < \theta_0 < 5\pi$, two pulses are to be expected according to the area theorem, and the first two equations in (7.12.51) with $N=2$ are employed. The two-soliton amplitudes may be determined once $\gamma$ has been obtained. To evaluate $\gamma$, and thus the parameter $\epsilon$ introduced in (7.12.41), we note that for $\theta_0 = 3\pi$ one of the roots of the two equations should be zero. This result will be assured if we choose $\gamma$ so that $x_2 = 0$ for $\theta_0 = 3\pi$. A numerical solution (Schnack and Lamb, 1973) shows that this requires $\gamma = 0.673$. Since $\gamma$ is related to the population that is left inverted at the interface where the pulse enters the medium, we could expect it to depend upon the strength of the incident pulse. However, this dependence appears to be quite weak. To see this we use the same value of $\gamma$ in the solution of all three of the equations in (7.12.52) when $5\pi < \theta_0 < 7\pi$ and observe the value of $\theta_0$ at which the smallest of the three roots vanishes. The value is found to be $\theta_0 = 5.05\pi$, which is quite close to the value of $5\pi$ required by the area theorem.

The resulting locus of roots, along with points that represent the result of numerical solutions of the Maxwell–Bloch equations, are shown in Figure 7.6. Since $\gamma = 0.673$, the value of $\epsilon \tau_0 / \sqrt{\pi}$ to be used in obtaining (7.12.42) is approximately 0.6. While this value may be considered to be too large to justify the approximation used in obtaining (7.12.42), a more meaningful measure of the validity of the procedure is provided by the agreement with numerical results shown in Figure 7.6.

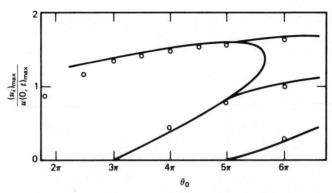

**Figure 7.6**   Pulse amplitudes obtained from conservation laws when energy transfer to medium is accounted for. Circles represent results of numerical solutions provided by M. O. Scully and F. A. Hopf.

## 7.13   LEVEL DEGENERACY

Self-induced transparency (i.e., the propagation of optical solitons) requires the existence of a pulse profile $\mathcal{E}(x,t)$ *and* a matrix element $p_0$ such that

$$\frac{p_0}{\hbar}\int_{-\infty}^{\infty} dt\, \mathcal{E}(x,t) = 2n\pi, \qquad n = 0, 1, 2, \ldots \qquad (7.13.1)$$

Unfortunately, most atomic systems possess level degeneracy. This is the term used to describe the situation in which there is more than one wave function associated with one or both of the levels $a$ and $b$. The matrix element $\int d^3 r\, \psi_a^* r \psi_b$, which has the single value $p_0$ in the previous development in Section 7.5, can now have a number of different values, depending upon the various wave functions associated with the levels $a$ and $b$. If there are different matrix elements connecting the resonant levels, then in general there will be no one pulse profile that can satisfy condition (7.13.1) for all different matrix elements. Thus self-induced transparency cannot be expected to occur in general. There are two ways of attempting to circumvent this limitation. The first way is to make $\int \mathcal{E}\, dt = 0$. Analytical expressions for field profiles satisfying this restriction are easily obtained for the nondegenerate case in the sine–Gordon limit. They are the breather solutions. For the degenerate case they would satisfy (7.13.1) for $n = 0$ for any value of the matrix element. At present, no zero area pulse solutions of the equations with level degeneracy have been obtained analytically. However, numerical and experimental considerations have shown the feasibility of this approach (Hopf et al., 1971; Grieneisen et al., 1972).

We now consider a second way of overcoming the restrictions of level degeneracy. Although the situation is quite restrictive experimentally, it leads to an interesting extension of the sine–Gordon equation. To have a simple example in mind, let us consider a degenerate two-level system of the type shown in Figure 7.7. More specifically, let us assume that the wave functions are of the form

$$\psi_{am}(r,\theta,\varphi) = u_a(r) P_l^m(\cos\theta) e^{im\varphi}, \qquad -l \leq m \leq l \qquad (7.13.2)$$

where the $P_l^m(\cos\theta)$ are the associated Legendre polynomials and similarly for the state $b, l', m'$. Only some one $l$ value is assumed to lead to resonance with the incident wave, so this quantum number will be ignored in specifying the wave function. Spin eigenfunctions could also be included, but their effect is easily appended and will be ignored for the present.

If the coordinate system with respect to which the angles $\theta$ and $\varphi$ are measured is taken to have the $z$ axis along the direction of polarization of the incident (plane-polarized) electric field, the calculation of the matrix elements is quite simple. We must consider integrals of the form

$$P_{ab} = \int \psi_a^*(-e\mathbf{r}\cdot\mathbf{E})\psi_b r^2 \sin\theta \, dr \, d\theta \, d\varphi \qquad (7.13.3)$$

Since $\mathbf{r}\cdot\mathbf{E} = rE\cos\theta$, the $\varphi$ integration requires $m' = m$ while the $\theta$ integration requires $l' = l \pm 1$. This latter result follows from the recurrence relations and orthogonality properties of Legendre polynomials. (Abramowitz and Stegun, 1964, p. 332). (When the spin quantum number $s$ is included, the result is $j = l + s$ with $j' = j \pm 1$ or 0.) The total wave function for the system is now

$$\psi = \sum_m (a_m \psi_{am} + b_m \psi_{bm}) \qquad (7.13.4)$$

where

$$\psi_{am} = u_a(r) P_l^m(\cos\theta) e^{im\varphi}$$
$$\psi_{bm} = u_b(r) P_{l'}^m(\cos\theta) e^{im\varphi} \qquad (7.13.5)$$

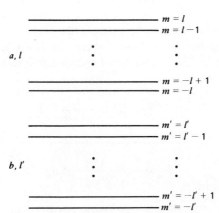

**Figure 7.7**  Example of energy levels for degenerate two-level system.

Only some *one* of the choices $l' = l \pm 1$ is assumed to lead to resonance with the incident light wave. Substituting the total wave function in (7.13.4) into the Schrödinger equation $i\hbar\psi_t = (H_0 + V)\psi$ and proceeding as in Section 7.6, we obtain the equations

$$i\hbar\dot{a}_m = E_a a_m + b_m p_m E$$
$$i\hbar\dot{b}_m = E_b b_m + a_m p_m E \tag{7.13.6}$$

where

$$p_m = \int d^3\mathbf{r}' \psi_{am}^*(-er\cos\theta)\psi_{bm} \tag{7.13.7}$$

Hence only the same $m$ values are connected and we find a number of essentially independent transitions taking place simultaneously.

The total polarization is again given by

$$P = \int d^3\mathbf{r}\,\psi_a^*(-e\mathbf{r})\psi_b \tag{7.13.8}$$

For the wave function given in (7.13.4), this becomes

$$P = \sum p_m(a_m^* b_m + b_m^* a_m) \tag{7.13.9}$$

which is an obvious extension of (7.6.6). For each of the sublevels there will be a pair of equations of the same form as those obtained for the nondegenerate case in (7.8.2). We thus obtain

$$\frac{\partial \mathcal{S}_m}{\partial t} = \frac{p_m}{\hbar}\mathcal{E}\,\mathfrak{N}_m$$
$$\frac{\partial \mathfrak{N}_m}{\partial t} = -\frac{p_m}{\hbar}\mathcal{E}\,\mathcal{S}_m \tag{7.13.10}$$

for each value of $m$. If we assume that all the degenerate levels are equally likely to occur, and that for a given $l$ there are $2l+1$ levels, which follows from the properties of the associated Legendre polynomials, then the first integral of (7.13.10) will be

$$\mathcal{S}_m^2 + \mathfrak{N}_m^2 = \frac{1}{2l+1} \tag{7.13.11}$$

Writing

$$\mathcal{S}_m = -\frac{1}{\sqrt{2l+1}}\sin\sigma_m$$
$$\mathfrak{N}_m = -\frac{1}{\sqrt{2l+1}}\cos\sigma_m \tag{7.13.12}$$

we then obtain

$$\frac{p_m}{\hbar}\, \mathcal{E} = \frac{\partial \sigma_m}{\partial t} \tag{7.13.13}$$

In place of (7.7.3a) we now have

$$\frac{\partial \mathcal{E}}{\partial t} + c\frac{\partial \mathcal{E}}{\partial x} = 2\pi n_0 \omega_0 \sum_{-l}^{l} p_m \mathcal{S}_m$$

$$= -\frac{2\pi n_0 \omega_0}{\sqrt{2l+1}} \sum_{-l}^{l} p_m \sin\left[\frac{p}{\hbar}\int_{-\infty}^{t} dt' \mathcal{E}(x,t')\right] \tag{7.13.14}$$

It is convenient to set $p_m = \kappa_m p$, where $p$ is the largest of the values of $p_m$. Then

$$\frac{\partial \mathcal{E}}{\partial t} + c\frac{\partial \mathcal{E}}{\partial x} = -\frac{2\pi n_0 \omega_0 p}{\sqrt{2l+1}} \sum_{-l}^{l} \kappa_m \sin(\kappa_m \sigma) \tag{7.13.15}$$

where

$$\sigma(x,t) = \frac{p}{\hbar}\int_{-\infty}^{t} dt'\, \mathcal{E}(x,t') \tag{7.13.16}$$

When spin is included, the summation on $m$ is from $-j$ to $+j$. Finally, setting $t' = t = x/c$, and recalling the definition of $\Omega^2$ in (7.8.3), we have

$$\frac{\partial^2 \sigma}{\partial x\, \partial t'} = -\frac{\Omega^2}{c\sqrt{2l+1}} \sum_{-j}^{j} \kappa_m \sin(\kappa_m \sigma) \tag{7.13.17}$$

For $\Delta j \equiv j - j' = \pm 1, 0$, the $\kappa_m$ are known to have the values (Condon and Shortley, 1957, p. 63)

$$\kappa_m = \begin{cases} \dfrac{(j^2 - m^2)^{1/2}}{j}, & \Delta j = -1 \\[2mm] \dfrac{m}{j}, & \Delta j = 0 \\[2mm] \dfrac{[(j+1)^2 - m^2]^{1/2}}{j}, & \Delta j = 1 \end{cases} \tag{7.13.18}$$

For our purposes the interesting case is $\Delta j = 0$ (known as $Q$ branch transitions), for then the various $\kappa_m$ are integral multiples of the lowest value. We can then have an optical field profile that inverts one degenerate state once, another twice, and so on, and thus obtain self-induced transparency in the presence of this type of level degeneracy.

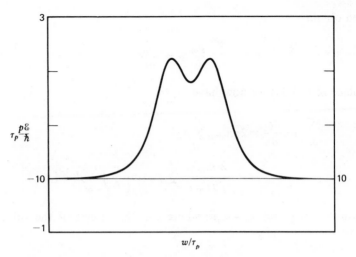

**Figure 7.8** Steady-state pulse profile (7.13.21) for $Q(2)$ transition.

The simplest case is for $j=2$, the $Q(2)$ transition. Then $\kappa_{-2}=-2$, $\kappa_{-1}=-1$, $\kappa_0=0$, $\kappa_1=1$, $\kappa_2=2$ and (7.13.17) becomes

$$\frac{\partial^2 \sigma}{\partial x\, \partial t'} = -\frac{2\Omega^2}{c\sqrt{5}}\left(\sin\sigma + \tfrac{1}{2}\sin\tfrac{1}{2}\sigma\right) \tag{7.13.19}$$

The steady-state solution $\sigma = \sigma(t-x/v)$ is readily found to be (Seeger, 1955, p. 566)

$$\sigma\left(t-\frac{x}{v}\right) = -4\tan^{-1}\left[\sqrt{5}\ \operatorname{csch}\left(\frac{t-x/v}{\tau_p}\right)\right] \tag{7.13.20}$$

where $cv = 1 + \tfrac{1}{2}\sqrt{5}\ \Omega^2\tau_p^2$. The associated electric field envelope is

$$\frac{p}{\hbar}\mathscr{E} = \frac{\partial\sigma}{\partial t} = \frac{4\sqrt{5}}{\tau_p}\ \frac{\operatorname{sech}(w/\tau_p)}{1+4\operatorname{sech}^2(w/\tau_p)}$$

$$= \frac{2\sqrt{5}}{\tau_p}\operatorname{sech} a\left[\operatorname{sech}\left(\frac{w}{\tau_p}+a\right) + \operatorname{sech}\left(\frac{w}{\tau_p}-a\right)\right] \tag{7.13.21}$$

where $\cosh 2a = 9$. A graph of this result is shown in Figure 7.8. Higher values of $j$ lead to steady-state solutions with more peaks in the pulse profile (Rhodes et al., 1968). Numerical solution of the interaction of two such pulses (Bullough and Caudrey, 1978) shows that they can experience fluctuations in the amplitudes of the peaks. Exact analytical expressions that describe these interactions have not yet been obtained. They hold out the hope of finding solitons with internal degrees of freedom.

## CHAPTER 8

## Bäcklund Transformations

We saw in Chapter 2 that the solution of the Schrödinger equation for the potential $V(x) = -v \operatorname{sech}^2 x$ can be expressed in terms of elementary functions when $v = n(n+1)$, where $n$ is a positive integer. These special values of $n$ correspond to reflectionless potentials and also provide the *initial* pulse profiles that evolve into the pure multisoliton solutions of the Korteweg–deVries equation. Starting with $n = 0$, so that $V(x) = 0$ these special potentials and the solutions of the associated Schrödinger equations can be constructed by a simple technique that was developed in Chapter 2.

It seems natural to inquire whether or not there is a simple transformation theory that relates these solutions at *later* times in their evolution. We now develop such a transformation theory not only for the Korteweg–deVries equation but also for some of the other more common soliton equations. For the sine–Gordon equation, which arose long ago in differential geometry (Eisenhart, 1909, p. 284), the appropriate transformation equations were obtained by A. V. Bäcklund and transformations of the type that we shall encounter have come to be known as Bäcklund transformations. Much of the early literature in this field was summarized in a paper by Goursat (1925) and has been reproduced in Miura (1976, p. 77). The expectation that other soliton equations would also have Bäcklund transformations was realized only more recently after Wahlquist and Estabrook (1973) obtained a Bäcklund transformation for the Korteweg–deVries equation.

One of the most interesting results of this transformation theory is that it leads to a simple superposition formula—the theorem of permutability—by which multisoliton solutions may be constructed from single soliton solutions by purely algebraic means.

## 8.1 BÄCKLUND TRANSFORMATION FOR THE KORTEWEG–deVRIES EQUATION

In Section 2.6 we considered the relation between the potentials in two Schrödinger equations that were related in such a way that the solution of one equation is a linear combination of the solution and first derivative of the solution of the other equation. The potentials that arose were the initial conditions for the pure multisoliton solutions. Let us now extend these considerations to the case in which the potentials depend upon a parameter while the eigenvalue term is required to be independent of this parameter (Wadati et ál., 1975). We know from our past considerations that the relation between two such potentials implies a corresponding relation between two solutions of the Korteweg–deVries equation. We shall find that this relation is of a special kind known as a Bäcklund transformation.

We consider the two Sturm–Liouville equations

$$y_{xx} = [\lambda + \varphi(x,t)] y \tag{8.1.1}$$

and

$$w_{xx} = [\lambda + \psi(x,t)] w \tag{8.1.2}$$

in which $w$ and $y$ are related according to

$$w = A(x,t,\lambda)y + y_x \tag{8.1.3}$$

If $\lambda_t = 0$, the functions $\varphi$ and $\psi$ can, of course, be two solutions of the Korteweg–deVries equation. Since $t$ is merely parametric in these equations, we may follow exactly the same procedure as that used in Section 2.6. Noting that integration now introduces an arbitrary function of time we find that $A(x,t,\lambda)$ must satisfy

$$2A_x + \varphi - \psi = 0 \tag{8.1.4}$$

$$A^2 - A_x - \varphi = \tilde{\lambda}(t) \tag{8.1.5}$$

which correspond to (2.6.5) and (2.6.6), respectively. When we again introduce the linearizing transformation $A = -\tilde{y}_x/\tilde{y}$, we find that $\tilde{y}$ satisfies (8.1.1) with $\lambda$ replaced by $\tilde{\lambda}(t)$. Since we are imposing the restriction that the eigenvalue parameter be time-independent, we see that $\tilde{\lambda}(t)$ must be a constant.

If we now introduce potential functions through the definitions $\varphi = z'_x$ and $\psi = z_x$, we may write (8.1.4) as

$$A = \tfrac{1}{2}(z - z') \tag{8.1.6}$$

Ultimately we shall be interested in spatial derivatives of $z$ and $z'$ which yield $\varphi$ and $\psi$. Hence the arbitrary function of time in (8.1.6) has been discarded. Introducing the notation $p = z_x$, we find that (8.1.5) may be written as

$$p + p' = m + \tfrac{1}{2}(z - z')^2 \tag{8.1.7}$$

where $m = -2\tilde{\lambda}$ and, of course, $p = \psi, p' = \varphi$. We thus obtain a first-order nonlinear (Riccati) differential equation that relates two functions that are closely related to solutions of the Korteweg–deVries equation. If $\psi$ is assumed to satisfy the Korteweg–deVries equation in the form $\psi_t - 6\psi\psi_x + \psi_{xxx} = 0$, then $z$ satisfies

$$z_t - 3(z_x)^2 + z_{xxx} = 0 \tag{8.1.8}$$

and similarly for $z'$.

In (8.1.7) we have a relation between the spatial derivatives of $z$ and $z'$. To obtain an expression relating their time derivatives we may first differentiate (8.1.7) with respect to time and then integrate with respect to $x$ to obtain

$$z_t + z_t' = \int dx (z - z')(z_t - z_t') \tag{8.1.9}$$

Using (8.1.8) and introducing the notation $u = z + z', v = z - z'$, we can write

$$z_t + z_t' = \int dx \left[ \tfrac{3}{2}(v^2)u_x - vv_{xxx} \right] \tag{8.1.10}$$

With the help of (8.1.7) and a partial integration, we obtain

$$z_t + z_t' = \tfrac{3}{2}(u_x)^2 - vv_{xx} + \tfrac{1}{2}(v_x)^2$$
$$= 2\left[ p^2 + pp' + (p')^2 \right] - (z - z')(z_{xx} - z_{xx}') \tag{8.1.11}$$

The final result in (8.1.11), coupled with (8.1.7.), constitute the Bäcklund transformation for the Korteweg–deVries equation. If a solution $z'$ to the third order equation (8.1.8) is known, then another solution $z$ may be obtained by solving the Bäcklund transformation equations which are the first-order equation (8.1.7) and the second-order equation (8.1.11).

As a simple example of the use of this transformation theory, we may begin with the trivial solution $z' = 0$ in (8.1.7) and find that $z$ satisfies

$$z_x = m + \tfrac{1}{2}z^2 \tag{8.1.12a}$$

$$z_t = 2p^2 - zz_{xx} = 2mz_x \tag{8.1.12b}$$

where we have used $z_{xx} = zp$ from (8.1.12a) in obtaining the last form in (8.1.12b). After integrating (8.1.12a) and determining the arbitrary function of time by using (8.1.12b), we have

$$z = -2\kappa \tanh(\kappa x - 4\kappa^3 t), \qquad |z| < 2\kappa \tag{8.1.13a}$$

and

$$z^* = -2\kappa \coth(\kappa x - 4\kappa^3 t), \qquad |z^*| < 2\kappa \tag{8.1.13b}$$

where $m = -2\kappa^2$. The function $\varphi = z_x$ is seen to be the single-soliton of the Korteweg–deVries equation. The solution $z^*$ is not of interest as a single-soliton solution since it diverges at $k_x = 4\kappa^3 t$, but it will be useful later.

The solution obtained in (8.1.13a) could now play the role of $z'$, in (8.1.7) and another solution for $z$ could be obtained. This procedure would, of course, entail the integration of a more complicated differential equation than that in (8.1.12a). Fortunately, there is a much simpler way to construct more elaborate solutions. They may be obtained by means of a simple superposition formula that requires no additional integration at all. To obtain such a formula, which, for the sine–Gordon equation, is known as the theorem of permutability, (Eisenhart, 1909, p. 286), we first consider two solutions $z_1$ and $z_2$ that are obtained from (8.1.7) by starting from a solution $z_0$ with two different values of $m$. We then have

$$p_1 + p_0 = m_1 + \tfrac{1}{2}(z_1 - z_0)^2 \qquad (8.1.14a)$$

$$p_2 + p_0 = \dot{m}_2 + \tfrac{1}{2}(z_2 - z_0)^2 \qquad (8.1.14b)$$

If we again use (8.1.7) but with $p_1$ substituted for $p'$ and $m_2$ for $m$ and denote the new solution by $z_{12}$, we have

$$p_{12} + p_1 = m_2 + \tfrac{1}{2}(z_{12} - z_1)^2 \qquad (8.1.15a)$$

Similarly, we can use $p_2$ for $p'$ and $m_1$ for $m$ to obtain a solution that we shall denote by $z_{21}$. We then have

$$p_{21} + p_2 = m_1 + \tfrac{1}{2}(z_{21} - z_2)^2 \qquad (8.1.15b)$$

As we shall see below, we may set $z_{12} = z_{21}$. Thus, if we subtract (8.1.14b) from (8.1.14a) and also subtract (8.1.15b) from (8.1.15a), we obtain two different expressions for $p_1 - p_2$. Equating the two expressions for this difference and setting $z_{12} = z_{21} = z_3$, we arrive at the purely algebraic relation,

$$z_3 - z_0 = \frac{2(m_2 - m_1)}{z_1 - z_2} \qquad (8.1.16)$$

This result leads quite directly to the two-soliton formula for the Korteweg–deVries equation if we begin with the solution $z_0 = 0$. More specifically, assume that $\kappa_1 > \kappa_2$ and then use $z^*$ from (8.1.13b) for $z_1$ and use (8.1.13a) for $z_2$. This choice of solutions precludes the possibility of a zero in the denominator of the superposition formula (8.1.16). In particular, the choice of constants $m_1 = -8$ and $m_2 = -2$ yields

$$z_3 = -6[2\coth(2x - 32t) + \tanh(x - 4t)]^{-1} \qquad (8.1.17)$$

The spatial derivative of this solution is the two-soliton formula given in (1.3.25). This method will also be used for generating multisoliton solutions of the sine–Gordon equation in Section 8.2.

## Validity of the Theorem of Permutability

Any proof of the theorem of permutability hinges upon the equality $z_{12} = z_{21}$. This equality is readily established if we interpret the solutions $y$ and $w$ as representing solutions to scattering problems. If we can show equality of certain scattering data, we can then rely upon inverse scattering theory to justify the claim that the scattering potentials themselves are also equal (Flaschka and McLaughlin, 1976).

Let us assume that $y(x)$, a solution of (8.1.1), is normalized as in (2.2.1) so that

$$y_< = e^{ikx} + R_0(k)e^{-ikx}, \qquad x \to -\infty$$
$$y_> = T_0(k)e^{ikx}, \qquad x \to +\infty \tag{8.1.18}$$

where $k^2 = -\lambda$. For localized potentials, (8.1.5) becomes $A^2 - A_x = \tilde{\lambda}$ as $|x| \to \infty$. Writing $\tilde{\lambda} = k_1^2$, a solution for $A$ is $A = -k_1 \tanh[k_1 x + C(t)]$. Thus

$$A_< = k_1, \qquad x \to -\infty$$
$$A_> = -k_1, \qquad x \to +\infty \tag{8.1.19}$$

A similarly normalized solution for $w$, the corresponding solution of (8.1.2), may be obtained by writing $w = K(Ay + y_x)$ where $K$ is to be chosen so as to again provide an incident wave of unit amplitude. From (8.1.3) we obtain

$$w_< = K[(k_1 + ik)e^{ikx} + (k_1 - ik)R_0 e^{-ikx}], \qquad x \to -\infty \tag{8.1.20}$$

which shows that to obtain unit incident amplitude we must choose $K = (k_1 + ik)^{-1}$. As $x \to +\infty$, we then have

$$w_> = -\frac{k_1 - ik}{k_1 + ik} T_0(k)e^{ikx} = T_1(k)e^{ikx} \tag{8.1.21}$$

Clearly, a subsequent transformation from the solution $w$ to a new solution with $\tilde{k} = k_2$ will yield the transmission coefficient

$$T_{12} = \frac{k_2 - ik}{k_2 + ik}\left(\frac{k_1 - ik}{k_1 + ik}\right)T_0 \tag{8.1.22}$$

Since the algebraic factors commute, the same result would be obtained if the terms $k_1$ and $k_2$ had been used in reverse order to yield $T_{21}$. The zeros of the two transmission coefficients also occur at the same locations in the $k$ plane (with the same residues), and hence we may conclude that the scattering potentials $(z_{12_x})$ and $(z_{21_x})$ that could be constructed from these scattering data are equal.

## 8.2 BÄCKLUND TRANSFORMATIONS FOR SOME OTHER EVOLUTION EQUATIONS

Bäcklund transformations for nonlinear evolution equations related to the linear system

$$v_{1x} + i\zeta v_1 = qv_2$$
$$v_{2x} - i\zeta v_2 = - qv_1 \tag{8.2.1}$$

where $q$ is real may be obtained by a simple extension to two components of the procedure used in the previous section for the Korteweg–deVries equation. We could also introduce a one-component formulation with complex potential (Wadati et al., 1975). Let us introduce the additional set of linear equations

$$w_{1x} + i\zeta w_1 = \varphi w_2$$
$$w_{2x} - i\zeta w_2 = - \varphi w_1 \tag{8.2.2}$$

and assume that $w$ and $v$ are related according to

$$\begin{pmatrix} w_1 \\ w_2 \end{pmatrix} = \begin{pmatrix} v_1 \\ v_2 \end{pmatrix}_x + \mathcal{C} \begin{pmatrix} v_1 \\ v_2 \end{pmatrix} \tag{8.2.3}$$

where $\mathcal{C}$ is a $2 \times 2$ matrix having elements $a_{ij}$. Using (8.2.1), we may rewrite (8.2.3) as

$$w_1 = Av_1 + Bv_2$$
$$w_2 = Cv_1 + Dv_2 \tag{8.2.4}$$

where $A = a_{11} - i\zeta$, $B = a_{12} + q$, $C = a_{21} - q$, and $D = a_{22} + i\zeta$. If we consider the case $a_{21} = - a_{12}$ so that $C = - B$, substitution of (8.2.4) into (8.2.2) yields

$$A_x = (q - \varphi)B$$
$$B_x = - 2i\zeta B - qA + \varphi D$$
$$B_x = 2i\zeta B + \varphi B - qD \tag{8.2.5}$$
$$D_x = (q - \varphi)B$$

These equations have first integrals $AD + B^2 = f^2(t)$ and $A - D = 2g(t)$, which imply that $B^2 = f^2 + g^2 - (A - g)^2$. Hence the relation $(A + D)_x = 2B(q - \varphi)$ may be written

$$\frac{A_x}{\sqrt{h^2 - (A - g)^2}} = q - \varphi \tag{8.2.6}$$

Setting $q = z_x$, $\varphi = z'_x$ and integrating we obtain $A = g + h \cos(z - z')$, $B = h \sin(z - z')$, and $D = -g + h \cos(z - z')$. The constant of integration has been chosen so that $B$ vanishes when $z$ and $z'$ vanish. For simplicity we set $g = h = \frac{1}{2}$ and obtain

$$(z + z')_x = -2i\zeta \sin(z - z') \tag{8.2.7}$$

from the second and third of (8.2.5). This first-order equation that relates $z$ and $z'$ is one of the Bäcklund transformation equations. The other Bäcklund transformation equation that relates the time dependence of $z$ and $z'$ depends upon the particular evolution equation being considered. This corresponds to the role played by the two pairs of linear equations in the two-component inverse method. In developing the inverse method we found that (5.4.2) for the space dependence is the same for all equations while (5.4.1) for the time dependence contains coefficients $A$, $B$, and $C$ that depend upon the particular evolution equation.

Before proceeding to a specific case, we note that the theorem of permutability follows from (8.2.7) by writing out a set of four equations similar to those in (8.1.14) and (8.1.15). After employing trigonometric identities, we obtain

$$\tan\left(\frac{z_3 - z_0}{z}\right) = \pm\left(\frac{a_1 + a_2}{a_1 - a_2}\right)\tan\left(\frac{z_2 - z_1}{2}\right) \tag{8.2.8}$$

where $a_i = -2i\zeta_i$. The choice of signs has been introduced into this result, since linear equations (8.2.1) are related to evolution equations for which the functions $-z$ is a solution whenever $z$ is a solution. The standard form of the theorem of permutability for the sine–Gordon equation is obtained when the lower sign in (8.2.8) is chosen. The solution $q_3 = z_{3x}$ that may be obtained from the theorem of permutability (8.2.8) applies to solutions of all evolution equations for which (8.7.1) are the appropriate linear equations. As noted above, the differences among the various soliton equations is governed by the transformation equation relating the time derivatives.

### Example—The Sine–Gordon Equation

We saw in Section 5.2 that the sine–Gordon equation $\sigma_{xt} = \sin \sigma$ is associated with the two-component inverse method when we set $q = -\frac{1}{2}\sigma_x$ in (8.2.1). In our present notation, where $q = z_x$, this implies that $z = -\frac{1}{2}\sigma$ and $z' = -\frac{1}{2}\sigma'$. We can now write the Bäcklund transformation equation (8.2.7) as

$$\frac{1}{2}(\sigma_x + \sigma'_x) = -2i\zeta \sin\left(\frac{\sigma - \sigma'}{2}\right) \tag{8.2.9}$$

Differentiating this relation with respect to time and noting that $\sigma$ and $\sigma'$ satisfy the sine–Gordon equation, we obtain

$$\sin\sigma + \sin\sigma' = -2i\zeta(\sigma - \sigma')_t \cos\left(\frac{\sigma - \sigma'}{2}\right) \qquad (8.2.10)$$

Using the identity $\sin\sigma + \sin\sigma' = 2\sin\left(\frac{\sigma + \sigma'}{2}\right)\cos\left(\frac{\sigma - \sigma'}{2}\right)$ and setting $2i\zeta = -a$, we obtain the other Bäcklund transformation equation

$$\tfrac{1}{2}(\sigma_t - \sigma'_t) = \frac{1}{a}\sin\left(\frac{\sigma + \sigma'}{2}\right) \qquad (8.2.11a)$$

This result together with (8.2.9), which now reads,

$$\tfrac{1}{2}(\sigma_x + \sigma'_x) = a\sin\left(\frac{\sigma - \sigma'}{2}\right) \qquad (8.2.11b)$$

are the Bäcklund transformation for the sine–Gordon equation. It is these equations that were obtained by Bäcklund in treating a problem in differential geometry (Eisenhart, 1909, p. 284).

Use of the transformation equations (8.2.11), as well as the theorem of permutability (8.2.8) with $z_i = -\tfrac{1}{2}\sigma_i$, to obtain what are now referred to as multisoliton solutions of the sine–Gordon equation may be traced to the work of Seeger et al. (1953). In particular, the two-soliton solution of the sine–Gordon equation is obtained from the theorem of permutability (8.2.8) by following the method used in the previous section for the Korteweg–de Vries equation. Divergent solutions do not occur in this case, however.

To proceed beyond the two-soliton solution it is sometimes convenient to introduce a diagrammatic representation of the Bäcklund transformation (8.2.11). A transformation from a solution $\sigma_i$ to a solution $\sigma_j$ with a parameter $a_k$ is represented as shown in Figure 8.1a. The diagram for the theorem of permutability is shown in Figure 8.1b. These diagrams may be compounded to yield analytical expressions that describe the interaction of more than two solitons. The diagram corresponding to a solution $\sigma_f$ that is the three-soliton solution of the sine–Gordon equation is shown in Figure 8.2. An example of the pulse profile is shown in figure 8.3.

The same diagrams may be also used for constructing solutions of the modified Korteweg–deVries equation, since the theorem of permutability (8.2.8) also applies to this equation. We need only start from the solution $z = 0$ and generate the single-soliton solutions of the form given in (5.1.22). The diagram for the interaction of two breather solutions is shown in Figure 8.4. Note that the constants in the upper half of the figure have been chosen to be

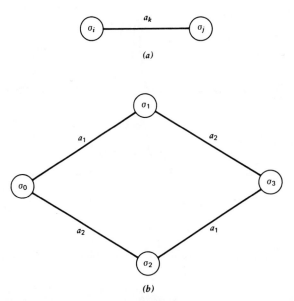

**Figure 8.1** Diagrammatic representations of Bäcklund transformations (*a*) single transformation given by (8.2.11). (*b*) Sequence of transformations for theorem of permutability (8.2.8).

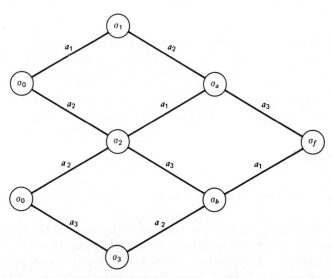

**Figure 8.2** Diagrammatic representations of sequence of Bäcklund transformations leading to three-soliton interaction.

**251**

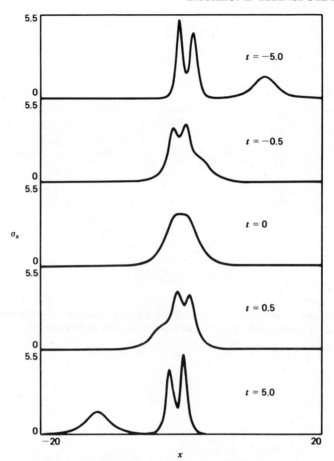

**Figure 8.3**   Three-soliton interaction for sine–Gordon equation.

the complex conjugates of those at the image locations in the lower half of the figure. This technique was used to obtain the two breather solution of the modified Korteweg–deVries equation shown in Figure 5.2. The method proceeds in a stepwise fashion and avoids the evaluation of large determinants.

## 8.3   MORE GENERAL BÄCKLUND TRANSFORMATIONS

The transformations considered in the two previous sections, in which the original function $z'$ and the transformed function $z$ satisfy the *same* partial differential equation, are a special class of Bäcklund transformations. In general, the functions $z$ and $z'$ may be governed by different equations. Furthermore, there is no need for there to be any close association with a

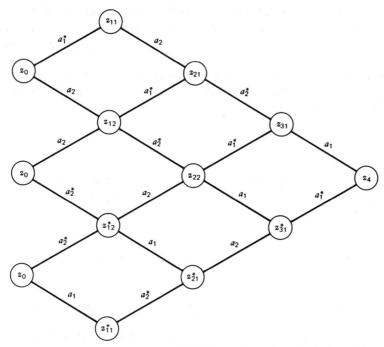

**Figure 8.4** Diagrammatic representation of Bäcklund transformations for the interaction of two breather solutions of the modified Korteweg–deVries equation (cf. Fig. 5.2).

linear Sturm–Liouville equation. In this section we briefly consider a technique developed by Clairin (1903) for constructing either the previously considered Bäcklund transformations or the more general type in which $z$ and $z'$ satisfy different partial differential equations.

Let us consider Bäcklund transformations relating second-order partial differential equations of the form $Rr + Ss + Tt + V = 0$, where $r = z_{xx}$, $s = z_{xy}$, $t = z_{yy}$ and $R, S, T$, and $V$ are functions of $x, y$, and $z$ as well as $p$ and $q$ where $p = z_x, q = z_y$. Equations of this form are frequently referred to as Monge-–Ampère equations. A Bäcklund transformation between two such second-order equations is a pair of first-order partial differential equations of the form

$$p = f(z, z', p' q')$$
$$q = \varphi(z, z', p', q')$$

(8.3.1)

A classification of the various types of transformations that can arise requires extensive consideration. It has been carried out by Clairin (1903) and summarized by Goursat (1925). A brief survey is included in Forsyth (1906, Vol. VI, p. 433). In certain instances the transformation is merely a simpler transformation known as a contact transformation. Otherwise, it is referred to as a Bäcklund transformation. Forsyth (1906, p. 441) has also shown that an arbitrarily chosen Monge–Ampère equation cannot be associated with a Bäcklund transformation. (It must be possible to satisfy four equations with three unknowns.)

To determine $f$ and $\varphi$ we begin by noting that the integrability condition (i.e., equality of mixed second derivatives) requires that the functions in (8.3.1) satisfy the relation

$$\Omega \equiv \frac{\partial p}{\partial y} - \frac{\partial q}{\partial x} = 0 \tag{8.3.2}$$

Noting that each of the four variables $z$, $z'$, $p'$, and $q'$ depends upon $x$ and $y$ and indicating partial derivatives by subscripts, we have

$$p_y = f_z q + f_{z'} q' + f_{p'} s' + f_{q'} t' \tag{8.3.3}$$

A similar expression holds for $q_x$. Using (8.3.1) to eliminate $p$ and $q$ and rearranging terms, we find that

$$\Omega = -\varphi_{p'} r' + (f_{p'} - \varphi_{q'}) s' + f_{q'} t' - \varphi_{z'} p' + f_{z'} q' + f_z \varphi - \varphi_z f = 0 \tag{8.3.4}$$

If the function $z$ is considered known, we see that (8.3.4) is a second-order partial differential equation for the function $z'$ as long as at least one of the coefficients $f_{q'}$, $\varphi_{p'}$, or $f_{p'} - \varphi_{q'}$ is unequal to zero.

### Example—Liouville's Equation

As an example of the construction of a Bäcklund transformation between two different equations by Clairin's method, we derive a transformation between Liouville's equation $s' = e^{z'}$, which is nonlinear, and the linear equation $s = 0$. The latter equation is the form taken by the wave equation when characteristic coordinates are employed. The general solution of the wave equation in terms of two arbitrary functions is, of course, known. We will see that the Bäcklund transformation may be used to obtain the general solution of Liouville's equation.

Since Liouville's equation contains $s'$ but not $r'$ or $t'$, we expect from the compatibility relation (8.3.4) that $\varphi_{p'} = f_{q'} = 0$ and $f_{p'} - \varphi_{q'} \neq 0$. We thus begin with the transformation equations in the form

$$p = f(z, z', p') \tag{8.3.5a}$$

$$q = \varphi(z, z', q') \tag{8.3.5b}$$

The $y$ derivative of (8.3.5a) yields

$$s = f_z q + f_{z'} q' + f_{p'} s' \tag{8.3.6}$$

Using (8.3.5b) plus the fact that $s=0$ and $s'=e^{z'}$, we have

$$0 = f_z \varphi + f_z q' + f_{p'} e^{z'} \tag{8.3.7}$$

Two derivatives with respect to $q'$ yield

$$f_z \varphi_{q'q'} = 0 \tag{8.3.8}$$

which implies (since $f_z$ presumably depends upon $z$) that $\varphi$ is a linear function of $q'$. Similarly, $f$ may be shown to be a linear function of $p'$ and we may write

$$\begin{aligned} p &= a(z,z')p' + m(z,z') \\ q &= \alpha(z,z')q' + \mu(z,z') \end{aligned} \tag{8.3.9}$$

As noted above, we must impose the restriction $f_{p'} - \varphi_{q'} = a - \alpha \neq 0$. We may avoid a relatively tedious analysis and obtain a simple transformation quite readily if we merely set $a=1, \alpha=-1$. The transformation equations are then

$$p = p' + m(z,z') \tag{8.3.10a}$$

$$q = -q' + \mu(z,z') \tag{8.3.10b}$$

and the compatibility relation (8.3.4) is

$$\Omega(z,\varepsilon',p',q') = 2e^{z'} - p'\mu_{z'} + q'm_{z'} + m_z(-q'+\mu) - \mu_z(p'+m) \tag{8.3.11}$$

where we have introduced the fact that $z'$ satisfies Liouville's equation. The unknown functions $m$ and $\mu$ are now determined by examining various derivatives of $\Omega$. In particular, we have

$$\begin{aligned} \Omega_{p'} &= -(\mu_{z'} + \mu_z) = 0 \\ \Omega_{q'} &= m_{z'} - m_z = 0 \end{aligned} \tag{8.3.12}$$

With these relations $\Omega$ reduces to

$$\Omega = \mu m_z - m\mu_z + 2e^{z'} = 0 \tag{8.3.13}$$

In addition (8.3.12) are partial differential equations having solutions

$$\begin{aligned} m(z,z') &= m(z+z') \\ \mu(z,z') &= \mu(z-z') \end{aligned} \tag{8.3.14}$$

From (8.3.13) we may obtain

$$\Omega_z = \mu m_{zz} - m\mu_{zz} = 0 \tag{8.3.15}$$

When the new independent variables

$$u = z + z'$$
$$v = z - z'$$

(8.3.16)

are introduced, (8.3.15) may be written

$$\frac{1}{m}\frac{d^2m}{du^2} = \frac{1}{\mu}\frac{d^2\mu}{dv^2} = -k^2$$

(8.3.17)

where the variables have been separated and $k^2 > 0$ is the separation constant. (The choice of sign is appropriate for the real exponentials appearing in Liouville's equation.) The expressions for $m$ and $\mu$ are thus of the form

$$m(u) = be^{ku} + ce^{-ku}$$
$$\mu(v) = \beta e^{kv} + \gamma e^{-kv}$$

(8.3.18)

The integration constants $b$, $c$, $\beta$, and $\gamma$ may now be determined by employing (8.3.7). We find

$$k\left( \beta b e^{2kz} - \beta c e^{-2kz'} + b\gamma e^{-2kz'} - \gamma c e^{-2kz} \right) + e^{z'} = 0$$

(8.3.19)

which may be satisfied by choosing either $k = \frac{1}{2}$, $\beta = c = 0$, and $\alpha = -2b$ or $k = -\frac{1}{2}$, $b = \gamma = 0$, and $\beta = -2/c$. With the former choice we obtain the Bäcklund transformation in the form

$$p = p' + be^{(1/2)(z+z')}$$
$$q = -q' - \frac{2}{b}e^{-(1/2)(z-z')}$$

(8.3.20)

The latter choice of constants yields the same one-parameter Bäcklund transformation.

The transformation equations (8.3.20) may be used to obtain the general solution of Liouville's equation. To this end we first note that the general solution of the wave equation $s = 0$ may be written $z = X(x) + Y(y)$. Also, on setting $\zeta = e^{-z'/2}$, the transformation equations are linearized and we obtain

$$\zeta_x + \frac{1}{2}X'\zeta = \frac{1}{2}be^{(X+Y)/2}$$
$$\zeta_y - \frac{1}{2}Y'\zeta = \frac{1}{b}e^{-(X+Y)/2}$$

(8.3.21)

The general solution of these linear equations is readily determined. If we set $e^X = (2/b)\theta'(x)$ and $e^{-Y} = b\chi'(y)$, then the general solution of Liouville's equation is found to be

$$e^{z'} = 2\frac{\theta'(x)\chi'(y)}{[\theta(x) + \chi(y)]^2}$$

(8.3.22)

In this example, where there is a Bäcklund transformation to a simple linear equation, the general solution of the nonlinear equation has been obtained. For transformations of a nonlinear equation into itself, as is the case for the soliton equations, the method only yields particular solutions.

Certain other transformations between different partial differential equations have been known for some time and they may be shown to be examples of Bäcklund transformations. The nonlinear Burgers equation $z'_t + z'z'_x + z'_{xx} = 0$ is related to the linear diffusion equation $z_t + z_{xx} = 0$ by the transformation (Forsyth, 1906, Vol. VI, p. 101; Hopf, 1950; Cole, 1950).

$$z' = 2\frac{\partial}{\partial x}(\ln z) \qquad (8.3.23)$$

In the notation of the present section this relation is equivalent to

$$p = \tfrac{1}{2}zz' \qquad (8.3.24)$$

As outlined in Ex. 1, the technique used above may be employed to obtain the additional equation

$$q = -\tfrac{1}{2}(zp' + pz') \qquad (8.3.25)$$

Equations 8.3.24 and 8.3.25 constitute a Bäcklund transformation that relates the Burgers equation to the diffusion equation.

We saw in Chapter 5 that the Korteweg–deVries and modified Korteweg–deVries equations are related by the Miura transformation. An additional equation may be adjoined to the Miura transformation to obtain a Bäcklund transformation. The procedure is outlined in Ex. 5.

*Exercise 1*

Use Clairin's method to determine the form of the relation $q = \varphi(p', z, z')$ that must be adjoined to the transformation $p = \tfrac{1}{2}zz'$ to obtain a Bäcklund transformation relating the Burgers equation to the diffusion equation. Show that the result may be put in the form given in (8.3.25).

*Exercise 2*

The Bäcklund transformation for the modified Korteweg–deVries equation is composed of (8.2.7) and an equation for the time derivative that may be obtained in a manner similar to that employed in deriving (8.1.11) for the Korteweg–deVries equation. Show that the result is (Wadati, 1974)

$$(z + z')_t = -a\{(z_{xx} - z'_{xx})\cos v + [(z_x)^2 + (z'_x)^2]\sin v\}$$

where $a = -2i\zeta$ and $v = z - z'$. Show that if $z' = 0$, then $z = 2\tan^{-1}\exp(ax - a^3t)$.

*Exercise 3*

Use Clairin's method to derive the Bäcklund transformation (8.2.11) that transforms the sine–Gordon equation $s = \sin z$ into itself. For simplicity, assume that the transformation is in the form of (8.3.10).

*Exercise 4*

Show that the substitution $\Gamma = \tan[(z + z')/4]$ converts the relation $p + p' = 2a \sin[(z - z')/2]$ into the Riccati equation

$$\Gamma_x + a\Gamma - \tfrac{1}{2}p(1 + \Gamma^2) = 0$$

With the further transformation $\Gamma = w_1/w_2$, this equation may be replaced by the pair of linear equations

$$w_{1x} + \tfrac{1}{2}aw_1 = \tfrac{1}{2}pw_2$$
$$w_{2x} - \tfrac{1}{2}aw_2 = -\tfrac{1}{2}pw,$$

Similar relations between Bäcklund transformations and the linear equations of inverse scattering theory may be obtained for other evolution equations (Miura, 1976).

*Exercise 5*

Write the Korteweg and modified Korteweg–deVries equations as $z_t + 6z'z'_x + z'_{xxx} = 0$ and $z_t - 6z^2 z_x + z_{xxx} = 0$, respectively, and consider a Bäcklund transformation between these two equations in the form $z_x = f(z, z')$ and $z_t = \varphi(z, z', z'_x, z'_{xx})$. Use Clairin's method to obtain the Bäcklund transformation

$$z_x = z' + z^2$$
$$z_t = z'_{xx} - 2(zz'_x + z'z_x)$$

The first of these relations is a form of the Miura transformation.

# CHAPTER 9

# Perturbation Theory

The physical situations that give rise to the standard soliton equations tend to be highly idealized. Inclusion of effects that are present in more realistic experimental situations, especially various forms of dissipation, leads to equations that differ from the standard soliton equations. If the additional terms in the equations are small in some sense, we may expect that at least for some initial interval of time, their effect on the various soliton phenomena considered previously may be small. In this chapter we develop a perturbation theory that may be applied to equations that differ slightly from the Korteweg–deVries equation and the cubic Schrödinger equation. The presentation is based on the recent results of Karpman and Maslov (1978), in which perturbation methods have also been applied to some of the other soliton equations. In this regard the paper by Kaup and Newell (1978) and a more recent paper by Karpman (1979) may also be consulted.

## The Korteweg–deVries Equation

### 9.1 BASIC EQUATIONS

We shall consider a perturbed Korteweg–deVries equation in the form

$$u_t - 6uu_x + u_{xxx} = \epsilon R(u) \tag{9.1.1}$$

where $\epsilon \ll 1$ and $R(u)$ is some specified function of the solution $u(x,t)$. Typical examples would be $\epsilon R(u) = \gamma u$, which provides a simple description of a dissipative process, and $\epsilon R(u) = \gamma \partial^2 u / \partial x^2$, which introduces the diffusion effects associated with the Burgers equation (cf. Sections 6.5 and 8.3).

From our previous considerations in Chapters 1 and 4 we know that the expression $-6uu_x + u_{xxx}$ may be written as the commutator of the operators $L = D^2 - u$ and $B = -4D^3 + 6uD + 3u_x$, where $D = \partial / \partial x$. From those previous developments it follows that (9.1.1) may be written

$$u_t - [L, B] = \epsilon R(u) \tag{9.1.2}$$

The operator $L$ still satisfies the Sturm–Liouville equation

$$Ly = \lambda y \qquad (9.1.3)$$

except that the eigenvalue parameter $\lambda$ is no longer time-independent. Also, the expression $By$ is no longer equal to $y_t$. In fact, the difference $\Phi \equiv y_t - By$ will play a fundamental role in the subsequent perturbation analysis. We shall find that the determination of $\Phi$ leads to expressions for the time dependence of $\lambda$ as well as of the reflection and transmission coefficients due to the perturbation $\epsilon R(u)$.

An equation governing $\Phi$ may be obtained from the time derivative of (9.1.3). Recalling that $L_t = -u_t$ and using (9.1.2) to eliminate $u_t$, we obtain

$$(L - \lambda)\Phi = \epsilon R(u)y + \lambda_t y \qquad (9.1.4)$$

We shall consider this equation for both bound states and scattering solutions. For the bound-state solutions we are primarily interested in the time dependence of $\lambda(t)$. The scattering solutions, however, involve the reflection and transmission of a wave of specified wave number that is incident upon the potential $u(x,t)$. The incident and scattered waves are at the same wave number and hence no $t$ dependence occurs in $\lambda$. Thus $\lambda_t = 0$ for the scattering solutions. There is, of course, $t$ dependence in the reflection and transmission coefficients.

Equation 9.1.4 may be viewed as an inhomogeneous differential equation for $\Phi$. A solution for $\Phi$ may be constructed from solutions of the homogeneous equation $(L - \lambda)\Phi = 0$ by the standard method of variation of parameters. Let us introduce fundamental solutions $f_1(x,k; t)$ and $f_2(x,k; t)$ that are solutions of the homogeneous counterpart of (9.1.4) when $u$ is a solution of (9.1.1.). The asymptotic properties of $f_1$ and $f_2$ are the same as those given in (2.8.2). Since $u$ is unknown, the fundamental solutions are also unknown but will be obtained approximately when the perturbation procedure is developed.

Let us consider in detail the solution of (9.1.4) for a function $\Phi$ that is related to the fundamental solution $f_2$. Writing $y = h(t)f_2$ and recalling that $B$ reduces to $-4D^3$ when $u$ vanishes, we find that

$$\Phi = y_t - By \underset{x \to -\infty}{\to} \left(h_t + 4ik^3 h\right)e^{-ikx} \qquad (9.1.5)$$

If we again set $h(t) = h(0)e^{4ik^3 t}$, we see that $\Phi$ vanishes as $x \to -\infty$. The solution $y$ then reduces to the unperturbed result when the perturbation vanishes.

On the other hand, as $x \to +\infty$, the relation (2.8.7a) between $f_2$ and $f_1$ leads to

$$\Phi \underset{x \to +\infty}{\to} h(t)\left[c_{12t}e^{-ikx} + \left(c_{11t} - 8ik^3 c_{11}\right)e^{ikx}\right] \qquad (9.1.6)$$

We now proceed to relate the time dependence of $c_{11}$ and $c_{12}$ to the perturbation $\epsilon R(u)$. This may be done by considering the solution $\Phi$ for intermediate values of $x$.

The solution of (9.1.4) for arbitrary values of $x$ may be written in terms of $f_1$ and $f_2$ as the linear combination

$$\Phi = \alpha(x,k,t)f_1(x,k;t) + \beta(x,k,t)f_2(x,k;t) \tag{9.1.7}$$

A standard application of the method of variation of parameters to (9.1.4) yields

$$\alpha = \frac{h(t)}{2ikc_{12}} \int_{-\infty}^{x} dx'\, G(u)[\, f_2(x',k;\, t)\,]^2 \tag{9.1.8a}$$

$$\beta = -\frac{h(t)}{2ikc_{12}} \int_{-\infty}^{x} dx'\, G(u)f_1(x',k;\, t)f_2(x',k;\, t) \tag{9.1.8b}$$

where $2ikc_{12} = W[\,f_1;\, f_2\,]$ and

$$G(u) = \begin{cases} \epsilon R(u) + \lambda_t & \text{bound states} \\ \epsilon R(u) & \text{scattering solutions} \end{cases} \tag{9.1.9}$$

The solution for $\Phi$ is seen to vanish both as $x \to -\infty$ and as $\epsilon \to 0$. As $x \to +\infty$, (9.1.7) and (9.1.8) yield

$$\Phi \xrightarrow[x \to +\infty]{} \frac{h(t)}{2ikc_{12}} \left[ e^{ikx} \int_{-\infty}^{\infty} dx'\, G(u)f_2(f_2 - c_{11}f_1) \right.$$

$$\left. - c_{12}e^{-ikx} \int_{-\infty}^{\infty} dx'\, G(u)f_1 f_2 \right] \tag{9.1.10}$$

Equations governing the time dependence of $c_{11}$ and $c_{12}$ may be obtained by equating coefficients of $e^{\pm ikx}$ in (9.1.6) and (9.1.10). The result is

$$c_{11t} - 8ik^3 c_{11} = -\frac{i}{2k} \int_{-\infty}^{\infty} dx'\, G(u)f_2(x',k;\, t)f_1(x', -k;\, t) \tag{9.1.11a}$$

$$c_{12t} = \frac{i}{2k} \int_{-\infty}^{\infty} dx'\, G(u)f_2(x',k;\, t)f_1(x',k;\, t) \tag{9.1.11b}$$

For the scattering solution, in which $G(u) = \epsilon R(u)$, we are interested in the reflection coefficient which is the ratio $c_{11}/c_{12}$. From (9.1.11) we find that

$$\left(\frac{c_{11}}{c_{12}}\right)_t - 8ik^3 \left(\frac{c_{11}}{c_{12}}\right) = -\frac{i\epsilon}{2k^2 c_{12}^2} \int_{-\infty}^{\infty} dx'\, R(u)f_2^2 \tag{9.1.12}$$

For the bound-state solutions, where $G(u) = \epsilon R(u) + \lambda_t$, we must still impose $c_{12}(i\kappa) = 0$ so that $f_2(x, i\kappa) = c_{11}(i\kappa) f_1(x, i\kappa)$. The second of (9.1.11), with $\lambda = \kappa^2$ and $k = i\kappa$, now yields

$$\kappa_t = -\frac{\epsilon}{2\kappa} \frac{\int_{-\infty}^{\infty} dx' R(u) f_2^2(x, i\kappa; t)}{\int_{-\infty}^{\infty} dx' f_2^2(x, i\kappa; t)} \tag{9.1.13}$$

In treating the first of (9.1.11), the function $f_1(x, -k; t)$ is singular at the pole locations associated with the bound states. [This is evident in the single pole solution given in Ex. 15 of Chapter 2 and in (9.2.2a).] However, by using $f_1(x, -k) = [f_2(x, k) - c_{11} f_1(x, k)]/c_{12}$ and evaluating the resulting indeterminate form as $k \to i\kappa$, we obtain

$$c_{11t} - 8\kappa^3 c_{11} = -\frac{\epsilon}{2\kappa \dot{c}_{12}(i\kappa)} \int_{-\infty}^{\infty} dx' R(u) f_2(i\kappa) \frac{d}{dk} \left[ f_2(k) - c_{11}(i\kappa) f_1(k) \right]_{k=i\kappa} \tag{9.1.14}$$

where the dot refers to a derivative with respect to $k$. The term involving $\dot{c}_{11}(i\kappa)$ does not appear since it is multiplied by the integral in (9.1.11b), which vanishes since $c_{12t} = 0$ for bound states. Also, the term proportional to $\lambda_t$ does not appear since it is proportional to $\int_{-\infty}^{\infty} dx(f_2 \dot{f}_1 - f_1 \dot{f}_2)$. This latter integral, as noted in Ex. 17 of Chapter 2, is proportional to $W(\dot{f}_1; \dot{f}_2)|_{x=-\infty}^{x=+\infty}$ which vanishes for bound-state solutions.

The results in (9.1.12) to (9.1.14) are the expressions upon which the perturbation analysis is based. It is clear from their structure that we may begin a perturbation expansion in powers of $\epsilon$ by using unperturbed expressions for $f_1$ and $f_2$ under the integrals. We now carry out this procedure for the single-soliton solution.

## 9.2  PERTURBATION OF THE SINGLE-SOLITON SOLUTION

The fundamental solutions associated with the unperturbed single soliton are solutions of the equation

$$\frac{d^2 f}{d^2 x} + \left[ k^2 - u_s(x) \right] f = 0 \tag{9.2.1}$$

with $u_s(z) = -2\kappa^2 \operatorname{sech}^2 z$ and $z = \kappa(x - \xi)$. The soliton parameters $\kappa$ and $\xi$ must now have a time dependence that is determined by the perturbation $\epsilon R(u)$. From Ex. 4 of Chapter 3 we have

$$f_1(x, k; t) = \frac{e^{ikx}(k + i\kappa \tanh z)}{k + i\kappa} \tag{9.2.2a}$$

$$f_2(x, k; t) = \frac{e^{-ikx}(k - i\kappa \tanh z)}{k + i\kappa} \tag{9.2.2b}$$

In addition, since the perturbed solution will not in general be reflectionless, we expect that the continuous spectrum will be excited by the perturbation.

## Perturbation of Bound-State Parameters

For the pole at $k = i\kappa$ we have

$$f_1(i\kappa) = \tfrac{1}{2} e^{-\kappa\xi} \operatorname{sech} z \tag{9.2.3a}$$

$$f_2(i\kappa) = \tfrac{1}{2} e^{\kappa\xi} \operatorname{sech} z \tag{9.2.3b}$$

and thus

$$c_{11}(i\kappa) = \frac{f_2(i\kappa)}{f_1(i\kappa)} = e^{2\kappa\xi} \tag{9.2.4}$$

The change in the eigenvalue parameter is obtained from (9.1.13), which becomes

$$\kappa_t = -\frac{\epsilon}{4\kappa} \int_{-\infty}^{\infty} dz\, R(u_s) \operatorname{sech}^2 z \tag{9.2.5}$$

The time dependence of the phase term $\xi(t)$ is derived from that of $c_{11}(ik)$ and relation (9.2.4). We first use (9.2.2) to obtain

$$\frac{d}{dk}\left[ f_2(k) - c_{11}(i\kappa) f_1(k) \right]_{k=i\kappa} = -\frac{i}{\kappa} e^{\kappa\xi} \operatorname{sech} z \left( \kappa x + \tfrac{1}{2}\sinh 2z \right) \tag{9.2.6}$$

Then (9.1.14) plus the relation $c_{11t} = (2k_t\xi + 2\kappa\dot{\varsigma}_t) e^{2\kappa\xi}$, which follows from (9.2.4) yields

$$\xi_t = 4\kappa^2 - \frac{\epsilon}{4\kappa^3} \int_{-\infty}^{\infty} dz\, R(u_s) \operatorname{sech}^2 z \left( z + \tfrac{1}{2}\sinh 2z \right) \tag{9.2.7}$$

We have also set $\dot{c}_{12}(i\kappa) = (2i\kappa)^{-1}$ since $c_{12}(k) = (k - i\kappa)/(k + i\kappa)$.

As an example of the use of these results, let us consider the effect of damping on the single-soliton solution. This is done by setting $\epsilon R(u_s) = -\gamma u_s = 2\gamma\kappa^2 \operatorname{sech}^2 z$. The relation $\int_{-\infty}^{\infty} dz \operatorname{sech}^4 z = \tfrac{4}{3}$ reduces (9.2.5) to the form $\kappa_t = -2\gamma\kappa/3$, or

$$\kappa(t) = \kappa_0 e^{-2\gamma t/3} \tag{9.2.8}$$

Since $\kappa$ is proportional to both the amplitude and the reciprocal of the soliton width, we have determined the rate at which the soliton spreads out and decays with increasing time.

Since the integrand in (9.2.7) is an odd function, we have $\xi_t = 4\kappa^2$. Integration yields

$$\xi(t) = \frac{3}{\gamma}\left[\kappa_0^2 - \kappa^2(t)\right] + \xi_0 \qquad (9.2.9)$$

where $\kappa(t)$ is given by (9.2.8). Note that as $\gamma \to 0$ we recover $\xi = 4\kappa_0^2 t + \xi_0$, the result for the unperturbed soliton.

### Perturbations in the Continuous Spectrum

We may expect that the solution to the perturbed single-soliton solution of the Korteweg–deVries equation will be of the form

$$u(x,t) = -2\kappa^2 \operatorname{sech}^2 z + \delta u \qquad (9.2.10)$$

where $\kappa$ and $\zeta$ have the time dependence obtained above and $\delta u$ is due to excitation of the continuous spectrum. To obtain $\delta u$ we may return to the Marchenko equation (3.2.7):

$$\Omega_R(x+y; t) + A_R(x,y; t) + \int_x^\infty dx' \Omega_R(x'+y; t) A_R(x,x'; t) = 0$$

$$(9.2.11)$$

where, for a single soliton,

$$\Omega_R(z; t) = \int_{-\infty}^\infty \frac{dk}{2\pi} \frac{c_{11}(k,t)}{c_{12}(k,t)} e^{ikz} - i\frac{c_{11}(i\kappa,t)}{\dot{c}_{12}(i\kappa,t)} e^{-\kappa t} \qquad (9.2.12)$$

We now set $\Omega_R = \Omega_S + \delta\Omega$, where $\Omega_S$ and $\delta\Omega$ refer to the second and first terms in (9.2.12), respectively. If we also write $A_R(x,y) = A_S(x,y) + \delta A(x,y)$ and neglect the higher-order product $\delta A\, \delta\Omega$ in the integral in the Marchenko equation, we find

$$\delta A(x,y; t) + \int_x^\infty dx' \Omega_S(x'+y; t)\delta A(x,x'; t) = \Psi(x,y; t) \qquad (9.2.13)$$

where

$$\Psi(x,y; t) = -\delta\Omega(x+y; t) - \int_x^\infty dx' A_S(x,x'; t)\delta\Omega(x',y; t) \qquad (9.2.14)$$

From Ex. 4 of Chapter 3 we have

$$\Omega_S(z) = 2\kappa e^{\kappa(2\xi - z)} \qquad (9.2.15a)$$

$$A_S(x,y) = -\frac{2\kappa e^{\kappa(2\xi - x - y)}}{1 + e^{2\kappa(\xi - x)}} \qquad (9.2.15b)$$

We now solve (9.2.13) for $\delta A(x,x)$ and use the relation

$$\delta u(x,t) = -2\frac{d}{dx}\delta A(x,x;\,t) \tag{9.2.16}$$

To obtain $\delta A(x,x;\,t)$ we first multiply (9.2.13) by $e^{-\kappa y}$ and integrate from $x$ to $\infty$. This yields

$$\int_x^\infty dy\,e^{-\kappa y}\delta A(x,y;\,t) = \frac{\int_x^\infty dy\,e^{-\kappa y}\Psi(x,y;\,t)}{1+e^{2\kappa(\xi-x)}} \tag{9.2.17}$$

Using the definition of $A_s(x,y)$ we then have

$$\delta A(x,y;\,t) = e^{\kappa x}A_s(x,y)\int_x^\infty dy'\,e^{-\kappa y'}\Psi(x,y';\,t)+\Psi(x,y;\,t) \tag{9.2.18}$$

The term $\Psi(x,y;\,t)$ may be expressed in terms of $c_{11}/c_{12}$. When (9.2.15b) and the definition

$$\delta\Omega(x,t) = \int_{-\infty}^\infty \frac{dk}{2\pi}e^{i\kappa x}\frac{c_{11}(k,t)}{c_{12}(k,t)} \tag{9.2.19}$$

are used in (9.2.14), we find that

$$\Psi(x,y;\,t) = \int_{-\infty}^\infty \frac{dk}{2\pi}\frac{c_{11}}{c_{12}}e^{ik(x+y)}\left[i-\frac{A_s(x+y)}{k+i\kappa}\right] \tag{9.2.20}$$

Then (9.2.18) yields $\delta A(x,x;\,t)$ in the form

$$\delta A(x,x;\,t) = \int_{-\infty}^\infty \frac{dk}{2\pi}\frac{c_{11}}{c_{12}}e^{2ikx}\left[i-\frac{A_s(x,x)}{k+i\kappa}\right]^2 \tag{9.2.21}$$

Noting that

$$i-\frac{A_s(x,x)}{k+i\kappa} = i\left(\frac{k+i\kappa\tanh z}{k+i\kappa}\right) \tag{9.2.22}$$

we may use (9.2.16) and (9.2.21) to obtain the final result

$$\delta u = 2\kappa\frac{d}{dz}\int_{-\infty}^\infty \frac{dk}{2\pi}\frac{c_{11}}{c_{12}}e^{2ik\xi+2izk/\kappa}\left(\frac{k+i\kappa\tanh z}{k+i\kappa}\right)^2 \tag{9.2.23}$$

The ratio $\rho = c_{11}/c_{12}$ is to be determined to first order in $\epsilon$ by solving (9.1.12). On the right-hand side of this equation we use the zero-order expression* $c_{12} = (k - i\kappa)/(k + i\kappa)$ and (9.2.2b) for $f_2$ and find that

$$\rho_t - 8ik^3\rho = \frac{i\kappa}{k(k - i\kappa)^2} \int_{-\infty}^{\infty} dz \, \epsilon R(u) e^{-2ikz/\kappa} (k - i\kappa \tanh z)^2 \quad (9.2.24)$$

For the example considered above, namely $\epsilon R = -\gamma u$, we have

$$\rho_t - 8ik^3\rho = -\frac{i\gamma k^3 e^{-2ik\xi(t)}}{k(k - i\kappa)^2} I\left(\frac{k}{\kappa}\right) \quad (9.2.25)$$

where $\kappa(t)$ and $\xi(t)$ are given by (9.2.8) and (9.2.9), respectively, and

$$I(p) \equiv \int_{-\infty}^{\infty} dz \, e^{-2ipz} \operatorname{sech}^2 z (p - i \tanh z)^2 \quad (9.2.26)$$

This integral may be written as a sum of three integrals as $I = p^2 I_1 - 2p I_2 - I_3$, where

$$I_1 = 2 \int_0^{\infty} dz \cos 2pz \operatorname{sech}^2 z = 2\pi p \operatorname{csch} \pi p$$

$$I_2 = 2 \int_0^{\infty} dz \sin 2pz \operatorname{sech}^2 z \tanh z = p I_1 \quad (9.2.27)$$

$$I_3 = 2 \int_0^{\infty} dz \cos 2pz \operatorname{sech}^2 z \tanh^2 z = \tfrac{1}{3}(1 - 2p^2) I_1$$

The relations among these integrals are readily established through partial integration. We find that

$$I(p) = -\frac{2\pi}{3} p(1 + p^2) \operatorname{csch} \pi p \quad (9.2.28)$$

The integration of (9.2.4) for $\rho$ with the time dependence of $\kappa$ and $\xi$ as given in (9.2.8) and (9.2.9) would be quite involved. Instead, we shall assume that[†] $\gamma t \ll 1$ and write $\kappa = \kappa_0$ and $\xi = 4\kappa_0^2 t + \xi_0$. Determination of $\rho(t)$ is then immediate and for the initial condition $\rho(0) = 0$ we find that

$$\rho(t) = \frac{c_{11}(t)}{c_{12}(t)} = \frac{\pi \gamma e^{-2ik\xi_0}}{12k(k^2 + \kappa_0^2)\sinh(\pi k/\kappa_0)} \left[ e^{8ik^3 t} - e^{-8ik\kappa_0^2 t} \right] \quad (9.2.29)$$

*Recently it has been shown that a correction term must be included (Karpman, 1979, equation 5.23). However, the qualitative results to be obtained here are unaffected by this correction.
[†]For a detailed consideration of various time scales involved in this problem, see Karpman (1979).

It should first be noted that this approximate expression for $\rho(t)$ violates the general result $c_{11}(0,t)/c_{12}(0,t) \to -1$ that was obtained in Ex. 15 of Chapter 2. However, we shall find that the $z$ derivative in (9.2.23) serves to introduce an additional factor of $k$ into the integrand in (9.2.23) and thus render the integrand sufficiently insensitive to small values of $k$ that the foregoing expression for $\rho(t)$ can provide a satisfactory description of $\delta u$. For a more thorough discussion of this point, the reader may refer to the results of Karpman and Maslov (1978).

To obtain the approximate form of $\delta u$, we first note that outside the vicinity of the soliton, which is where the contribution due to the perturbation is of most interest, we may neglect the slow $z$ dependence in $\tanh z$. Furthermore, we note that the integrand in (9.2.23) oscillates rapidly for large $k$. We thus expect that the dominant contribution to the integrand will come from the region near $k=0$. If we approximate the amplitude of the integrand of (9.2.23) in the region near $k=0$ and, as mentioned above, neglect the slow $z$ dependence of $\tanh z$ in carrying out the derivative with respect to $z$, we find that

$$\delta u = \frac{\gamma}{3k_0} F(x,t) \tag{9.2.30}$$

where

$$F(x,t) = \int_{-\infty}^{\infty} \frac{dk}{2\pi i} \frac{e^{2ik(x-\xi_0)}}{k} \left( e^{8ik^3 t} - e^{-8ik\kappa_0^2 t} \right) \tag{9.2.31}$$

and a factor $\tanh^2 z$ has been set equal to unity. The function $F(x,t)$ is readily understood by first considering the derivative

$$\frac{\partial F}{\partial x} = \frac{1}{2\pi} \int_{-\infty}^{\infty} d\zeta \, e^{i\zeta(x-x_0)} \left( e^{i\zeta^3 t} - e^{-4i\kappa_0^2 t \zeta} \right)$$

$$= \frac{1}{\pi} \int_{0}^{\infty} d\zeta \cos\left[ \zeta(x-\xi_0) + \zeta^3 t \right] - \delta\left(x-\xi_0-4\kappa_0^2 t\right)$$

$$= \frac{1}{(3t)^{1/3}} \text{Ai}\left[ \frac{x-\xi_0}{(3t)^{1/3}} \right] - \delta\left(x-\xi_0-4\kappa_0^2 t\right) \tag{9.2.32}$$

where $\text{Ai}(x)$ is the Airy function (Abramowitz and Stegun, 1964). From the first form (9.2.32) we see that as $t \to 0$ the expression involving the Airy function becomes a delta function. If we assume that the perturbation $\delta u$ given in (9.2.30) vanishes as $x \to -\infty$, that is, well behind the pulse, then

$$F(x,t) = \int_{-\infty}^{(x-\xi_0)/(3t)^{1/3}} d\eta \, \text{Ai}(\eta) - \theta\left(x-\xi_0-4\kappa_0^2 t\right) \tag{9.2.33}$$

where $\theta(x)$ is the usual step function. Tables of the integral over the Airy function are available (Abramowitz and Stegun, 1964, p. 478). The function $\int_{-\infty}^{\eta} d\eta' \, \text{Ai}(\eta')$ oscillates about 0 for $x < 0$ and reaches an amplitude of $\frac{2}{3}$ at $x = 0$. For $x > 0$ it increases monotonically to unity. Consequently, the function $F(x, t)$ has the shape shown in Figure 9.1. The perturbation term $-\gamma u$ in (9.1.1) is thus seen to produce a plateau behind the soliton. This effect has been observed in numerical solutions (Randall and Leibovich, 1973).

The results just obtained provide an illuminating example of the effects of a perturbation on both the bound-state parameters (the soliton) and the continuous spectrum (the plateau behind the soliton). An indication of the relative magnitude of each of these contributions to the total solution may be obtained by first rewriting the evolution equation in the form

$$u_t + \left(-3u^2 + u_{xx}\right)_x = -\gamma u \tag{9.2.34}$$

Thus the area under a localized solution must satisfy

$$\frac{\partial}{\partial t} \int_{-\infty}^{\infty} dx \, u(x, t) = -\gamma \int_{-\infty}^{\infty} dx \, u(x, t) \tag{9.2.35}$$

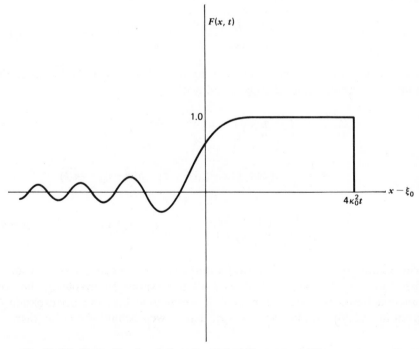

**Figure 9.1**   Shelf behind soliton in solution of perturbed Korteweg–deVries equation $u_t - 6uu_x + u_{xxx} = \gamma u$.

For the single-soliton solution we have found that

$$u(x,t) = -2\kappa^2 \operatorname{sech}^2 z + \frac{\gamma}{3\kappa_0} F(x,t) \qquad (9.2.36)$$

Noting from Figure 9.1 that the area under $F(x,t)$ is approximately equal to $4\kappa_0^2 t$, we find that

$$\int_{-\infty}^{\infty} dx\, u = -4\kappa(t) + \tfrac{4}{3}\gamma\kappa_0 t \qquad (9.2.37)$$

To first order in $\gamma$ the right-hand side of (9.2.34) is thus $4\gamma\kappa_0$, while the left-hand side reduces to

$$\frac{\partial}{\partial t}\left( -4\kappa_0 e^{-2\gamma t/3} + \tfrac{4}{3}\gamma\kappa_0 t \right) = \tfrac{8}{3}\gamma\kappa_0 + \tfrac{4}{3}\gamma\kappa_0 = 4\gamma\kappa_0 \qquad (9.2.38)$$

We thus see that one-third of the time rate of change of the area under the disturbance is contributed by the plateau behind the soliton. Further discussion of these results may be found in the research literature quoted at the beginning of this chapter.

## A Simple Procedure

We have found that the perturbation due to damping has a very simple effect on the shape of the soliton solution of the Korteweg–deVries equation. As might be expected, so simple a result may be obtained in a more direct way. We now show how the energy, as well as other conserved quantities for the *unperturbed* Korteweg–deVries equation, may be used to analyze the decay of the soliton solution when damping is present.

We begin by noting that the conserved quantity $g_4$ given in (2.8.51) is proportional to the energy contained in a solution of the Korteweg–deVries equation. To show this we use the procedure outlined in Section 2.1 and introduce a Lagrangian density from which the undamped Korteweg–deVries equation may be derived. The Hamiltonian density, which is the energy density, is related to this Lagrangian.

Setting $u = \varphi_x$, the Lagrangian density for the Korteweg–deVries equation (4.1.1) is

$$\mathcal{L} = -\tfrac{1}{2}\varphi_x \varphi_t + (\varphi_x)^3 + \tfrac{1}{2}(\varphi_{xx})^2 \qquad (9.2.39)$$

Since a second derivative appears, the equation of motion follows from

$$\frac{\partial}{\partial t}\left( \frac{\partial \mathcal{L}}{\partial \varphi_t} \right) + \frac{\partial}{\partial x}\left( \frac{\partial \mathcal{L}}{\partial \varphi_x} \right) - \frac{\partial^2}{\partial x^2}\left( \frac{\partial \mathcal{L}}{\partial \varphi_{xx}} \right) - \frac{\partial \mathcal{L}}{\partial \varphi} = 0 \qquad (9.2.40)$$

With the Lagrangian above, this yields

$$\varphi_{xt} - 6\varphi_x\varphi_{xx} + \varphi_{xxxx} = 0 \tag{9.2.41}$$

which is the undamped Korteweg–deVries equation (4.1.1) when we reintroduce $u = \varphi_x$.

The Hamiltonian density is

$$\mathcal{H} = \varphi_t \frac{\partial \mathcal{L}}{\partial \varphi_t} - \mathcal{L}$$

$$= -(\varphi_x)^3 - \tfrac{1}{2}(\varphi_{xx})^2 \tag{9.2.42}$$

The energy in the solution is then

$$\mathcal{E} = \int_{-\infty}^{\infty} dx\, \mathcal{H} = -\int_{-\infty}^{\infty} dx\left(u^3 + \tfrac{1}{2}u_x^2\right)$$

$$= \int_{-\infty}^{\infty} dx\, g_4 \tag{9.2.43}$$

In relating this result to $g_4$, a partial integration has been performed and perfect derivatives have been ignored.

For the undamped Korteweg–deVries equation, $\mathcal{E}_t = 0$. We now *define* the energy of the damped Korteweg–deVries equation to be the same expression given in (9.2.43) and determine how it changes with time. The procedure being used is thus quite similar to that employed to describe the damping of a mechanical oscillator that is subject to resistance.

For the single-soliton solution we use $u = -2\kappa^2 \operatorname{sech}^2 z$. Noting that $\kappa\, dx = dz$, we find

$$\mathcal{E} = -8\kappa^5 \int_{-\infty}^{\infty} dz(\operatorname{sech}^4 z - 2\operatorname{sech}^6 z)$$

$$= \tfrac{32}{5}\kappa^5 \tag{9.2.44}$$

so that $\mathcal{E}_t = 32\kappa^4 \kappa_t$.

On the other hand, from (9.2.43) we have

$$\mathcal{E}_t = -\int dx\left(3\varphi_x^2\varphi_{xt} + \varphi_{xx}\varphi_{xxt}\right)$$

$$= -\int dx\, \varphi_{xt}\left(3\varphi_x^2 - \varphi_{xxx}\right)$$

$$= \int dx\, \varphi_t\left(6\varphi_x\varphi_{xx} - \varphi_{xxxx}\right)$$

$$= \gamma \int dx\, \varphi_x\varphi_t \tag{9.2.45}$$

Since this result is already proportional to $\gamma$, we use the unperturbed solutions in evaluating the integral and obtain $\mathcal{E}_t = -\frac{64}{3}\gamma\kappa^5$. Equating the two expressions for $\mathcal{E}_t$, we have $\kappa_t = -\frac{2}{3}\gamma\kappa$, which was obtained previously and leads to the expression for $\kappa$ given in (9.2.8).

This same result is even more easily obtained by using the conserved quantity $g_2$. Further consideration of this approach in the context of a perturbed sine–Gordon equation may be found in McLaughlin and Scott (1978).

*Exercise 1*

A perturbation in the form of a diffusion term may be treated by setting $\epsilon R(u) = \gamma u_{xx}$. Show that the parameter determining the soliton width now satisfies $\kappa_t = -\frac{8}{15}\gamma\kappa^3$. Recover this result by using conserved quantities.

# The Cubic Schrödinger Equation

## 9.3  BASIC EQUATIONS

A perturbation procedure for the evolution equations related to the two-component eigenvalue problem considered in Chapter 5 can also be developed. The method involves only minor extensions of that used for the Korteweg–deVries equation in the two previous sections. As an example, we consider a perturbed cubic Schrödinger equation of the form

$$iu_t + u_{xx} + 2|u|^2 u = i\epsilon R(u) \tag{9.3.1}$$

We begin by constructing an operator form of this equation (and its complex conjugate) that corresponds to (9.1.2) for the Korteweg–deVries equation. We use the operator that was obtained in Ex. 3 of Chapter 5, namely

$$A = 2i\begin{pmatrix} 1 & 0 \\ 0 & -1 \end{pmatrix}D^2 - 2i\begin{pmatrix} 0 & u \\ u^* & 0 \end{pmatrix}D - i\begin{pmatrix} -|u|^2 & u_x \\ u_x^* & |u|^2 \end{pmatrix} \tag{9.3.2}$$

where $D = \partial/\partial x$. We also use the operator

$$L = \begin{pmatrix} 1 & 0 \\ 0 & -1 \end{pmatrix}D - \begin{pmatrix} 0 & u \\ u^* & 0 \end{pmatrix} \tag{9.3.3}$$

which follows from a rewriting of (5.3.2) in the form

$$Lv = \lambda v \tag{9.3.4}$$

with $v = \begin{pmatrix} v_1 \\ v_2 \end{pmatrix}$ and $\lambda = -i\zeta$. We then obtain

$$[L,A] = \begin{pmatrix} 0 & i(u_{xx} + 2u|u|^2) \\ -i(u_{xx}^* + 2u^*|u|^2) & 0 \end{pmatrix} \qquad (9.3.5)$$

Since $L_t = -U_t$ where

$$u = \begin{pmatrix} 0 & u \\ u^* & 0 \end{pmatrix} \qquad (9.3.6)$$

we may rewrite (9.3.1) and its complex conjugate as

$$U_t - [L,A] = \epsilon R \qquad (9.3.7)$$

where

$$R = \begin{pmatrix} 0 & R \\ R^* & 0 \end{pmatrix} \qquad (9.3.8)$$

In complete analogy with the derivation of (9.1.4), we find that a $t$ derivative of (9.3.4) leads to

$$(L - \lambda)\Phi = Gv \qquad (9.3.9)$$

where $\Phi = v_t - Av$,

$$G = \begin{cases} \epsilon R + \lambda_t I & \text{bound states} \\ \epsilon R & \text{scattering solutions} \end{cases} \qquad (9.3.10)$$

and $I$ is a $2 \times 2$ unit matrix. Let us consider a fundamental solution of (9.3.9) that is proportional to $\begin{pmatrix} 1 \\ 0 \end{pmatrix} e^{-ikx}$ as $x \to -\infty$. We thus set $v = h(t)\varphi(x,k; t)$, where $\varphi$ is a fundamental solution of the type defined in (2.11.15). Using the form of $A$ in the limit $x \to -\infty$, where $u = 0$, we find that

$$\Phi \underset{x \to -\infty}{\to} (h_t + 2ik^2h)\begin{pmatrix} 1 \\ 0 \end{pmatrix} e^{-ikx} \qquad (9.3.11)$$

and thus $h(t) = h(0)e^{-2ik^2t}$.

As $x$ approaches $+\infty$, we obtain

$$\Phi \underset{x \to \infty}{\to} \begin{pmatrix} 0 \\ 1 \end{pmatrix} e^{ikx}h(t)(c_{11t} - 4ik^2c_{11}) + \begin{pmatrix} 1 \\ 0 \end{pmatrix} e^{-ikx}h(t)c_{12t} \qquad (9.3.12)$$

where we have used the relation between $\varphi$ and $\psi$ given in (3.9.8a) and the limiting forms

$$\psi \underset{x \to \infty}{\to} \begin{pmatrix} 0 \\ 1 \end{pmatrix} e^{ikx}$$

$$\tilde{\psi} \underset{x \to \infty}{\to} \begin{pmatrix} 1 \\ 0 \end{pmatrix} e^{-ikx} \tag{9.3.13}$$

The time dependence of $c_{12}$ as well as the additional time dependence of $c_{11}$ due to the perturbation is now determined by solving (9.3.9) for intermediate values of $x$ as was done for the Korteweg–deVries equation in Section 9.1. Again we use a variation of parameters method and write

$$\Phi = \alpha\psi + \beta\varphi \tag{9.3.14}$$

where $\psi$ and $\varphi$ are fundamental solutions of the homogeneous conterpart of (9.3.9) while $\alpha$ and $\beta$ are functions to be determined. Substitution of (9.3.14) into (9.3.9) yields

$$K\alpha_x\psi + K\beta_x\varphi = Gv \tag{9.3.15}$$

where we have set

$$K \equiv \begin{pmatrix} 1 & 0 \\ 0 & -1 \end{pmatrix} \tag{9.3.16}$$

To solve (9.3.15) for $\alpha_x$ and $\beta_x$ it is useful to introduce

$$\hat{\varphi} \equiv \begin{pmatrix} \varphi_2 \\ \varphi_1 \end{pmatrix}, \quad \hat{\psi} \equiv \begin{pmatrix} \psi_2 \\ \psi_1 \end{pmatrix} \tag{9.3.17}$$

According to (9.3.16) and the definition of the Wronskian in (2.11.18), we have

$$\hat{\varphi}^T K\psi = (\varphi_2, \varphi_1) \begin{pmatrix} \psi_1 \\ -\psi_2 \end{pmatrix} = -W(\varphi, \psi) = -c_{12} \tag{9.3.18}$$

We also find $\hat{\psi}^T K\psi = 0$. Similarly, we find $\hat{\varphi}^T K\varphi = 0$ and $\psi^T K\varphi = c_{12}$. These relations provide a convenient means of determining $\alpha_x$ and $\beta_x$. Multiplying (9.3.15) by $\hat{\varphi}^T$ and $\hat{\psi}^T$ in turn, we immediately obtain

$$\alpha_x = \frac{-\hat{\varphi}^T Gv}{c_{12}}$$

$$\beta_x = \frac{\hat{\psi}^T Gv}{c_{12}} \tag{9.3.19}$$

The solution for $\Phi$ that vanishes as $x \to -\infty$ is thus

$$\Phi = \frac{1}{c_{12}} \left( \psi \int_{-\infty}^{x} dx' \, \hat{\varphi}^T G v + \varphi \int_{-\infty}^{x} dx' \, \hat{\psi}^T G v \right) \tag{9.3.20}$$

When we evaluate this result as $x \to +\infty$ and equate the coefficients of $\begin{pmatrix} 0 \\ 1 \end{pmatrix} e^{ikx}$ and $\begin{pmatrix} 1 \\ 0 \end{pmatrix} e^{-ikx}$ with those obtained in (9.3.12), we find that

$$c_{11t} - 4ik^2 c_{11} = -\int_{-\infty}^{\infty} dx' \, \hat{\tilde{\psi}}^T G \varphi \tag{9.3.21a}$$

$$c_{12t} = \int_{-\infty}^{\infty} dx' \, \hat{\psi}^T G \varphi \tag{9.3.21b}$$

In obtaining (9.3.21a) we have used (3.9.8a) to write

$$-\hat{\varphi}^T + c_{11} \hat{\psi}^T = -c_{12} (\tilde{\psi}_2, \tilde{\psi}_1) \tag{9.3.22}$$

and have defined

$$\hat{\tilde{\psi}}^T \equiv (\tilde{\psi}_2, \tilde{\psi}_1) \tag{9.3.23}$$

The time dependence of the eigenvalue $\zeta = k_n$ is obtained by setting $c_{12t} = 0$ in (9.3.21b) and using $G = \epsilon R(u) + \lambda_t \mathsf{I}$. Recalling the definition of $R$ given in (9.3.8) and noting that $\lambda = -i\zeta = -ik_n$, we may carry out the matrix multiplication in the integrand and reduce (9.3.21b) to the form

$$k_{nt} = -\frac{i\epsilon}{2} \frac{\displaystyle\int_{-\infty}^{\infty} dx' \left( R \varphi_2^2 + R^* \varphi_1^2 \right)}{\displaystyle\int_{-\infty}^{\infty} dx' \, \varphi_1 \varphi_2} \tag{9.3.24}$$

The functions $\varphi_1$ and $\varphi_2$ are evaluated at $\zeta = k_n$ where $\tilde{\psi}$ and $\varphi$ are proportional.

In evaluating (9.3.21a) at $\zeta = k_n$, the function $\hat{\tilde{\psi}}^T$ is singular at the pole locations in the upper half-plane that are associated with the bound states. This is evident in the single-pole solutions to be given subsequently in (9.4.5). However, by using (3.9.8a) and evaluating the resulting indeterminate form as $\zeta \to k_n$, we obtain

$$c_{11t} - 4ik_1^2 c_{11} = -\frac{\epsilon}{\dot{c}_{12}} \int_{-\infty}^{\infty} dx' \left[ R \varphi_2 (\dot{\varphi}_2 - c_{11} \dot{\psi}_2) + R^* \varphi_1 (\dot{\varphi}_1 - c_{11} \dot{\psi}_1) \right] \big|_{\zeta = k_1} \tag{9.3.25}$$

where the dot indicates a derivative with respect to $\zeta$. The procedure parallels that used in obtaining (9.1.14). The results expressed in (9.3.24) and (9.3.25) provide the basis for the perturbation theory for the bound-state parameters.

We now consider the use of these formulas for analyzing the change in the single-soliton solution due to the presence of dissipation.

## 9.4  DAMPING OF THE SINGLE SOLITON

The single-soliton solution due to a zero of $c_{12}(\zeta)$ at $\zeta = k_1 = \alpha + i\beta$ with $\beta > 0$ has the general form given in Ex. 5 of Chapter 5, namely

$$u_s(z) = 2\beta e^{-i\theta} \operatorname{sech} z \qquad (9.4.1)$$

where now

$$z = 2\beta(t)[x - \xi(t)] \qquad (9.4.2)$$

and

$$\theta = \frac{\alpha(t)z}{\beta(t)} + \delta(t) \qquad (9.4.3)$$

When no perturbation is present, we have $\alpha = \alpha_0, \beta = \beta_0, \xi = -4\alpha_0 t + \text{constant}$ and $\delta = -4(\alpha_0^2 + \beta_0^2)t + \text{constant}$, as given in Ex. 4 of Chapter 5. In addition, the fundamental solutions given there may be written

$$\varphi(x, k; t) = \frac{e^{-ik\sigma}}{k - k_1^*} \begin{pmatrix} k - \alpha - i\beta \tanh z \\ -i\beta e^{i\theta} \operatorname{sech} z \end{pmatrix} \qquad (9.4.4)$$

$$\psi(x, k; t) = \frac{e^{ik\sigma}}{k - k_1^*} \begin{pmatrix} -i\beta e^{-i\theta} \operatorname{sech} z \\ k - \alpha + i\beta \tanh z \end{pmatrix} \qquad (9.4.5)$$

where $\theta$ is defined in (9.4.3) and

$$\sigma = \frac{z}{2\beta(t)} + \xi(t) \qquad (9.4.6)$$

The time dependence of $\xi$ and $\delta$ as well as that of $\alpha$ and $\beta$ is determined by the perturbation. For a zero of $c_{12}(\zeta)$ located at $\zeta = k_1 = \alpha + i\beta$, the fundamental solutions reduce to

$$\varphi(x, k_1; t) = \tfrac{1}{2} e^{-ik_1\sigma} \operatorname{sech} z \begin{pmatrix} e^{-z} \\ -e^{i\theta} \end{pmatrix} \qquad (9.4.7)$$

$$\psi(x, k_1, t) = \tfrac{1}{2} e^{ik_1\sigma} \operatorname{sech} z \begin{pmatrix} -e^{-i\theta} \\ e^z \end{pmatrix} \qquad (9.4.8)$$

Then we find that

$$c_{11}(k_1) = \frac{\varphi_1(k_1)}{\psi_1(k_1)} = \frac{\varphi_2(k_1)}{\psi_2(k_1)} = -e^{-z - 2ik_1\sigma + i\theta} \qquad (9.4.9)$$

We also have

$$c_{12}(\zeta) = W(\varphi, \psi) = \frac{\zeta - k_1}{\zeta + k_1} \tag{9.4.10}$$

and therefore $\dot{c}_{12}(k_1) = (2k_1)^{-1}$. The time dependence of the eigenvalue is now given by (9.3.24), which reduces to

$$k_{nt} = \frac{i\epsilon}{4} \int_{-\infty}^{\infty} dz\, \text{sech}^2 z (Re^{z+i\theta} + R^* e^{-z-i\theta}) \tag{9.4.11}$$

Separating real and imaginary parts, we obtain

$$\alpha_t = -\tfrac{1}{2}\epsilon \, \text{Im} \int_{-\infty}^{\infty} dz\, Re^{i\theta} \, \text{sech}\, z \tanh z \tag{9.4.12a}$$

$$\beta_t = \tfrac{1}{2}\epsilon \, \text{Re} \int_{-\infty}^{\infty} dz\, Re^{i\theta} \, \text{sech}\, z \tag{9.4.12b}$$

As a simple example of the use of these results, we consider the effect of damping on a single soliton. The damping may be introduced by considering the equation

$$iu_t + u_{xx} + 2|u|^2 u = -i\gamma u \tag{9.4.13}$$

We thus have $\epsilon R = -\gamma u_s$ or $\epsilon Re^{i\theta} = -2\beta\gamma \, \text{sech}\, z$. We then find that $\beta_t = -2\gamma\beta$ and $\alpha_t = 0$, so that

$$\alpha = \alpha_0$$
$$\beta = \beta_0 e^{-2\gamma t} \tag{9.4.14}$$

The soliton thus spreads out and decays, as was the case for the Korteweg–deVries equation.

The change in the phase of the soliton is obtained from (9.3.25) by first using (9.4.9) to write

$$c_{11t} = e^{-2ik_1\xi + i\delta} [2i(k_{1t}\xi + k_1\xi_t) - i\delta_t] \tag{9.4.15}$$

In evaluating (9.3.25) we require

$$\varphi_2(\dot{\varphi}_2 - c_{11}\dot{\psi}_2) = -\frac{i}{2} e^{i\theta + z - 2ik_1\xi + i\delta} \, \text{sech}\, z \left( \sigma e^z \, \text{sech}\, z - \frac{1}{2\beta} \right)$$

$$\varphi_1(\dot{\varphi}_1 - c_{11}\dot{\psi}_1) = -\frac{i}{2} e^{-i\theta - 2ik_1\xi + i\delta} \, \text{sech}\, z \left( \sigma e^{-z} \, \text{sech}\, z + \frac{1}{2\beta} \right) \tag{9.4.16}$$

Separating real and imaginary terms in (9.3.25), we obtain

$$\xi_t = -4\alpha + \frac{\epsilon}{4\beta^2} \operatorname{Re} \int_{-\infty}^{\infty} dz\, z \operatorname{sech} z Re^{i\theta} \tag{9.4.17a}$$

$$\delta_t = 2\alpha\xi_t + 4(\alpha^2 - \beta^2) - \frac{\epsilon}{2\beta} \operatorname{Im} \int_{-\infty}^{\infty} dz \operatorname{sech} z (1 - z\tanh z) Re^{i\theta} \tag{9.4.17b}$$

For the perturbation $\epsilon R = -\gamma u_s$ introduced above, both integrals in (9.4.17) vanish. We then have $\xi_t = -4\alpha_0$ and $\delta_t = 2\alpha_0 \xi_t + 4[\alpha_0^2 - \beta^2(t)]$, where $\beta(t)$ is given by (9.4.14). Integration yields the result

$$\xi = -4\alpha_0 t + \xi_0 \tag{9.4.18}$$

$$\delta = -4\alpha_0^2 t - \frac{\beta_0^2}{\gamma}(1 - 4e^{-\gamma t}) + 2\alpha_0 \xi_0$$

As $\gamma \to 0$ this reduces to the unperturbed result given in Ex. 4 of Chapter 5.

The time dependence of the parameters $\alpha$ and $\beta$ that determine the shape of the single-soliton solution of the damped cubic Schrödinger equation may also be determined by using the conserved quantities associated with the undamped equation. The procedure is the same as that employed for the Korteweg–deVries equation in the previous section.

From (3.9.28) we have

$$2iC_1 = \int_{-\infty}^{\infty} g_1\, dx = \int_{-\infty}^{\infty} |u|^2\, dx$$

$$(2i)^3 C_3 = \int_{-\infty}^{\infty} g_3\, dx = \int_{-\infty}^{\infty} (|u|^4 - |u_x|^2)\, dx \tag{9.4.19}$$

The latter of these expressions corresponds to the energy of the solution, as may be seen by using the Lagrangian density $\mathcal{L} = u_x^* u_x - \frac{1}{2}(u^* u_t - u_t^* u) - (u^* u)^2$. The integral over all space of the associated Hamiltonian density $\mathcal{H} = u_t \partial\mathcal{L}/\partial u_t + u_t^* \partial\mathcal{L}/\partial u_t^* - \mathcal{L}$ is found to be proportional to $C_3$.

For the single-soliton solution $u = 2\beta e^{-i\theta} \operatorname{sech} z$ given in (9.4.1), we find that

$$2iC_1 = 8\beta$$

$$(2i)^3 C_3 = 32\beta(\tfrac{1}{3}\beta^2 - \alpha^2) \tag{9.4.20}$$

The time derivative of $C_1$ and $C_3$ when $u$ is assumed to satisfy the damped cubic Schrödinger equation (9.4.13) yields

$$2iC_{1t} = -2i\gamma \int_{-\infty}^{\infty} |u|^2\, dx$$

$$(2i)^3 C_{3t} = i\gamma \int_{-\infty}^{\infty} (u^* u_t - u_t^* u)\, dx \tag{9.4.21}$$

When the time derivative of $C_1$, as given by (9.4.20), is equated to that obtained from (9.4.21) we recover the relation $\beta_t = -2\gamma\beta$. A similar consideration of $C_{3_t}$ reduces to $\alpha_t = 0$ so that both results given in (2.4.14) are recovered.

The similarity with the perturbation theory for the Korteweg–deVries equation should now be evident. The consideration of the continuous spectrum is also similar to that for the Korteweg–deVries equation and will not be developed here. This topic as well as additional examples of perturbation calculations may be found in the research papers referred to at the beginning of this chapter.

# REFERENCES

Abella, J. D., N. A. Kurnit, and S. R. Hartmann. 1966. Photon echoes. *Phys. Rev.* **141**, 391–406.

Ablowitz, M. J., D. J. Kaup, A. C. Newell, and H. Segur. 1974a. The inverse scattering transform —Fourier analysis for nonlinear problems. *Stud. Appl. Math.* **53**, 249–315.

Ablowitz, M. J., D. J. Kaup, A. C. Newell, and H. Segur. 1974b. Coherent pulse propagation, a dispersive, irreversible phenomenon. *J. Math. Phys.* **15**, 1852–1858.

Abramowitz, M. and I. A. Stegun. 1964. *Handbook of mathematical functions.* National Bureau of Standards Applied Mathematics Series, No. 55. U.S. Department of Commerce.

Agranovich, Z. S. and V. A. Marchenko. 1964. *The inverse problem of scattering theory.* Gordon and Breach, New York.

Allen, L. and J. H. Eberly. 1975. *Optical resonance and two-level atoms.* Wiley, New York.

Ames, W. F. 1965. *Nonlinear partial differential equations in engineering.* Academic, New York.

Amsler, M. H. 1955. Des surfaces à courbure négative constante dans l'espace à trois dimensions et de leurs singularités. *Math. Ann.* **130**, 234–256.

Bäcklund, A. V. 1882. Zur Theorie der Flächentransformationen, *Math. Ann.* **19**, 387–422.

Balanis, G. N. 1972. The plasma inverse problem. *J. Math. Phys.* **13**, 1001–1005.

Bargmann, V. 1949. On the connection between phase shifts and scattering potential. *Rev. Mod. Phys.* **21**, 488–493.

Batchelor, G. K. 1967. *An introduction to fluid dynamics.* Cambridge University Press, New York.

Berezin, Yu. A. and V. I. Karpman. 1964. Theory of nonstationary finite amplitude waves in a low-density plasma. *Sov. Phys. JETP* **19**, 1265–1271.

Berezin, Yu. A. and V. I. Karpman, 1967. Nonlinear evolution of disturbances in plasmas and other dispersive media. *Sov. Phys. JETP* **24**, 1049–1056.

Bergmann, P. G. 1946. The wave equation in a medium with a variable index of refraction. *J. Acoust. Soc. Am.* **17**, 329–333.

Bloch, F. 1946. Nuclear induction. *Phys. Rev.* **70**, 460–474.

Brekhovskikh, L. M. 1960. *Waves in layered media.* Academic, New York.

Bullough, R. K. and P. J. Caudrey. 1978. The double-sine-Gordon equation: wobbling solitons? *Rocky Mt. J. Math.* **8**, 53–70.

Cajori, F. 1904. *An introduction to the theory of equations.* Macmillan, New York. Reprinted 1969, Dover, New York.

Calogero, F. and A. Degasperis. 1968. Values of the potential and its derivatives at the origin in terms of the $s$-wave phase shift and bound state parameters. *J. Math. Phys.* **9**, 90–116.

Calogero, F. and A. Degasperis. 1977. Nonlinear evolution equations solvable by the inverse spectral transform II. *Nuovo Cimento* **39B**, 1–54.

Churchill, R. V. 1948. *Introduction to complex variables and applications.* McGraw-Hill, New York.

Clairin, J. 1903. Sur quelques équations aux dérivées partielles du second ordre, *Ann. Fac. Sci. Univ. Toulouse* 2e Sér. **5**, 437–458.

Coddington, E. and N. Levinson. 1955. *Theory of ordinary differential equations*. McGraw-Hill, New York.

Cole, J. D. 1950. On a quasilinear parabolic equation occurring in aerodynamics. *Q. Appl. Math.* **9**, 225–236.

Condon, E. V. and G. H. Shortley. 1957. *The theory of atomic spectra*. Cambridge University Press, New York.

Crum, M. M. 1955. Associated Sturm–Liouville equations. *Q. J. Math.* **6**, 121–127.

Darboux, G. 1882. Sur une proposition relative aux équations linéaires. *Comptes Rendus.* **94**, 1456–1459.

Darboux, G. 1915. *Leçons sur la théorie générale des surfaces et les applications géometriques du calcul infinitésimal*, Vol. 2, 2nd ed. Gauthier-Villars, Paris.

Davey, A. 1972. The propagation of a weak nonlinear wave. *J. Fluid Mech.* **53**, 769–781.

Davis, H. T. 1960. *Introduction to nonlinear differential and integral equations*. United States Atomic Energy Commission. U.S. Government Printing Office, Washington, D.C. Reprinted Dover, New York.

Dodd, R. K., R. K. Bullough, and S. Duckworth. 1975. Multisoliton solutions of nonlinear dispersive wave equations not soluble by the inverse method. *J. Phys. A* **8**, L64–L68.

Eckart, C. 1930. The penetration of a potential barrier by electrons. *Phys. Rev.* **35**, 1303–1309.

Eisenhart, L. P. 1909. *A treatise on the differential geometry of curves and surfaces*. Ginn, Boston. Reprinted 1960, Dover, New York.

Faddeev, L. D. 1958. On the relation between the $s$-matrix and potential for the one-dimensional Schrödinger operation. *Sov. Phys. Dokl.* **3**, 747–751.

Faddeev, L. D. 1963. The inverse problem in the quantum theory of scattering. *J. Math. Phys.* **4**, 72–104.

Faddeev, L. D. 1967. Properties of the $s$-matrix of the one-dimensional Schrödinger equation. *Am. Math. Soc. Transl. Ser.* 2 **65**, 139–166.

Flaschka, H. and D. W. McLaughlin. 1976. Some comments on Bäcklund transformations and inverse scattering problems. In *Bäcklund transformations*. Vol. 515 in Lecture Notes in Mathematics, A. Dold and B. Eckmann, eds. Springer-Verlag, Berlin.

Forsyth, A. R. 1906. *Theory of differential equations*, Vol. 6. Cambridge University Press, New York. Reprinted 1959, Dover, New York.

Frank, F. C. and J. H. van der Merwe. 1949. One dimensional dislocations. I. Static theory. *Proc. R. Soc. Lond. A* **198**, 205–216.

Frank, F. C. and J. H. van der Merwe. 1950. One dimensional dislocations. IV. Dynamics. *Proc. R. Soc. Lond. A* **201**, 261–268.

Gardner, C. S., J. M. Greene, M. D. Kruskal, and R. M. Miura. 1974. Korteweg–deVries equation and generalizations. VI. Methods for exact solution. *Commun. Pure Appl. Math.* **27**, 97–133.

Gibbon, J. D., P. J. Caudrey, R. K. Bullough, and J. C. Eilbeck. 1973. An $N$-soliton solution of a nonlinear optics equation derived by a general inverse method. *Lett. Nuovo Cimento* **8**, 775–779.

Gibbs, H. M. and R. E. Slusher. 1972. Sharp line self-induced transparency. *Phys. Rev. A.* **6**, 2326–2334.

Goldstein, H. 1950. *Classical mechanics*. Addison-Wesley, Reading, Mass.

Goursat, E. 1925. Le problème de Bäcklund. *Mem. Sci. Math. Fasc.* **6**, Gauthier-Villars, Paris.

Grieneisen, H. P., J. Goldhar, N. A. Kurnit, A. Javan, and H. R. Schlossberg. 1972. Observation of the transparency of a resonant medium to zero-degree optical pulses. *Appl. Phys. Lett.* **21**, 559–562.

Hahn, E. 1950. Spin echoes. *Phys. Rev.* **80**, 580–594.

Hammack, J. L. and H. Segur. 1974. The Korteweg–deVries equation and water waves. Part 2. Comparison with experiments. *J. Fluid Mech.* **65**, 289–314.

Hasimoto, H. and H. Ono. 1972. Nonlinear modulation of gravity waves. *J. Phys. Soc. Japan* **33**, 805–811.

Hasimoto, H. 1972. A soliton on a vortex filament. *J. Fluid Mech.* **51**, 477–485.

Hildebrand, F. B. 1976. *Advanced calculus for applications*, 2nd ed. Prentice-Hall, Englewood Cliffs, N.J.

Hirota, R. 1973. Exact envelope-soliton solutions of a nonlinear wave equation. *J. Math. Phys.* **14**, 805–809.

Hopf, E. 1950. The partial differential equation $u_t + uu_x = \mu u_{xx}$. *Commun. Pure Appl. Math.* **3**, 201–230.

Hopf, F. A. and M. O. Scully. 1969. Theory of an inhomogeneously broadened laser amplifier. *Phys. Rev.* **179**, 399–416.

Hopf, F. A., G. L. Lamb, Jr., C. K. Rhodes, and M. O. Scully. 1971. Some results on coherent radioactive phenomena with $0\pi$ pulses. *Phys. Rev. A* **3**, 758–766.

Icsevgi, A. and W. E. Lamb, Jr. 1969. Propagation of light pulses in a laser amplifier. *Phys. Rev.* **185**, 517–545.

Ikezi, H. 1973. Experiments on ion-acoustic solitary waves. *Phys. Fluids* **16**, 1668–1675.

Ince, E. L. 1969. *Ordinary differential equations*. Longmans, New York. Reprinted 1956, Dover, New York.

Indenbom, V. L. 1958. Mobility of dislocations in the Frenkel–Kontorova model. *Sov. Phys. Crystallogr.* **3**, 193–201.

Infeld, L. and T. E. Hull. 1951. Factorization method. *Rev. Mod. Phys.* **23**, 21–68.

Jackson, J. D. 1962. *Classical electrodynamics*. Wiley, New York.

Jeffrey, A. and T. Kakutani. 1972. Weak nonlinear dispersive waves: a discussion centered around the Korteweg–deVries equation. *SIAM Rev.* **14**, 582–643.

Jeffreys, H. and B. S. Jeffreys, 1956. Methods of mathematical physics, 3rd ed. Cambridge University Press, New York.

Kamke, E. 1971. *Differentialgleichungen Losungsmethoden und Losungen*. Chelsea, New York.

Karpman, V. I. 1975. *Non-linear waves in dispersive media*. Pergamon, Elmsford, N.Y.

Karpman, V. I. 1979. Soliton evolution in the presence of perturbation. *Phys. Scr.* **20** 462–478.

Karpman, V. I. and V. P. Sokolov. 1968. On solitons and the eigenvalues of the Schrödinger equation. *Sov. Phys. JETP* **27**, 839–845.

Karpman, V. I. and E. M. Maslov. 1978. Perturbation theory for solitons. *Sov. Phys. JETP* **46**, 281–291.

Kaup, D. J. 1977. Coherent pulse propagation: a comparison of the complete solution with the McCall–Hahn theory and others. *Phys. Rev. A* **16**, 704–719.

Kaup, D. J. and A. C. Newell, 1978. Solitons as particles, oscillators, and in slowly varying media: a singular perturbation theory. *Proc. R. Soc. Lond. A* **361**, 413–446.

Kay, I. and H. E. Moses. 1956. Reflectionless transmission through dielectrics and scattering potentials. *J. Appl. Phys.* **27**, 1503–1508.

Kay, I. 1960. The inverse scattering problem when the reflection coefficient is a rational function. *Commun. Pure Appl. Math.* **13**, 371–393.

Korteweg, D. J. and G. deVries. 1895. On the change of form of long waves advancing in a rectangular channel, and on a new type of long stationary waves. *Phil. Mag.* (5) **39**, 422–443.

Lake, B. M., H. C. Yuen, H. Rungaldier, and W. E. Ferguson. 1977. Nonlinear deep water waves: theory and experiment. Part 2. Evolution of a continuous wave train. *J. Fluid Mech.* **83**, 49–74.

Lakshmanan, M. 1979. Rigid body motions, space curves, prolongation structures, connection forms, fiber bundles and solitons. *J. Math Phys.* **20**, 1667–1672.

Lax, P. 1968. Integrals of nonlinear equations of evolution and solitary waves. *Commun. Pure Appl. Math.* **21**, 467–490.

Leibovich, S. and A. R. Seebass, eds. 1974. *Nonlinear waves*. Cornell University Press, Ithaca, N.Y.

Liouville, J. 1853. Sur l'équation aux différences partielles: $\partial^2 \log\lambda / \partial u\, \partial v \pm \lambda/2a^2 = 0$. *J. Math.* **18**, 71–72.

Love, A. E. H. 1927. *A treatise on the mathematical theory of elasticity*, 4th ed. Cambridge University Press, New York. Reprinted 1944, Dover, New York.

Lovitt, W. V. 1924. *Linear Integral Equations*. McGraw-Hill, New York. Reprinted 1950, Dover, New York.

McCall, S. L. and E. L. Hahn. 1969. Self-induced transparency. *Phys. Rev.* **183**, 457–485.

McLaughlin, D. W. and A. C. Scott. 1978. Perturbation analysis of fluxon-dynamics. *Phys. Rev. A* **18**, 1652–1680.

Magnus, W. and F. Oberhettinger. 1954. *Formulas and theorems for the functions of mathematical physics*. Chelsea, New York.

Miura, R. M. 1968. Korteweg–deVries equation and generalization. I. A remarkable explicit nonlinear transformation. *J. Math. Phys.* **9**, 1202–1204.

Miura, R. M., ed. 1976. *Bäcklund transformations*. Vol. 515 in Lecture Notes in Mathematics, A. Dold and B. Eckmann, eds. Springer-Verlag, Berlin.

Miura, R. M., C. S. Gardner, and M. D. Kruskal. 1968. Korteweg–deVries equation and generalizations. II. Existence of conservation laws and constants of motion. *J. Math. Phys.* **9**, 1204–1209.

Morse, P. M. 1948. *Vibration and sound*, 2nd ed. McGraw-Hill, New York.

Morse, P. M. and H. Feshbach. 1953. *Methods of theoretical physics*. McGraw-Hill, New York.

Newton, R. 1960. Analytic properties of radial wave functions. *J. Math. Phys.* **1**, 319–347.

Nussenzweig, H. M. 1959. The poles of the $s$-matrix of a rectangular potential well or barrier. *Nucl. Phys.* **11**, 499–521.

Patel, C. K. N. and R. E. Slusher. 1968. Photon echoes in gases. *Phys. Rev. Lett.* **20**, 1087–1089.

Randall, J. D. and S. Leibovich. 1973. The critical state; a trapped wave model of vortex breakdown. *J. Fluid Mech.* **58**, 495–515.

Rhodes, C. K., A. Szöke, and A. Javan. 1968. The influence of level degeneracy on the self-induced transparency effect. *Phys. Rev. Lett.* **21**, 1151–1155.

Reiter, G. 1980. Lie groups, spin equations and the geometrical interpretation of solitons. *J. Math. Phys.* To be published.

Schiff, L. I. 1949. *Quantum mechanics*. McGraw-Hill, New York.

Schnack, D. D. and G. L. Lamb, Jr. 1973. Higher conservation laws and coherent pulse propagation. In *Coherence and quantum optics*, L. Mandel and E. Wolf, eds., pp. 23–33. Plenum, New York.

Seeger, A., H. Donth, and A. Kochenforfer. 1953. Theorie der Versetzungen in eindimensionalen Atomreihen. *Z Phys.* **134**, 173–193.

Seeger, A. 1955. Theorie der Gitterfehlstellen. In *Handbuch der Physik*, Vol. 7, p. 1, pp. 383–665, p. 566. Springer-Verlag, Berlin.

Seeger, A. and P. Schiller. 1966. Kinks and dislocation lines and their effects on the internal friction in crystals. In *Physical Acoustics*, Vol. 3, P. A. W. P. Mason, Ed., pp. 361–495. Academic, New York.

Shimizu, K. and Y. H. Ichikawa. 1972. Automodulation of ion oscillation modes in plasma. *J. Phys. Soc. Japan* **33**, 789–792.

Slusher, R. E. and H. M. Gibbs. 1972. Self-induced transparency in atomic rubidium. *Phys. Rev. A* **5**, 1634–1659.

Spitzer, L., Jr. 1956. *Physics of fully ionized gases.* Interscience, New York.

Steuerwald, R. 1936. Über enneper'sche Flächen und Bäcklund'sche Transformation. *Abh. Bayer. Akad. Wiss. (Muench.)* **40**, 1–105.

Struik, D. J. 1961. *Lectures on classical differential geometry*, 2nd ed. Addison-Wesley, Reading, Mass.

Su, C. H. and C. S. Gardner. 1969. Korteweg–deVries equation and generalizations. III. Derivation of the Korteweg–deVries equation and Burgers' equation. *J. Math. Phys.* **10**, 536–539.

Tanaka, S. 1972. Modified Korteweg–deVries equation and scattering theory. *Proc. Japan Acad.* **48**, 466–489.

Taniuti, T. and C-C. Wei. 1968. Reductive perturbation method in nonlinear wave propagation I. *J. Phys. Soc. Japan* **24**, 941–946.

Tappert, F. D. and N. J. Zabusky. 1971. Gradient-induced fission of solitons. *Phys. Rev. Lett.* **27**, 1774–1776.

Tidman, D. A. and N. A. Krall. 1971. *Shock waves in collisionless plasma.* Wiley-Interscience, New York.

Verde, M. 1955. Asymptotic expansion of phase shifts at high energies. *Nuovo Cimento* **2**, 1001–1014.

Wadati, M. 1973. The modified Korteweg–deVries equation. *J. Phys. Soc. Japan* **34**, 1289–1296.

Wadati, M. 1974. Bäcklund transformation for solutions of the modified Korteweg–deVries equation. *J. Phys. Soc. Japan* **36**, 1498.

Wadati, M. and M. Toda. 1972. The exact $N$-soliton solution of the Korteweg–deVries equation. *J. Phys. Soc. Japan* **32**, 1403–1411.

Wadati, M., H. Sanuki, and K. Konno. 1975. Relationships among inverse method, Bäcklund transformation and an infinite number of conservation laws. *Prog. Theor. Phys.* **53**, 419–436.

Wahlquist, H. and F. B. Estabrook. 1973. Bäcklund transformation for solutions of the Korteweg–deVries equation. *Phys. Rev. Lett.* **31**, 1386–1390.

Whitham, G. B. 1974. *Linear and nonlinear waves.* Wiley, New York.

Zabusky, N.J. 1968. Solitons and bound states of the time-independent Schrödinger equation. *Phys. Rev.* **168**, 124–128.

Zakharov, V. E. and A. B. Shabat. 1972. Exact theory of two-dimensional self-focusing and one-dimensional self-modulation of waves in nonlinear media. *Sov. Phys. JETP* **34**, 62–69.

# INDEX

Airy function, 129, 267
Area theorem, 214, 227

Bäcklund transformations, 243
  Burger's equation and diffusion equation, 257
  Korteweg-deVries equation, 243, 245, 258
  Liouville's equation, 256
  modified Korteweg-deVries equation, 257, 258
  sine-Gordon equation, 250
Bargmann potential, 1, 8, 96
  for Schrödinger equation:
    linear case, 8, 97
    quadratic case, 9, 97
  for two-component equations, 75
Bloch equations, 210
  similarity to Serret-Frenet equations, 214
Boundary, locally reacting, 14
Bound states:
  and Schrödinger equation, 33, 50
  and two-component system, 73, 77
Breather solutions:
  interaction between, 250
  for modified Korteweg-deVries equation, 138
  for sine-Gordon equation, 149
  as soliton-antisoliton pair, 151
Burgers equation, 189, 257

Circular cylinder, 13
Circulation of vortex, 191
Clairin's method, 253
Coefficients $c_{ij}$:
  normalization of bound state wave functions, 53
  relations among:
    for Schrödinger equation, 49

  for two-component system, 72
    with complex potential, 108
  temporal variation:
    for cubic Schrödinger equation, 157
    for Korteweg-deVries equation, 119
    for modified Korteweg-deVries equation, 136
    for sine-Gordon equation, 152
Conserved quantities:
  asymptotic solution:
    Schrödinger equation, 57
    two-component system:
      complex potential, 111
      real potential, 78
  Born approximation, 58
  for Korteweg-deVries equation, 126
  in optical pulse propagation, 234
  Schrödinger equation, 58
  soliton amplitudes, 127
  symmetric functions of roots, algebraic equations, 128
  two-component system:
    complex potential, 111
    real potential, 78
  use in perturbation theory:
    cubic Schrödinger equation, 277
    Korteweg-deVries equation, 269
  WKB method, 58
Conservation law, energy on string, 25
Cubic Schrödinger equation, 55, 133, 155, 177, 194, 197
  from general inverse method, 167
  Hamiltonian density, 277
  Hirota equation, 202
  and Korteweg-deVries equation, 174
  Lagrangian density, 277
  linear equations, 156
    in matrix form, 158, 271
  and modified Korteweg-deVries equation, 141

perturbation, single soliton solution,
    275
perturbed, 271
shallow water waves, 174
and sine-Gordon equation, 149
single soliton solution, 157
vortex filament, 197
Cutoff frequency, 30

Darboux method, 38
    for $\operatorname{csch}^2 x$ potential, 94
    for $\operatorname{sech}^2 x$ potential, 40
Darboux vector, 203
Debye shielding length, 179
Delta function, 32
Delta function potential:
    derivative, 32
    reflection and transmission coefficients,
        33
    from $\operatorname{sech}^2 x$ potential, 37
Delta function pulse, integral representa-
        tion, 28
Diffusion equation, 257
Dislocation theory, 182
Dispersion:
    and Korteweg-deVries equation, 19
    waves on elastically braced string, 30
Dispersion relation, 13, 16
Doppler shift, 208

Eckart potential, 8
Eigenfunctions, squared, 165
Eigenvalue:
    constant in time, 4, 5
    constant in space, 228
    reflectionless potential, 37
Elastic rings, 13
Energy flow on string, 23
Energy in single soliton, 186
Error function, complex, 223
Expansion parameters, choice, 186
Euler-Lagrange equation, 25

Fluid, incompressible, 191
Frenet equations, see Serret-Frenet equa-
        tions
Fundamental solution, 46
    Schrödinger equation:
        alternative representation, 63
        analytic in upper half plane, 47, 64
        with delta function potential, 54
        linear combinations, 48

with potential $-2 \operatorname{sech}^2 x$, 89
with potential $-6 \operatorname{sech}^2 x$, 54
variation of parameters, 47
single soliton solution, cubic Schrödinger
        equation, 159
two-component system:
    with complex potential, 107
        single soliton result, 110
    linear combination of solutions, 71
    with real potential, single soliton result,
        76, 77
    variation of parameters, 70

Hamiltonian density, 25
    cubic Schrödinger equation, 277
    Korteweg-deVries equation, 270
    sine-Gordon equation, 150
Hirota equation, 202

Integral equation, Volterra type, 47
Inverse method:
    for Schrödinger equation, 84
    two-component system:
        complex potential, 107
        real potential, 99
Inverse scattering, 4, 84
Ion plasma waves, 178

Jacobian elliptic function, 114

Korteweg-deVries equation, 1, 4, 17, 189
    Bäcklund transformation, 243, 245
    and cubic Schrödinger equation, 174
    fluid in elastic walled cylinder, 18
    form under rescaling, 115
    from general inverse method, 168
    Hamiltonian density, 270
    ion plasma waves, 178, 181
    Lagrangian density, 269
    multi-soliton solutions, 121
    numerical results, 115
    perturbation, continuous spectrum, 264
    perturbed, 259
    perturbed single solition solution, 262
    shallow water waves, 169, 173
    similarity solution, 128
    single soliton solution, 7, 115
    solution by inverse scattering, 118
    solution by two-component inverse
        method, 131
    steady state solution, 7, 19, 113
    two-solition solution, 7, 11, 122, 246

Lagrangian density, 25
  cubic Schrödinger equation, 277
  Korteweg-deVries equation, 269
  sine-Gordon equation, 150
Level degeneracy, 238
Levinson's theorem, 59
Line source, inhomogeneous medium,
  43
Liouville's equation, 254
  Bäcklund transformation, 256
  general solution, 256

Marchenko equation:
  and inverse scattering, 66
  and scattering of pulses, 62
  for Schrödinger equation, 66
    attractive delta function potential, 88
    criterion for validity of procedure, 85
    presence of bound states, 87
    repulsive delta function potential, 86
    sech$^2$ x potential, 89
  for two-component system, 101
Maxwell equations, 205, 227
Miura transformation, 135, 258
  as Bäcklund transformation, 257
Modified Korteweg-deVries equation, 20,
    133, 134
  Bäcklund transformation, 257
  breather solution, 138
  from general inverse method, 168
  and Hirota equation, 202
  linear equations, 135
    in matrix form, 158
  multisoliton solution, 137
  similarity solution, 142
  single solition solution, 138
  solution by inverse scattering, 136
Multisoliton solution:
  cubic Schrödinger equation, 157
  Korteweg-deVries equation, 121
  modified Korteweg-deVries equation,
    137
  sine-Gordon equation, 152

Nonlinear Schrödinger equation, see Cubic
    Schrödinger equation
Normal mode, 2

Operator:
  linear differential, 5
  linear integral and sine-Gordon equation,
    160

Optical pulse propagation, 204
  amplifier, 213, 221, 226
  conservation laws, 234
  two-component method, 227
Oscillator-string system, 26
  damping, radiation, 27
  scattering of delta function pulse, 28

Painlevé transcendents, 142, 154
Perfect transmission:
  oscillator-string system, 27
  square well potential, 31
  sech$^2$ x potential, 37
Peremutability theorem, 246, 249
Perturbation expansion:
  cubic Schrödinger equation, 177,
    271
  elastic walled cylinder, 16
  ion plasma waves, 180
  Korteweg-deVries equation, 16, 259
  parameters, 186
  shallow water waves, 171, 174
Perturbation theory:
  cubic Schrödinger equation, 271
  Korteweg-deVries equation, 259
Phase terms, two-soliton solution, 123
Plasma frequency, 179
Polarization, 206
Poles of transmission coefficient:
  for Schrödinger equation:
    residue, 51
    simple for real potentials, 52
  location, real potentials, Schrödinger
    equation, 50
  location, two-component system, real
    potential, 73
  residues, two-component system, 74
Potential:
  attractive, one bound state, 62
  delta function, 62, 64
  parameter dependent, 13
  reflectionless:
    Schrödinger equation, 90
    general expression, 92
  two-component system:
    complex potential, 110
    real potentials, 103
  for Schrödinger equation, quantities
    $A_R$, $A_L$, 85
  for two-component system:
    from $A_1$, $A_2$, 102
    from $B_1$, $B_2$, 102

Potential, truncated:
  Schrödinger equation, 60
  two-component system, 78
Pulse propagation, reduction of en-
    velope velocity, 231
Pulse scattering, delta function
    potential, 62, 64

Rectangular potential, 29, 34
  and two-component system, 77
Reflected wave, 3, 26, 30, 36
Reflection coefficient:
  coefficients $c_{ij}$:
    for Schrödinger equation, 48
    for two-component system, 72
  exponential time dependence,
    162
  Fourier transform, 53
  for oscillator attached to string, 27
  phase change, 26
  rational function of k:
    Schrödinger equation, 93
    two-component system, 105
  for $sech^2 x$ potential, 37
  singularities, 27
  spatial dependence, 231
  time dependence, Korteweg-deVries
    equation, 120
  and transmission coefficient,
    Schrödinger equation, 55
  vanishing, and soliton propagation,
    218
  at zero frequency, 50
Reflectionless potential, eigenvalues,
    37
Refractive index, 3, 42
Relaxation time:
  homogeneous, 210
  inhomogeneous, 223
Reynolds number, 187
Riccati equation:
  associated Sturm-Liouville equations,
    39
  cross ratio theorem, 83
  general solution, 81
  relation between Schrödinger and
    two-component equations, 80
  sech x potential, 82
  standard form, 81
Rigid body motion, 203

Scattering, 3, 22
  by oscillator, 26

Scattering theory, one-dimensional, 22
Scattering, two-component:
  fundamental solutions:
    complex potential, 107
    real potential, 69
  and semi-permeable membrane, 61
Schrödinger equation, 33
  asymptotic solution, conserved
    quantities, 57
  complex potential, 134, 216
  and particle flux density, 33
  time dependent, 207
$Sech^2 x$ potential, 3
  eigenfunction, associated Legendre
    function, 38
  reflectionless, 37
  in Schrödinger equation, 34
  solution for n eigenvalues, 40
Self-induced transparency, 204
Serret-Frenet equations, 190, 194
  similarity to Bloch equations, 214
Shallow water waves:
  cubic Schrödinger equation, 174
  Korteweg-deVries equation, 169
Sine-Gordon equation, 20, 133, 143
  antisolitons, 146
  Bäcklund transformation, 250
  breather solution, 149
  and cubic Schrödinger equation,
    149
  in differential geometry, 144
  in dislocation theory, 182, 185
  from general inverse method, 167
  Hamiltonian density, 150
  Lagrange density, 150
  linear equations, 144
  multisoliton solution, 152
  in optical pulse propagation, 210
  similarity solution, 153
  single soliton solution, 146
  twisted curves, 203
  two-soliton solution, 147, 153
Soliton, 1
  as balance between nonlinearity and
    dispersion, 19
  in optical pulse propagation, 204, 213,
    220
  on vortex filament, 190
Soliton velocity, 18
Spectrum, 2
Square well, attractive potential, 34
Stability, 212
Stationary atoms, 210

Sturm-Liouville equation, 2
Sturm-Liouville equation, associated,
    Darboux's method for treating,
    38
String:
  attached to oscillator, 26
  elastically braced, 2, 29
  wave motion, 22

Transmission coefficient:
  coefficients $c_{ij}$:
    for Schrödinger equation, 48
    for two-component system, 72
  Fourier transform, 53
  number of poles, 59
  oscillator attached to string, 27
  and reflection coefficient:
    Schrödinger equation, 55
    two-component system, 105
  $\text{sech}^2 x$ potential, 37
  singularities, 27
Transmitted wave, 3, 26, 30, 34, 37
Two-level atom, 204
  degenerate, 238

Two-soliton solution:
  as determinant in symmetric form,
    122
  phase shift, 13, 123
  for sine-Gordon equation, 153, 213
  and soliton trajectory, 13

Vibrating string, see String
Viscosity, 186
Vortex filament, 190
Vorticity vector, 191

Wave motion:
  dispersive, 13
  localized, 30
Waves in two dimensions, 41
Wave velocity:
  group, 13
  phase, 13
Wronskian relations:
  Schrödinger equation, 47
  two-component system, 71

Zero area pulses, 213